Lecture Notes in Statistics **159**

Springer

New York
Berlin
Heidelberg
Barcelona
Hong Kong
London
Milan
Paris
Singapore
Tokyo

Marc Moore (Editor)

Spatial Statistics: Methodological Aspects and Applications

Marc Moore
Département de mathématiques et de génie industriel
École Polytechnique de Montréal
C.P. 6079 Succursale Centre-ville
Montréal, Québec H3C 3A7
Canada
marc.moore@courriel.polymtl.ca

Library of Congress Cataloging-in-Publication Data

Spatial statistics : methodological aspects and applications / editor, Marc Moore.
p. cm.—(Lecture notes in statistics ; 159)
Includes bibliographical references.
ISBN 0-387-95240-3 (softcover : alk. paper)
1. Spatial analysis (Statistics) I. Moore, Marc. II. Lecture notes in statistics (Springer-Verlag) ; v. 159.

QA278.2.S64 2001
519.5—dc21 00-067925

Printed on acid-free paper.

Camera-ready copy provided by the Centre de Recherches Mathématiques.
Printed and bound by Edwards Brothers, Inc., Ann Arbor, MI.
Printed in the United States of America.

9 8 7 6 5 4 3 2 1

ISBN 0-387-95240-3 SPIN 10793817

Springer-Verlag New York Berlin Heidelberg
A member of BertelsmannSpringer Science+Business Media GmbH

Preface

In 1997-98 the theme of the year at the CRM (Centre de recherches mathématiques, Université de Montréal) was Statistics. Among the many workshops organized four were on topics related to Spatial Statistics: Statistical Inference for Spatial Processes; Image Analysis; Applications of Spatial Statistics in Earth, Environmental and Health Sciences; Statistics of Brain Mapping. This volume contains papers based on some of the contributions to these four workshops. They range from asymptotic considerations for spatial processes to practical considerations related to particular applications including important methodological aspects. These contributions can be divided into two main categories: inference for spatial processes and image analysis, mainly images related to brain mapping.

Statistical inference for spatial processes has been an active field for many years and, as illustrated here, important progress are still made.

Zhu, Lahiri and Cressie consider the important concept of Spatial Cumulative Distribution Function (SCDF) associated to a random field. This function offers an effective summary for the distribution of the process over a region. A functional central limit theorem is provided for the empirical version of the SCDF when the random field has a nonstationary structure. The importance of the sampling design is clearly shown.

Kriging is a widely used spatial prediction procedure. An important ingredient in this procedure is the variogram that must be selected and fitted from the data. Genton presents a survey of his contributions toward the robustification of the three steps leading to the modeling of an appropriate variogram. This provides a fitted variogram that remains closed to the true underlying variogram even if outliers are present in the data, ensuring that kriging supplies informative predictions.

Hardouin and Guyon introduce the concept of spatial coordination among the state taken by each point on an array. Its relation to correlation is used to establish and study tests to detect such coordination.

Perera presents a survey of recent and deep results about the asymptotic behaviour of additive functionals of random fields over irregular sets.

Statistical techniques and considerations in image analysis are more recent. They have been made possible by the increasing availability of computing facilities. Also, image analysis has stimulated the development of sophisticated statistical methods and stochastic models.

Motivated by command and control problems, Huang and Cressie propose multiscale graphical Markov models over acyclic directed graphs for a true scene. They derive the generalized Kalman filter algorithm to obtain an optimal prediction of the true scene from noisy data.

Idier, Goussard and Ridolfi propose a nonsupervised three-dimensional segmentation method. Representing the 3D unknown image by an unilateral Markov random field (Pickard type) and assuming a particular Gaussian observation process, they obtain a feasible segmentation method. A compromise is made between numerical costs and coarseness of the image model, that still produces good results.

Gelpke and Künsch study the estimation of the motion in a velocity field from the observation of a sequence of images. The approach is based on physical considerations leading to a so called continuity equation, and on a discretization of it. A statistical approach produces estimates of the displacement field parameters. The proposed methodology is applied to successive images of measurements of total ozone on the northern hemisphere.

Cao and Worsley present a survey of the pioneer work of Worsley and his collaborators regarding the utilization of random field theory, and appropriate statistical tools, to test for activations in brain mapping applications.

Bullmore, Suckling and Brammer remind the origin of permutation tests. Fisher admitted that "the main disadvantage of this procedure is that it would be intorelably boring to do." With nowadays computing facilities this disadvantage has disappeared. These authors illustrate some of the traditional strengths of permutation testing and show that it can be an appropriate and useful tool in brain mapping.

Coulon and his collaborators present a new method for cerebral activation detection. The analysis is made through the comparison of multiscale object-based description of the individual maps. This comparison is made using a graph on which a labeling process is performed; this process is represented by a Markovian random field.

Lütkenhöner illustrates how new technologies generate the necessity to construct appropriate models and statistical analysis. The case of magnetoencephalography which is a noninvasive functional imaging technique, allowing almost unlimited temporal resolution, is presented.

Schorman and Dabringhaus explain the importance to detect nonlinear deformation resulting from histological preparation when the resulting micro information has to be integrated to the macro-structural information derived from magnetic resonance images. A procedure is proposed and its utility illustrated.

Bookstein offers an essay proposing techniques (morphometrics) and arguing for their utility, to take into account the explicit geometric normalization of Cartesian shifts as an integral part of any quantitative analysis in brain mapping.

We thank all these contributors who made this volume possible. Gratitude is also expressed to all the speakers and participants to these workshops. We express deep appreciation to X. Guyon, R. Lockhart and K. Worsley for their important contribution in the organization of these events. Thanks are due to X. Guyon, B. Rémillard and K. Worsley for their help in the refereeing process. We are most grateful to the CRM for having made possible this theme year and to all its staff for their excellent cooperation in bringing out this volume.

Marc Moore
Montreal, October 2000

Contents

Contributors

Isabelle Bloch Département Traitement du signal et de l'image, École nationale supérieure des télécommunications, 46, rue Barrault, 75634 Paris Cedex 13, France; Isabelle.Bloch@enst.fr

Fred L. Bookstein Institute of Gerontology, University of Michigan, Ann Arbor, MI 48109, USA; fred@brainmap.med.umich.edu

Michael Brammer Department of Biostatistics & Computing, Institute of Psychiatry, De Crespigny Park, London SE5 8AF, UK; m.brammer@iop.kcl.ac.uk

Edward Bullmore Department of Psychiatry, University of Cambridge, Addenbrooke's Hospital, Cambridge CB2 2QQ, UK; etb23@cam.ac.uk

Jin Cao Bell Laboratories, Lucent Technologies, 700 Mountain Avenue, Room 2C-260, Murray Hill, NJ 07974-2070, USA; cao@research.bell-labs.com

Olivier Coulon Department of Computer Sciences, University College London, Gower St., London WC1E 6BT, UK; O.Coulon@cs.ucl.ac.uk

Noel A.C. Cressie Department of Statistics, The Ohio State University, Columbus, OH 43210-1247, USA; ncressie@iastate.edu

Andreas Dabringhaus Refrather Straße 3, 51069 Köln, Germany; ad@online.de

Vincent Frouin Service Hospitalier Frédéric-Joliot, CEA, 91401 Orsay Cedex, France; frouin@shfj.cea.fr

Verena Gelpke Seminar für Statistik, ETH-Zentrum, 8092 Zürich, Switzerland; gelpke@stat.math.ethz.ch

Marc G. Genton Department of Statistics, North Carolina State University, Box 8203, Raleigh, NC 27695-8203, USA; genton@stat.ncsu.edu

Yves Goussard Département de génie électrique et de génie informatique and Institut de génie biomédical, École Polytechnique , C.P. 6079, Succ. Centre-Ville, Montréal (Québec) H3C 3A7, Canada; Yves@grbb.polymtl.ca

Xavier Guyon Statistique Appliquée, Modélisation Stochastique, Université Paris 1, 90, rue de Tolbiac, 75634 Paris Cedex 13, France; guyon@univ-paris1.fr

Cécile Hardouin Statistique Appliquée, Modélisation Stochastique, Université Paris 1, 90, rue de Tolbiac, 75634 Paris Cedex 13, France; hardouin@univ-paris1.fr

Hsin-Cheng Huang Academia Sinica, Taiwan

Jérôme Idier Laboratoire des signaux et systèmes, Supélec, plateau de Moulon, 91190 Gif-sur-Yvette, France; Jerome.Idier@lss.supelec.fr

Hans R. Künsch Seminar für Statistik, ETH-Zentrum, 8092 Zürich, Switzerland; kuensch@stat.math.ethz.ch

Soumendra N. Lahiri Department of Statistics, Iowa State University, Ames, IA 50011, USA; snlahiri@iastate.edu

Bernd Lütkenhöner Institute of Experimental Audiology, University of Münster, Münster, Germany; Lutkenh@uni-muenster.de

Jean-François Mangin Service Hospitalier Frédéric-Joliot, CEA, 91401 Orsay Cedex, France; mangin@shfj.cea.fr

Luis Gonzalo Perera Ferrer Centro de Matemática, Facultad de Ciencias, Universidad de la República, Iguá 4225, Montevideo 11400, Uruguay; gperera@cmat.edu.uy

Jean-Baptiste Poline Service Hospitalier Frédéric-Joliot, CEA, 91401 Orsay Cedex, France; poline@shfj.cea.fr

Andrea Ridolfi Laboratoire des signaux et systèmes, Supélec, plateau de Moulon, 91190 Gif-sur-Yvette, France; Andrea.Ridolfi@lss.supelec.fr

Thorsten Schormann C. und O. Vogt Institut für Hirnforschung, Heinrich-Heine-Universität, Postfach 101007, 40001 Düsseldorf, Germany; thorsten@hirn.uni-duesseldorf.de

John Suckling Department of Biostatistics & Computing, Institute of Psychiatry, De Crespigny Park, London SE5 8AF, UK; j.suckling@iop.kcl.ac.uk

Keith J. Worsley Department of Mathematics and Statistics, McGill University, 805 Sherbrooke St. West, Montreal, Quebec, H3A 2K6, Canada; worsley@math.mcgill.ca

Jun Zhu Department of Statistics, Iowa State University, Ames, IA 50011, USA; jzhu@iastate.edu

1

Asymptotic Distribution of the Empirical Cumulative Distribution Function Predictor under Nonstationarity

Jun Zhu, S. N. Lahiri, and Noel Cressie

ABSTRACT In this paper, we establish a functional central limit theorem for the empirical predictor of a spatial cumulative distribution function for a random field with a nonstationary mean structure. The type of spatial asymptotic framework used here is somewhat nonstandard; it is a mixture of the so called "infill" and "increasing domain" asymptotic structures. The choice of the appropriate scaling sequence for the empirical predictor depends on certain characteristics of the spatial sampling design generating the sampling sites. A precise description of this dependence is given. The results obtained here extend a similar result of Lahiri (1999) who considered only the stationary case.

1 Introduction

A spatial cumulative distribution function (SCDF) is a random distribution function that provides a statistical summary of a spatially distributed random process over a given region of interest. Let $\{Z(\mathbf{s}) : \mathbf{s} \in \mathbb{R}^d\}$ be a random field (r.f.) and let R be a region of interest in $\mathbb{R}^d$. Then, the SCDF of $\{Z(\cdot)\}$ over R is defined as

$$F_\infty(z; R) \equiv \int_R I(Z(\mathbf{s}) \leq z)\, d\mathbf{s}/|R|; \quad z \in \mathbb{R}, \tag{1}$$

where $I(A)$ is the indicator function, equal to 1 if A is true and equal to 0 otherwise, and $|R| \equiv \int_R d\mathbf{s}$ denotes the volume of R. Note that for each realization of the r.f., the SCDF F_∞ is a (right) continuous, nondecreasing function with $\lim_{z\to-\infty} F_\infty(z; R) = 0$ and $\lim_{z\to\infty} F_\infty(z; R) = 1$, and hence possesses all the properties of a cumulative distribution function on the real line $\mathbb{R}$. As a spatial statistic, it provides an effective summary of the distribution of the values of the Z process over a given region R. Use of the SCDF was proposed by Overton (1989) in the context of analysis of survey data from the National Surface Water Surveys. Many commonly used summary measures about the behavior of the spatial process $\{Z(\cdot)\}$ over the region R can be recovered from the knowledge of the SCDF. For example, if $Z(\mathbf{s})$ denotes an air-pollution measurement at location $\mathbf{s}$, then one might be interested in the average air-pollution level over a region of interest R, that is, in the regional mean

$$Z(R) \equiv \int_R Z(\mathbf{s})\, d\mathbf{s}/|R|.$$

It is easy to see that $Z(R) = \int z dF_\infty(z; R)$, which is simply the mean of the random distribution function $F_\infty(z; R)$. For more on the properties and uses of the SCDF as a spatial statistic, see Majure et al. (1995), Lahiri et al. (1999).

In practice, the Z process is observed only at finitely many locations $\{\mathbf{s}_1, \ldots, \mathbf{s}_n\}$ lying in the (sampling) region of interest $R \equiv R_n$ and we wish to infer about the SCDF $F_\infty(\cdot; R_n)$ based on the data $\{Z(\mathbf{s}_1), \ldots, Z(\mathbf{s}_n)\}$. Note that the SCDF, being a functional of the Z process over the entire region R_n, is then *unobservable.* A basic predictor of $F_\infty(\cdot; R_n)$ is given by the empirical cumulative distribution function (ECDF),

$$F_n(z) \equiv n^{-1} \sum_{i=1}^{n} I(Z(\mathbf{s}_i) \leq z); \quad z \in \mathbb{R}. \tag{2}$$

Properties of the ECDF as a predictor of $F_\infty(\cdot; R_n)$ depend on the spatial sampling design generating the sampling sites $\{\mathbf{s}_1, \ldots, \mathbf{s}_n\}$ and on the properties of the r.f. $Z(\cdot)$. Assuming that the r.f. is *stationary*, Lahiri (1999) proved a functional central limit theorem (FCLT) for the process,

$$\xi_n(z) \equiv b_n(F_n(z) - F_\infty(\cdot; R_n)); \quad z \in \mathbb{R}, \tag{3}$$

where b_n is a suitable sequence of scaling constants. In this article, we extend his results and prove a similar FCLT for the process ξ_n, allowing the process $Z(\cdot)$ to have a nonstationary mean structure. Specifically, we assume that the observed process $Z(\cdot)$ is of the form,

$$Z(\mathbf{s}) = g(\mathbf{s}) + \epsilon(\mathbf{s}); \quad \mathbf{s} \in \mathbb{R}^d, \tag{4}$$

where $g(\cdot)$ is a possibly unknown (deterministic) regression function and $\{\epsilon(\mathbf{s}) : \mathbf{s} \in \mathbb{R}^d\}$ is a zero-mean, stationary r.f. If there is additional information available on certain covariates $X_1(\mathbf{s}), \ldots, X_k(\mathbf{s})$, associated with the Z process at location $\mathbf{s} \in \mathbb{R}^d$ (e.g., from remote sensing satellite data), then the regression function $g(\mathbf{s})$ may be taken as a nonparametric or parametric function of the covariate values $x_1(\mathbf{s}), \ldots, x_k(\mathbf{s})$, such as

$$g(\mathbf{s}) = g_1(x_1(\mathbf{s}), \ldots, x_k(\mathbf{s})); \quad \mathbf{s} \in \mathbb{R}^d$$

for some function $g_1 : \mathbb{R}^k \to \mathbb{R}$ or

$$g(\mathbf{s}) = \beta_0 + \beta_1 x_1(\mathbf{s}) + \cdots + \beta_k x_k(\mathbf{s}); \quad \mathbf{s} \in \mathbb{R}^d,$$

where $(\beta_0, \beta_1, \ldots, \beta_k)' \in \mathbb{R}^{k+1}$ are regression parameters. In either case, since the ECDF given by (2) requires only the knowledge of the observed variables $\{Z(\mathbf{s}_1), \ldots, Z(\mathbf{s}_n)\}$, it can be used as a predictor of the unobservable SCDF.

The main result of the paper establishes weak convergence of the process $\xi_n(\cdot)$ to a Gaussian process $W(\cdot)$ (say) as random elements of $D[-\infty, \infty]$, the space of right continuous functions on $[-\infty, \infty]$ with left limits (cf. Billingsley, 1968). The limiting Gaussian process has continuous sample paths with probability 1. The spatial asymptotic structure used for proving the FCLT is somewhat nonstandard. It is a combination of what are known as the "increasing domain asymptotics" and the "infill asymptotics" (cf. Cressie, 1993), and is similar to the asymptotic structures used by Lahiri (1999) for the SCDF in the stationary case, and by Härdle and Tuan (1986), Hall and Patil (1994) for certain other inference problems involving temporal and spatial processes. We describe the spatial sampling design and the asymptotic structure in more detail in Section 2. As in the stationary case, the sampling design plays a critical role in determining the accuracy of the predictor F_n. Indeed, as follows from the main result of the paper, the order of the scaling constant b_n in (3) depends on the sampling design. Since the asymptotic variance of $[F_n(\cdot) - F_\infty(\cdot; R_n)]$ is of the order b_n^{-2}, more accurate prediction of the SCDF $F_\infty(\cdot; R_n)$ is possible for a sampling design that produces a larger scaling constant b_n. We provide a precise description of the relation between the order of the scaling constant b_n and the relevant characteristics of the sampling design that determine it. It follows from this that F_n is a more accurate predictor under sampling designs that possess certain symmetry properties. More details are given in Section 3.

In proving the FCLT for ξ_n in the nonstationary case, we have to contend with some technical difficulties that are not encountered in the stationary case. For example, in the stationary case, the expected value of $\xi_n(z)$ is zero for all z. However, under model (4), this is no longer true for a general regression function $g(\mathbf{s})$. Nonetheless, under appropriate regularity conditions, the process ξ_n is shown to converge weakly to a *zero-mean* Gaussian process as in the stationary case. Thus, one may use the FCLT to construct large sample inference procedures (e.g., prediction intervals) for the unobservable SCDF $F_\infty(\cdot; R_n)$ for *nonstationary* r.f.s given by model (4).

The rest of the paper is organized as follows. In Section 2, we describe the details of the sampling design and spatial asymptotic structure. In Section 3, we state the assumptions and the main results of the paper. In Section 4, we provide the details of the proofs.

2 The Asymptotic Framework

There are different ways of studying asymptotic properties of estimators and predictors based on spatial data. However, they seem to stem from essentially two basic structures, known as the "increasing domain asymptotic structure" and the "infill asymptotic structure" (cf. Cressie, 1993). When all sampling sites are separated by a fixed positive distance, and the sampling region $R = R_n$ becomes unbounded as the sample size increases, the resulting structure leads to what is known as the "increasing domain asymptotics" (cf. Cressie, 1993). This is the most common framework used for asymptotics for spatial data observed on a given lattice. In contrast, when samples of increasing size are collected from within a sampling region R that does not become unbounded with the sample size, one obtains the "infill" structure. Various asymptotic structures for spatial data arise from a varying degree of combination of these two basic structures. As in Lahiri (1999), we assume here a "mixed" asymptotic structure where we let the sampling region $R = R_n$ grow, and at the same time, allow "infilling" of any fixed bounded subregion of R_n. For a discussion of why such a structure is natural for the present problem, see Remark 1 below. A similar "mixed" structure has been used by Härdle and Tuan (1986) in the context of smoothing time-series data and by Hall and Patil (1994) in the context of nonparametric estimation of the auto-covariance function of a r.f.

We begin with a description of the sampling region R_n which specifies the "increasing domain" component of our asymptotic structure. Let R_0 be an open connected subset of $(-\frac{1}{2}, \frac{1}{2}]^d$ containing the origin such that the following regularity condition holds:

Condition 1. For any sequence of positive real numbers, $\{a_n\}$, with $\lim_{n\to\infty} a_n = 0$, the number of cubes generated by the lattice $a_n\mathbb{Z}^d$ that intersect both R_0 and R_0^c is $O\big((a_n^{-1})^{d-1}\big)$ as $n \to \infty$. Here, $\mathbb{Z}$ denotes the set of all integers.

Next, let $\{\lambda_n\}$ be a sequence of real numbers going to infinity with n. Then, the sampling region R_n is obtained by "inflating" the set R_0 by the scaling sequence λ_n; that is,

$$R_n = \lambda_n R_0.$$

Since the origin is assumed to lie inside R_0, the shape of the sampling region is preserved for different values of n. Furthermore, Condition 1 on R_0 guarantees that the effect of the data points lying on the boundary of R_n is negligible compared to the totality of all data values. This formulation is similar to that of Sherman and Carlstein (1994) and Hall and Patil (1994), and allows the sampling region R_n to have a fairly irregular shape. Some common examples of such regions are spheres, ellipsoids, polyhedrons, and star-shaped regions (which can be nonconvex sets with irregular boundaries).

Next we describe the "infill" component of our asymptotic framework. Let $\{h_n\}$ be a sequence of real numbers such that $h_n \downarrow 0$ as $n \to \infty$, and let $\mathbf{c}$ be an arbitrary point in the interior of the unit cube $\Delta_0 \equiv (0,1]^d$. We assume that the sampling sites $\{\mathbf{s}_1, \ldots, \mathbf{s}_n\}$ are given by the points on the "shifted and scaled" integer grid $\{(\mathbf{i}+\mathbf{c})h_n : \mathbf{i} \in \mathbb{Z}^d\}$ that lie within the sampling region R_n; that is,

$$\{\mathbf{s}_1, \ldots, \mathbf{s}_n\} \equiv \{(\mathbf{i}+\mathbf{c})h_n : \mathbf{i} \in \mathbb{Z}^d\} \cap R_n.$$

Since h_n tends to zero with n, the sampling sites "fill in" any given subregion of R_n with an increasing density. Thus, the asymptotic framework we consider here is a mixture of "infill" and "increasing domain" structures, with the sampling sites generated by a nonstochastic uniform sampling design whose "starting" point $\mathbf{c}h_n$ is an arbitrary point in the cube $(0, h_n]^d$. Indeed, as explained in the next section, it is the choice of the "design parameter" $\mathbf{c}$ that critically determines the order of the scaling constant b_n in (3) and, hence, the accuracy of the predictor F_n.

Remark 1. The mixed asymptotic structure described in the paper appears to be the natural one for investigating large-sample properties of the statistic $F_n(z)$ as a *predictor* of the unobservable SCDF $F_\infty(z; R_n)$. Since the SCDF $F_\infty(z; R_n)$ is defined in terms of an integral on R_n, it cannot be *predicted consistently* without infilling. On the other hand, if the region R_n remains bounded, then there is not enough information to allow *consistent estimation* of population quantiles (cf. Lahiri, 1996) for constructing prediction bands for the SCDF $F_\infty(z; R_n)$. Thus, both the "infill" and the "increasing domain" components must be present in the spatial asymptotic structure for constructing meaningful inference procedures based on the FCLT of the paper.

3 Main Results

For clarity of exposition, we divide this section into three parts. In Section 3.1, we collect the necessary notation. In Section 3.2, we state the assumptions, and in Section 3.3, we state the main result of the paper.

3.1 Notation

For a vector $\mathbf{x} = (x_1, \ldots, x_k)' \in \mathbb{R}^k$ $(k \geq 1)$, let $\|\mathbf{x}\| \equiv (\sum_{i=1}^k x_i^2)^{1/2}$ and $|\mathbf{x}| \equiv \sum_{i=1}^k |x_i|$ denote the ℓ^2 and ℓ^1 norm of $\mathbf{x}$. For a countable set J, we let $|J|$ denote the cardinality of J, and for an uncountable set $A \subset \mathbb{R}^k$, $|A|$ denotes the volume, that is, the Lebesgue measure of A. Let $\mathbb{Z}^+$ be the set of all nonnegative integers. For $\boldsymbol{\alpha} = (\alpha_1, \ldots, \alpha_k)' \in (\mathbb{Z}^+)^k$, $\mathbf{x} = (x_1, \ldots, x_k)' \in \mathbb{R}^k$, and for a function $f\colon \mathbb{R}^k \to \mathbb{R}$, we define

$\mathbf{x}^{\boldsymbol{\alpha}} \equiv \prod_{i=1}^{k} x_i^{\alpha_i}$, $\boldsymbol{\alpha}! \equiv \prod_{i=1}^{k} \alpha_i!$, and $D^{\boldsymbol{\alpha}} f \equiv D_1^{\alpha_1} \ldots D_k^{\alpha_k}$, where D_j denotes partial derivative with respect to x_j. For notational simplicity, we shall also write $D_{\mathbf{x}}^{\boldsymbol{\alpha}} D_{\mathbf{y}}^{\boldsymbol{\beta}} f(\mathbf{x}, \mathbf{y}) \equiv D^{(\boldsymbol{\alpha}', \boldsymbol{\beta}')'} f(\mathbf{x}, \mathbf{y})$ for $\mathbf{x} \in \mathbb{R}^p$, $\boldsymbol{\alpha} \in (\mathbb{Z}^+)^p$, $\mathbf{y} \in \mathbb{R}^q$, $\boldsymbol{\beta} \in (\mathbb{Z}^+)^q$, and $f: \mathbb{R}^{p+q} \to \mathbb{R}$, $p \geq, q \geq 1$.

Under the assumption of stationarity of the r.f. $\{\epsilon(\mathbf{s}) : \mathbf{s} \in \mathbb{R}^d\}$, let the location-invariant marginal distribution function of $\epsilon(\mathbf{s})$ and the joint bivariate distribution function of $\epsilon(\mathbf{0})$ and $\epsilon(\mathbf{s})$ be denoted as,

$$F(z) \equiv P(\epsilon(\mathbf{0}) \leq z),$$
$$G(z_1, z_2; \mathbf{s}) \equiv P(\epsilon(\mathbf{0}) \leq z_1, \epsilon(\mathbf{s}) \leq z_2),$$

where $z, z_1, z_2 \in \mathbb{R}$, and $\mathbf{s} \in \mathbb{R}^d$. Also, let

$$G_1(z_1, z_2, z_3, z_4; \mathbf{s}) \equiv P(z_1 < \epsilon(\mathbf{0}) \leq z_2, z_3 < \epsilon(\mathbf{s}) \leq z_4);$$

$z_1, z_2, z_3, z_4 \in \mathbb{R}$, $\mathbf{s} \in \mathbb{R}^d$. Then, for any *given* $z_1, z_2 \in \mathbb{R}$,

$$P(Z(\mathbf{x}) \leq z_1, Z(\mathbf{y}) \leq z_2) = G(z_1 - g(\mathbf{x}), z_2 - g(\mathbf{y}); \mathbf{y} - \mathbf{x}),$$

which is abbreviated to $\mathcal{G}(\mathbf{x}, \mathbf{y})$ to simplify notation and, similarly,

$$P(z_1 < Z(\mathbf{x}) \leq z_2, z_1 < Z(\mathbf{y}) \leq z_2)$$
$$= G_1(z_1 - g(\mathbf{x}), z_2 - g(\mathbf{x}), z_1 - g(\mathbf{y}), z_2 - g(\mathbf{y}); \mathbf{y} - \mathbf{x}),$$

which is abbreviated to $\mathcal{G}_1(\mathbf{x}, \mathbf{y})$. It is easy to see that the partial derivatives of the function $\mathcal{G}$ with respect to $\mathbf{x}$ and $\mathbf{y}$ are of the form $f(\mathbf{x}, \mathbf{y}; \mathbf{x} - \mathbf{y})$ for some function f depending on z_1, z_2. For example, for $|\boldsymbol{\alpha}| = 1$,

$$D_{\mathbf{x}}^{\boldsymbol{\alpha}} \mathcal{G}(\mathbf{x}, \mathbf{y}) = D_1 G(z_1 - g(\mathbf{x}), z_2 - g(\mathbf{y}); \mathbf{y} - \mathbf{x}) \cdot [-D^{\boldsymbol{\alpha}} g(\mathbf{x})]$$
$$+ D^{(0,0,\boldsymbol{\alpha}')'} G(z_1 - g(\mathbf{x}), z_2 - g(\mathbf{y}); \mathbf{y} - \mathbf{x}) \cdot [-1].$$

In view of this observation, we shall also use the alternative notation

$$L_{\boldsymbol{\alpha}, \boldsymbol{\beta}}(z_1, z_2; \mathbf{x}, \mathbf{y}, \mathbf{x} - \mathbf{y}) \equiv D_{\mathbf{x}}^{\boldsymbol{\alpha}} D_{\mathbf{y}}^{\boldsymbol{\beta}} \mathcal{G}(\mathbf{x}, \mathbf{y});$$

$\mathbf{x}, \mathbf{y} \in \mathbb{R}^d$, $\boldsymbol{\alpha}, \boldsymbol{\beta} \in (\mathbb{Z}^+)^d$.

Let $\mathcal{L}_2'(A)$ be the collection of all random variables with zero mean and finite second moment that are measurable with respect to the σ-field generated by $\{\epsilon(\mathbf{s}) : \mathbf{s} \in A\}$; $A \subset \mathbb{R}^d$. For $A, B \subset \mathbb{R}^d$, write

$$\rho_1(A, B) \equiv \sup \left\{ |E\xi\eta| / (E\xi^2)^{1/2} (E\eta^2)^{1/2} : \xi \in \mathcal{L}_2'(A), \eta \in \mathcal{L}_2'(B) \right\}.$$

Then, define the ρ-mixing coefficient of the r.f. $\{\epsilon(\mathbf{s}) : \mathbf{s} \in \mathbb{R}^d\}$ by

$$\rho(k; m) \equiv \sup\{\rho_1(A, B) : |A| \leq m, |B| \leq m, d(A, B) \geq k\}, \qquad (5)$$

where the supremum is taken over all rectangles A, B in $\mathbb{R}^d$, and where $d(A, B)$ denotes the distance between the sets $A, B \subset \mathbb{R}^d$ in the $|\cdot|$-norm, given by $d(A, B) \equiv \inf\{|\mathbf{x}-\mathbf{y}| : \mathbf{x} \in A, \mathbf{y} \in B\}$. We recall that $\mathbf{c}$ determines the starting-sampling site and we let $\mathbf{c}_0$ denote the center of $\Delta_0 = (0, 1]^d$. We define the index κ as a function of the vector $\mathbf{c}$, where $\kappa \equiv 4$ if $\mathbf{c} = \mathbf{c}_0$ and $\kappa \equiv 2$ if $\mathbf{c} \neq \mathbf{c}_0$. Finally, for $\boldsymbol{\alpha}_1, \boldsymbol{\alpha}_2 \in (\mathbb{Z}^+)^d$, let

$$a(\boldsymbol{\alpha}_1, \boldsymbol{\alpha}_2) \equiv \int_{\Delta_0} \int_{\Delta_0} (\mathbf{x} - \mathbf{c})^{\boldsymbol{\alpha}_1} (\mathbf{y} - \mathbf{c})^{\boldsymbol{\alpha}_2} \, d\mathbf{x} \, d\mathbf{y}.$$

We are now ready to state the assumptions to be used to obtain the main result of the paper.

3.2 Assumptions

Assumption 1. There exist positive real numbers C, ζ, θ with $\zeta > 3d$, $\theta d < \zeta$ such that the ρ-mixing coefficient of the random field $\{\epsilon(\mathbf{s}) : \mathbf{s} \in \mathbb{R}^d\}$, given by (5), satisfies,

$$\rho(k; m) \leq C k^{-\zeta} m^{\theta}. \tag{6}$$

Assumption 2. The random field $\{\epsilon(\mathbf{s}) : \mathbf{s} \in \mathbb{R}^d\}$ is stationary and the marginal distribution function F is $(\kappa/2 + 1)$ times differentiable with bounded derivatives. In addition, there exists a constant K such that

$$|F(F^{-1}(u_1) - a) - F(F^{-1}(u_2) - a)| \leq K |u_1 - u_2|,$$

for all $u_1, u_2 \in (0, 1)$ and $a \in \mathbb{R}$.

Assumption 3. The function g is $(\kappa+1)$ times differentiable with bounded partial derivatives, and $\int |D^{\boldsymbol{\alpha}} g(\mathbf{x})| \, d\mathbf{x} < \infty$, for all $|\boldsymbol{\alpha}| = 2, \ldots, 1 + \kappa/2$.

Assumption 4. (i) The function G is $(\kappa + 1)$ times differentiable on $\mathbb{R}^{d+2}$ with bounded partial derivatives, and $\sup_{u,v} D_{\mathbf{y}}^{\boldsymbol{\alpha}} G(u, v; \mathbf{y})$ is Lebesgue integrable on $\mathbb{R}^d$ for all $|\boldsymbol{\alpha}| = \kappa + 1$.

(ii) For all $z_1, z_2 \in \mathbb{R}$, as $n \to \infty$,

$$|R_n|^{-1} \int_{R_n - \mathbf{y}} L_{\boldsymbol{\alpha}_1, \boldsymbol{\alpha}_2}(z_1, z_2; \mathbf{x}, \mathbf{y} + \mathbf{x}, \mathbf{y}) \, d\mathbf{x} \to \mathcal{L}_{\boldsymbol{\alpha}_1, \boldsymbol{\alpha}_2}(z_1, z_2; \mathbf{y}),$$

for all $\mathbf{y} \in \mathbb{R}^d$ and $|\boldsymbol{\alpha}_1| = |\boldsymbol{\alpha}_2| = \kappa/2$.

Assumption 5. There exist constants $K > 0$, $\frac{1}{2} < \gamma < 1$, such that

$$\sum_{|\boldsymbol{\alpha}_1| + |\boldsymbol{\alpha}_2| = 2}^{\kappa+1} D_{\mathbf{x}}^{\boldsymbol{\alpha}_1} D_{\mathbf{y}}^{\boldsymbol{\alpha}_2} \mathcal{G}_1(\mathbf{x}, \mathbf{y}) \leq K |F(z_2) - F(z_1)|^{\gamma},$$

for all $z_1, z_2 \in \mathbb{R}$, $\mathbf{x}, \mathbf{y} \in \mathbb{R}^d$.

Assumption 6. The growth rate and the infill rate satisfy:

$$\left(h_n^{(2\gamma+1)\kappa+1}\lambda_n^d\right)^{-1} + (h_n\lambda_n/\log\lambda_n)^{-1} \to 0, \quad \text{as } n \to \infty,$$

where γ is as in Assumption 5.

We now briefly discuss Assumptions 1–6 and their implications. Assumptions similar to Assumptions 1, 5, 6 have been used by Lahiri (1999) in the stationary case (i.e., where the function g is constant) and are discussed in detail therein. Hence, we concentrate on the remaining assumptions. The first part of Assumption 2 is a smoothness condition on the marginal CDF $F(\cdot)$ that can be verified directly. The second part is a type of uniform Hölder's condition and it holds if, for example, F has a density that is compactly supported and is bounded away from zero. Assumption 3 exclusively relates to the regression function g. The integrability conditions in Assumptions 3 and 4(i) jointly imply the uniform integrability of the partial derivatives of the function $G(\cdot,\cdot;\mathbf{s})$ over R_n, which is crucial for the establishment of the convergence of finite-dimensional distributions of the process ξ_n. Assumption 4(ii) is a version of the well-known Grenander's condition (Grenander, 1954, Anderson, 1971) that is commonly assumed in the context of inference problems involving regression models with dependent errors.

We are now ready to state the main result of the paper.

3.3 The Main Result

Define the scaling constants $\{b_n : n \geq 1\}$ by

$$b_n \equiv \lambda_n^{d/2} h_n^{-2} \quad \text{if } \mathbf{c} = \mathbf{c}_0 \tag{7}$$

and

$$b_n \equiv \lambda_n^{d/2} h_n^{-1} \quad \text{if } \mathbf{c} \neq \mathbf{c}_0. \tag{8}$$

Since h_n tends to zero with n, it follows that b_n has a larger order of magnitude in the case $\mathbf{c} = \mathbf{c}_0$ than in the case $\mathbf{c} \neq \mathbf{c}_0$. Also, note that being the scaled difference of two cumulative distribution functions, the sample paths of the process ξ_n lie in $D[-\infty,\infty]$ with probability one. The following result (proved in Section 4) demonstrates the weak convergence of $\{\xi_n : n \geq 1\}$ as random elements of the space $D[-\infty,\infty]$, equipped with the Skorohod metric (cf. Billingsley, 1968).

Theorem 1. *Suppose that Condition* 1 *and Assumptions* 1–6 *hold. Then,*

$$\xi_n \xrightarrow{\mathcal{D}} W,$$

where $\xrightarrow{\mathcal{D}}$ *denotes weak convergence of random elements in* $D[-\infty,\infty]$ *as* $n \to \infty$, *and* $W(\cdot)$ *is a zero-mean Gaussian process with covariance function,*

$$\sigma(z_1, z_2) \equiv |R_0|^{-1} \sum_{\substack{|\boldsymbol{\alpha}_1|=\kappa/2 \\ |\boldsymbol{\alpha}_2|=\kappa/2}} (\boldsymbol{\alpha}_1!\boldsymbol{\alpha}_2!)^{-1} a(\boldsymbol{\alpha}_1, \boldsymbol{\alpha}_2) \int_{\mathbb{R}^d} \mathcal{L}_{\boldsymbol{\alpha}_1\boldsymbol{\alpha}_2}(z_1, z_2; \mathbf{y})\, d\mathbf{y}.$$

Moreover, $W(-\infty) = W(\infty) = 0$ *a.s., and* $W(z)$ *has continuous sample paths with probability* 1.

Proof. See Section 4. □

Theorem 1 shows that the process ξ_n converges weakly to a Gaussian process as random elements of $D[-\infty,\infty]$ even under *nonstationarity* of the Z process. This extends a result of Lahiri (1999), who proved a similar FCLT for ξ_n under stationarity. Although the limiting Gaussian processes have mean zero in both cases, they have different covariance functions. In the stationary case, the covariance function depends only on the bivariate distribution function $G(z_1, z_2; \mathbf{s})$. In contrast, as Theorem 1 shows, the covariance function of the limiting Gaussian process in the nonstationary case depends in a nontrivial way on the mean function g as well as on the bivariate distribution function $G(z_1, z_2; \mathbf{s})$. Although it may be somewhat counter-intuitive, Theorem 1 provides an instance where second-order properties (viz., the covariance function $\sigma(z_1, z_2)$) of the limit distribution W depends on the first-order (mean) structure of the observed r.f. $\{Z(\mathbf{s})\}$.

Theorem 1 also shows how the scaling constant b_n is determined by the design parameter $\mathbf{c}$. When $\mathbf{c} = \mathbf{c}_0$, the sampling sites under the uniform spatial design are located at the *mid-points* of the rectangular tessellation $\{(\mathbf{i} + \Delta_0)h_n : \mathbf{i} \in \mathbb{Z}^d\}$ of $\mathbb{R}^d$. In this case, the scaling sequence $\{b_n\}$ is of a larger order of magnitude and, hence, the ECDF is a more accurate predictor of the SCDF $F_\infty(\cdot; R_n)$. For any other choice of $\mathbf{c} \in \Delta_0$, the symmetry of the sampling sites with respect to the rectangular tessellation is lost, resulting in less accurate predictions. Thus, for a *given rate of infilling* h_n, if the statistician has the option of choosing the starting point $\mathbf{c}h_n$ of the rectangular sampling grid, he or she should choose $\mathbf{c} = \mathbf{c}_0$ to ensure a better prediction of $F_\infty(\cdot; R_n)$. Note that this implicitly requires the smoothness conditions of Assumptions 2–5 to hold with $\kappa = 4$, which are more stringent than the smoothness conditions with $\kappa = 2$. Thus, for a given rate of infilling h_n, $\mathbf{c} = \mathbf{c}_0$ is the best choice, provided the functions F, g, G have enough smoothness to satisfy Assumptions 2–5 with $\kappa = 4$.

On the other hand, note that for a given growth rate λ_n, Assumption 6 allows a more dense infilling (i.e., a smaller h_n) when $\mathbf{c} \neq \mathbf{c}_0$. This seems to imply that the *maximal* scaling sequence b_n in (8) for the case $\mathbf{c} \neq \mathbf{c}_0$ may be "larger" than that in (7) for the case $\mathbf{c} = \mathbf{c}_0$. However, it is easy to see that for a given sequence λ_n, the *maximal* scaling sequence for $\mathbf{c} \neq \mathbf{c}_0$

is $\lambda_n^{[d/2]+[d/\{1+2(2\gamma+1)\}]}$, as compared to $\lambda_n^{[d/2]+[2d/\{1+4(2\gamma+1)\}]}$ for the case $\mathbf{c} = \mathbf{c}_0$, and that the second sequence grows to infinity at a *faster rate* for all $d \geq 1$ and all $\gamma > \frac{1}{2}$. Consequently, even when maximal amounts of infilling are used, the choice $\mathbf{c} = \mathbf{c}_0$ still yields a more accurate predictor, provided that the smoothness conditions of Assumptions 2–5 hold with $\kappa = 4$.

4 Proofs

We begin with a brief description of the steps used for proving Theorem 1. The proof of the FCLT in Theorem 1 follows the standard approach of establishing (i) tightness of the process ξ_n and (ii) weak convergence of its finite-dimensional distributions. We separate out the lengthy steps in the proofs of (i) and (ii) into a few preparatory lemmas (viz., Lemmas 1–5). Lemmas 1 and 4 give certain moment bounds. Note that unlike the constant-mean stationary case, $E\xi_n(z)$ is *not* necessarily zero for a general regression function g. For convergence to a *zero-mean* Gaussian process, we need to show that the bias of F_n vanishes sufficiently fast. This is done in Lemma 3, which establishes asymptotic negligibility of the bias of $\xi_n(z)$. The (co-)variance of $[F_n(z) - F_\infty(z; R_n)]$ is found in Lemma 2. Indeed, the proof of Lemma 2 reveals the effect of the vector $\mathbf{c}$ on the scaling sequence $\{b_n\}$. Together with Lemma 3, it yields the expression for the covariance function of the limiting Gaussian process in Theorem 1. Lemma 5 proves weak convergence of its finite-dimensional distributions of ξ_n (i.e., part (ii)) using Lemmas 1–4. Finally, tightness of the process ξ_n (i.e., part (i)) is established by verifying a known sufficient condition for tightness of $D[0,1]$-valued processes from Billingsley (1968).

For proving the lemmas and the theorem, we need to introduce some more notation at this point. Let $\mathbf{s_i} \equiv (\mathbf{i} + \mathbf{c})h_n$; $\mathbf{i} \in \mathbb{Z}^d$, and let $J_n \equiv \{\mathbf{i} \in \mathbb{Z}^d : (\mathbf{i} + \mathbf{c})h_n \in R_n\}$ denote the subset of indices in $\mathbb{Z}^d$ corresponding to the sampling sites in R_n. Let $\Gamma(\mathbf{i}) \equiv (\mathbf{i} + \Delta_0)h_n$; $\mathbf{i} \in \mathbb{Z}^d$. Next, define the sets $J_{1n} \equiv \{\mathbf{i} \in J_n : \Gamma(\mathbf{i}) \subset R_n\}$ and $J_{2n} \equiv \{\mathbf{i} \in J_n : \Gamma(\mathbf{i}) \cap R_n \neq \varnothing, \Gamma(\mathbf{i}) \cap R_n^c \neq \varnothing\}$ which, respectively, denote the collection of indices of unit cubes $\Gamma(\mathbf{i})$ that are completely contained in R_n and that are on the boundary of R_n. Note that $J_n = J_{1n} \cup J_{2n}$. In a similar manner, define the regions $R_{1n} \equiv \cup_{\mathbf{i} \in J_{1n}} \Gamma(\mathbf{i})$ and $R_{2n} \equiv \cup_{\mathbf{i} \in J_{2n}} \Gamma(1, \mathbf{i})$, where $\Gamma(1, \mathbf{i}) \equiv \Gamma(\mathbf{i}) \cap R_n$. For a real number a, let $a^+ \equiv \max\{a, 0\}$. Let C, $C(\cdot)$ to denote generic positive constants that depend on their arguments (if any). Also, unless otherwise specified, limits in order symbols are taken by letting n tend to ∞. For example, we would write $a_n = o(1)$ to mean $a_n \to 0$ as $n \to \infty$.

Lemma 1. *Let* $B_n \equiv t_n B_0$ *and* $Z(\mathbf{i}) \equiv \int f_{\mathbf{i}}(Z(\mathbf{s})) I(\mathbf{s} \in (\mathbf{i} + \Delta_0) \cap B_n)\, ds$; $\mathbf{i} \in \tilde{J}_n \equiv \{\mathbf{i} \in \mathbb{Z}^d : (\mathbf{i} + \Delta_0) \cap B_n \neq \phi\}$ *be random variables satisfying*

$$EZ(\mathbf{i}) = 0, \; |Z(\mathbf{i})| \leq 1, \; \textit{and } E|Z(\mathbf{i})|^2 \leq \delta_n, \quad \textit{for all } \mathbf{i} \in \tilde{J}_n,$$

where $t_n \to \infty$ and B_0 is a Borel subset of $(-\frac{1}{2}, \frac{1}{2}]^d$ that satisfies the same boundary condition (Condition 1) as the set R_0. Then, under Assumption 1,

$$E\Big(\sum_{\mathbf{i}\in\tilde{J}_n} Z(\mathbf{i})\Big)^4 \le C\big(d,\rho(\cdot)\big)\big(t_n^{2d}\delta_n^2 + t_n^d\delta_n\big).$$

Proof. In the stationary case (i.e., for $g(\mathbf{s}) \equiv 0$ for all $\mathbf{s} \in \mathbb{R}^d$), this lemma is proved in Lahiri (1999) (cf. Lemma 4.1, *op. cit.*). It may be checked that the arguments in the proof go through with minor changes even for the nonstationary case of model (4). We omit the details. □

Lemma 2. *Suppose that Condition 1 and Assumptions 1, 3, 4, and 6 hold. Then, for $z_1, z_2 \in \mathbb{R}$,*

$$\begin{aligned}
&E\Big(\sum_{\mathbf{i}\in J_n} Y_1(\mathbf{i})\Big)\Big(\sum_{\mathbf{i}\in J_n} Y_2(\mathbf{i})\Big)\\
&= |R_0|\lambda_n^d h_n^\kappa \sum_{\substack{|\boldsymbol{\alpha}_1|=\kappa/2\\ |\boldsymbol{\alpha}_2|=\kappa/2}} (\boldsymbol{\alpha}_1!\boldsymbol{\alpha}_2!)^{-1} a(\boldsymbol{\alpha}_1,\boldsymbol{\alpha}_2)\int_{\mathbb{R}^d}\mathcal{L}_{\boldsymbol{\alpha}_1\boldsymbol{\alpha}_2}(z_1,z_2;\mathbf{y})\,d\mathbf{y}\big(1+o(1)\big),
\end{aligned}$$

where $Y_j(\mathbf{i}) \equiv \int_{\Gamma(1,\mathbf{i})}\big(I(Z(\mathbf{s_i}) \le z_j) - I(Z(\mathbf{s}) \le z_j)\big)\,d\mathbf{s}$; $j = 1, 2$.

Proof. Here, we assume $\kappa = 4$. The proof for the case $\kappa = 2$ is similar. Let $I_n \equiv E\big(\sum_{\mathbf{i}\in J_n} Y_1(\mathbf{i})\big)\big(\sum_{\mathbf{i}\in J_n} Y_2(\mathbf{i})\big)$. Then, using Taylor's expansion, we have,

$$\begin{aligned}
I_n &= \sum_{\mathbf{i}\in J_n}\sum_{\mathbf{j}\in J_n}\int_{\Gamma(1,\mathbf{i})}\int_{\Gamma(1,\mathbf{j})} \mathcal{G}(\mathbf{s_i},\mathbf{s_j}) - \mathcal{G}(\mathbf{x},\mathbf{s_j}) - \mathcal{G}(\mathbf{s_i},\mathbf{y}) + \mathcal{G}(\mathbf{x},\mathbf{y})\,d\mathbf{x}\,d\mathbf{y}\\
&= \sum_{\mathbf{i}\in J_n}\sum_{\mathbf{j}\in J_n}\int_{\Gamma(1,\mathbf{i})}\int_{\Gamma(1,\mathbf{j})}\sum_{k=1}^{4}\Bigg[\sum_{|\boldsymbol{\alpha}|=k}\big(-D_{\mathbf{x}}^{\boldsymbol{\alpha}}\mathcal{G}(\mathbf{s_i},\mathbf{s_j})(\mathbf{x}-\mathbf{s_i})^{\boldsymbol{\alpha}}/\boldsymbol{\alpha}!\\
&\qquad\qquad - D_{\mathbf{y}}^{\boldsymbol{\alpha}}\mathcal{G}(\mathbf{s_i},\mathbf{s_j})(\mathbf{y}-\mathbf{s_j})^{\boldsymbol{\alpha}}/\boldsymbol{\alpha}!\big)\\
&\qquad + \sum_{|\boldsymbol{\alpha}_1|+|\boldsymbol{\alpha}_2|=k}\big(D_{\mathbf{x}}^{\boldsymbol{\alpha}_1}D_{\mathbf{y}}^{\boldsymbol{\alpha}_2}\mathcal{G}(\mathbf{s_i},\mathbf{s_j})(\mathbf{x}-\mathbf{s_i})^{\boldsymbol{\alpha}_1}(\mathbf{y}-\mathbf{s_j})^{\boldsymbol{\alpha}_2}/\boldsymbol{\alpha}_1!\boldsymbol{\alpha}_2!\big)\Bigg]\,d\mathbf{x}\,d\mathbf{y}\\
&\quad + \sum_{\mathbf{i}\in J_n}\sum_{\mathbf{j}\in J_n}\int_{\Gamma(1,\mathbf{i})}\int_{\Gamma(1,\mathbf{j})} r_n(\mathbf{s_i},\mathbf{s_j};\mathbf{x},\mathbf{y})\,d\mathbf{x}\,d\mathbf{y}\\
&\equiv I_{1n} + I_{2n},
\end{aligned}\tag{9}$$

where $r_n(\mathbf{s_i},\mathbf{s_j};\mathbf{x},\mathbf{y})$ are the remainder terms.

Next, we split the leading term I_{1n} into two parts. Let I_{11n} denote the term where the summations over both $\mathbf{i}$ and $\mathbf{j}$ extend over the index set J_{1n}, and let $I_{12n} \equiv I_{1n} - I_{11n}$ be defined by subtraction. Then, noting that the

$\Gamma(1,\mathbf{i}) = \Gamma(\mathbf{i})$ for all $\mathbf{i} \in J_{1n}$, and that the terms corresponding to $|\boldsymbol{\alpha}| \le 3$ integrate to 0, by Assumption 4, we obtain,

$$\begin{aligned}
I_{11n} &= \sum_{\mathbf{i}\in J_{1n}} \sum_{\mathbf{j}\in J_{1n}} \int_{\Gamma(\mathbf{i})} \int_{\Gamma(\mathbf{j})} \sum_{k=1}^{4} \Big[\sum_{|\boldsymbol{\alpha}|=k} (-D_{\mathbf{x}}^{\boldsymbol{\alpha}} \mathcal{G}(\mathbf{s_i},\mathbf{s_j})(\mathbf{x}-\mathbf{s_i})^{\boldsymbol{\alpha}}/\boldsymbol{\alpha}! \\
&\qquad\qquad - D_{\mathbf{y}}^{\boldsymbol{\alpha}} \mathcal{G}(\mathbf{s_i},\mathbf{s_j})(\mathbf{y}-\mathbf{s_j})^{\boldsymbol{\alpha}}/\boldsymbol{\alpha}!) \\
&\quad + \sum_{|\boldsymbol{\alpha}_1|+|\boldsymbol{\alpha}_2|=k} (D_{\mathbf{x}}^{\boldsymbol{\alpha}_1} D_{\mathbf{y}}^{\boldsymbol{\alpha}_2} \mathcal{G}(\mathbf{s_i},\mathbf{s_j})(\mathbf{x}-\mathbf{s_i})^{\boldsymbol{\alpha}_1}(\mathbf{y}-\mathbf{s_j})^{\boldsymbol{\alpha}_2}/\boldsymbol{\alpha}_1!\boldsymbol{\alpha}_2!) \Big] \, d\mathbf{x}\, d\mathbf{y} \\
&= h_n^{2d+4} \sum_{\substack{|\boldsymbol{\alpha}_1|=2\\|\boldsymbol{\alpha}_2|=2}} \Big[\sum_{\mathbf{i}\in J_{1n}} \sum_{\mathbf{j}\in J_{1n}} 4^{-1} D_{\mathbf{x}}^{\boldsymbol{\alpha}_1} D_{\mathbf{y}}^{\boldsymbol{\alpha}_2} \mathcal{G}(\mathbf{s_i},\mathbf{s_j}) \\
&\qquad\qquad \times \int_{\Delta_0}\int_{\Delta_0} (\mathbf{x}-\mathbf{c})^{\boldsymbol{\alpha}_1}(\mathbf{y}-\mathbf{c})^{\boldsymbol{\alpha}_2}\, d\mathbf{x}\, d\mathbf{y} \Big] \\
&= h_n^4 \sum_{\substack{|\boldsymbol{\alpha}_1|=2\\|\boldsymbol{\alpha}_2|=2}} \Big[4^{-1} a(\boldsymbol{\alpha}_1,\boldsymbol{\alpha}_2) \int_{R_n}\int_{R_n} D_{\mathbf{x}}^{\boldsymbol{\alpha}_1} D_{\mathbf{y}}^{\boldsymbol{\alpha}_2} \mathcal{G}(\mathbf{x},\mathbf{y})\, d\mathbf{x}\, d\mathbf{y} \Big] (1+o(1)) \\
&= h_n^4 |R_n| \sum_{\substack{|\boldsymbol{\alpha}_1|=2\\|\boldsymbol{\alpha}_2|=2}} \Big[4^{-1} a(\boldsymbol{\alpha}_1,\boldsymbol{\alpha}_2) \int_{\mathbb{R}^d} \mathcal{L}_{\boldsymbol{\alpha}_1\boldsymbol{\alpha}_2}(z_1,z_2;\mathbf{y}) d\mathbf{y} \Big] (1+o(1)). \quad (10)
\end{aligned}$$

In the last step, we need to use a version of the Lebesgue Dominated Convergence Theorem to conclude convergence of the integrals from pointwise convergence to the function $\mathcal{L}_{\boldsymbol{\alpha}_1\boldsymbol{\alpha}_2}(z_1,z_2;\mathbf{y})$. This can be done using Assumptions 3 and 4 as in the proof of (11) below.

Next, note that by Assumptions 3 and 4, there exists a Lebesgue integrable function $H_{\boldsymbol{\alpha}_1\boldsymbol{\alpha}_2}(\mathbf{s})$ such that

$$\sup_{z_1,z_2\in\mathbb{R}} |D_{\mathbf{x}}^{\boldsymbol{\alpha}_1} D_{\mathbf{y}}^{\boldsymbol{\alpha}_2} \mathcal{G}(\mathbf{x},\mathbf{y})| \le H_{\boldsymbol{\alpha}_1\boldsymbol{\alpha}_2}(\mathbf{y}-\mathbf{x}),$$

for all $\mathbf{x},\mathbf{y} \in \mathbb{R}^d$ and for all $1 \le |\boldsymbol{\alpha}_1|, |\boldsymbol{\alpha}_2| \le \kappa+1$. Hence, using the fact that $|J_{2n}| = O\big((\lambda_n h_n^{-1})^{d-1}\big)$, we obtain,

$$\begin{aligned}
|I_{12n}| = \Bigg| &\Big(\sum_{\mathbf{i}\in J_{1n}} \sum_{\mathbf{j}\in J_{2n}} + \sum_{\mathbf{i}\in J_{2n}} \sum_{\mathbf{j}\in J_{1n}} + \sum_{\mathbf{i}\in J_{2n}} \sum_{\mathbf{j}\in J_{2n}} \Big) \\
&\int_{\Gamma(1,\mathbf{i})} \int_{\Gamma(1,\mathbf{j})} \sum_{k=1}^{4} \Big[\sum_{|\boldsymbol{\alpha}|=k} (-D_{\mathbf{x}}^{\boldsymbol{\alpha}} \mathcal{G}(\mathbf{s_i},\mathbf{s_j})(\mathbf{x}-\mathbf{s_i})^{\boldsymbol{\alpha}}/\boldsymbol{\alpha}! \\
&\qquad\qquad - D_{\mathbf{y}}^{\boldsymbol{\alpha}} \mathcal{G}(\mathbf{s_i},\mathbf{s_j})(\mathbf{y}-\mathbf{s_j})^{\boldsymbol{\alpha}}/\boldsymbol{\alpha}!) \\
&+ \sum_{|\boldsymbol{\alpha}_1|+|\boldsymbol{\alpha}_2|=k} (D_{\mathbf{x}}^{\boldsymbol{\alpha}_1} D_{\mathbf{y}}^{\boldsymbol{\alpha}_2} \mathcal{G}(\mathbf{s_i},\mathbf{s_j})(\mathbf{x}-\mathbf{s_i})^{\boldsymbol{\alpha}_1}(\mathbf{y}-\mathbf{s_j})^{\boldsymbol{\alpha}_2}/\boldsymbol{\alpha}_1!\boldsymbol{\alpha}_2!) \Big] \, d\mathbf{x}\, d\mathbf{y} \Bigg|
\end{aligned}$$

$$\leq C(d)h_n^{2d+2}\sum_{k=2}^{4}\sum_{\substack{|\alpha_1|\geq 1\\|\alpha_2|\geq 1\\|\alpha_1|+|\alpha_2|=k}}\Big(\sum_{\mathbf{i}\in J_{1n}}\sum_{\mathbf{j}\in J_{2n}}+\sum_{\mathbf{i}\in J_{2n}}\sum_{\mathbf{j}\in J_{1n}}+\sum_{\mathbf{i}\in J_{2n}}\sum_{\mathbf{j}\in J_{2n}}\Big)$$

$$\left|D_{\mathbf{x}}^{\alpha_1}D_{\mathbf{y}}^{\alpha_2}\mathcal{G}(\mathbf{s_i},\mathbf{s_j})\int_{\Delta_0}\int_{\Delta_0}(\mathbf{x}-\mathbf{c})^{\alpha_1}(\mathbf{y}-\mathbf{c})^{\alpha_2}/\alpha_1!\alpha_2!\,d\mathbf{x}\,d\mathbf{y}\right|$$

$$\leq C(d)h_n^{2d+2}\sum_{k=2}^{4}\sum_{\substack{|\alpha_1|\geq 1\\|\alpha_2|\geq 1\\|\alpha_1|+|\alpha_2|=k}}\Big(\sum_{\mathbf{i}\in J_{1n}}\sum_{\mathbf{j}\in J_{2n}}+\sum_{\mathbf{i}\in J_{2n}}\sum_{\mathbf{j}\in J_{1n}}+\sum_{\mathbf{i}\in J_{2n}}\sum_{\mathbf{j}\in J_{2n}}\Big)|H_{\alpha_1\alpha_2}(\mathbf{s_j}-\mathbf{s_i})|$$

$$\leq C(d)h_n^{2d+2}|J_{2n}|h_n^{-d}$$

$$\leq C(d)h_n^3\lambda_n^{d-1}. \tag{11}$$

Also, by similar arguments, there exist Lebesgue integrable functions $H_{\alpha_1\alpha_2}(\cdot)$ and points $\mathbf{s_i}^* \in \Gamma(1,\mathbf{i})$ and $\mathbf{s_j}^* \in \Gamma(1,\mathbf{j})$ such that, uniformly in $z_1, z_2 \in \mathbb{R}$,

$$|I_{2n}| \leq \sum_{\mathbf{i}\in J_n}\sum_{\mathbf{j}\in J_n}\int_{\Gamma(1,\mathbf{i})}\int_{\Gamma(1,\mathbf{j})}|r_n(\mathbf{s_i},\mathbf{s_j};\mathbf{x},\mathbf{y})|\,d\mathbf{x}\,d\mathbf{y}$$

$$\leq C(d)\sum_{|\alpha_1|+|\alpha_2|=5}\sum_{\mathbf{i}\in J_n}\sum_{\mathbf{j}\in J_n}\int_{\Gamma(1,\mathbf{i})}\int_{\Gamma(1,\mathbf{j})}[|H_{\alpha_1,\alpha_2}(\mathbf{s_j}^*-\mathbf{s_i}^*)| \times(||\mathbf{x}-\mathbf{s_i}||^5+||\mathbf{y}-\mathbf{s_j}||^5)]\,d\mathbf{x}\,d\mathbf{y}$$

$$\leq C(d)h_n^5|R_n|\sum_{|\alpha_1|+|\alpha_2|=5}\int_{\mathbb{R}^d}H_{\alpha_1\alpha_2}(\mathbf{y})\,d\mathbf{y}$$

$$\leq C(d)h_n^5\lambda_n^d. \tag{12}$$

The lemma now follows from (9)–(12). □

Lemma 3. *Under Condition* 1 *and Assumptions* 2 *and* 3, *for any* $z \in \mathbb{R}$,

$$\left|E\sum_{\mathbf{i}\in J_n}Y(\mathbf{i})\right| = O(h_n^{\kappa/2}),$$

where $Y(\mathbf{i}) \equiv \int_{\Gamma(1,\mathbf{i})}\big(I(Z(\mathbf{s_i})\leq z)-I(Z(\mathbf{s})\leq z)\big)\,d\mathbf{s}$ *and* $\Gamma(1,\mathbf{i})$ *and* J_{1n} *are as in Lemma* 2.

Proof. We only consider the case $\kappa = 4$. Let $\tilde{F}(\mathbf{x}) \equiv F\big(z-g(\mathbf{x})\big)$; $\mathbf{x}\in\mathbb{R}^d$. Then, using Taylor's expansion and proceeding as in the proof of Lemma 2, by Assumptions 2 and 3, we obtain,

$$\left|E\sum_{\mathbf{i}\in J_n}Y(\mathbf{i})\right| = \left|\sum_{\mathbf{i}\in J_n}\int_{\Gamma(1,\mathbf{i})}\Big(F\big(z-g(\mathbf{s_i})\big)-F\big(z-g(\mathbf{x})\big)\Big)\,d\mathbf{x}\right|$$

$$= \left| \sum_{\mathbf{i} \in J_n} \int_{\Gamma(1,\mathbf{i})} \left(-\sum_{|\alpha|=1}^{2} (\alpha!)^{-1} D^{\alpha} \tilde{F}(\mathbf{s_i})(\mathbf{x}-\mathbf{s_i})^{\alpha} + r_n(\mathbf{s_i};\mathbf{x}) \right) d\mathbf{x} \right|$$

$$= h_n^{d+2} \sum_{|\alpha|=2} (2!)^{-1} \sum_{\mathbf{i} \in J_{1n}} \left| D^{\alpha} \tilde{F}(\mathbf{s_i}) \int_{\Delta_0} (\mathbf{x} - \mathbf{c})^{\alpha} \, d\mathbf{x} \right| (1 + o(1))$$

$$\leq C(d) h_n^2 \sum_{|\alpha|=2} \int_{\mathbb{R}^d} |D^{\alpha} \tilde{F}(\mathbf{x})| \, d\mathbf{x}$$

$$= O(h_n^{\kappa/2}). \qquad \square$$

Lemma 4. *Let $R_n^1 \equiv (a_{1n}, b_{1n}) \times \cdots \times (a_{dn}, b_{dn})$ be a rectangle in $\mathbb{R}^d$ with $b_{in} - a_{in} \geq c > 0$, for all $i = 1, \ldots, d$ and $n \geq 1$, and that $R_n^1 \cap R_n \neq \varnothing$. Suppose that Assumptions 1–6 hold. Then, for any $z_1, z_2 \in \mathbb{R}$, as $n \to \infty$,*

$$\text{(a)} \qquad E\Big(\sum_{\mathbf{i} \in J_{3n}} Y_3(\mathbf{i}) \Big)^2 \leq C(d) h_n^{\kappa} |R_n^1|,$$

$$\text{(b)} \qquad E\Big(\sum_{\mathbf{i} \in J_{3n}} Y_4(\mathbf{i}) \Big)^2 \leq C(d) h_n^{\kappa} |R_n^1| \cdot |F(z_1) - F(z_2)|^{\gamma},$$

where

$$Y_3(\mathbf{i}) \equiv \int_{\Gamma(2,\mathbf{i})} (I(Z(\mathbf{s_i}) \leq z_1) - I(Z(\mathbf{s}) \leq z_1)) \, d\mathbf{s},$$

$$Y_4(\mathbf{i}) \equiv \int_{\Gamma(2,\mathbf{i})} (I(z_1 < Z(\mathbf{s_i}) \leq z_2) - I(z_1 < Z(\mathbf{s}) \leq z_2)) \, d\mathbf{s},$$

$\Gamma(2, \mathbf{i}) \equiv \Gamma(\mathbf{i}) \cap R_n \cap R_n^1$, *and* $J_{3n} \equiv \{\mathbf{i} \in \mathbb{Z}^d : \Gamma(2, \mathbf{i}) \neq \varnothing\}$.

Proof. The proof is similar to that for Lemma 2 and is omitted. $\square$

Lemma 5. *Suppose that Condition 1 and Assumptions 1–6 hold. Then, for any $a_1, \ldots, a_r \in \mathbb{R}$, $z_1, \ldots, z_r \in R$, and $r \geq 1$,*

$$\sum_{j=1}^{r} a_j \xi_n(z_j) \xrightarrow{\mathcal{D}} N\Big(0, \sum_{i=1}^{r} \sum_{j=1}^{r} \sigma(z_i, z_j)\Big),$$

where $\xrightarrow{\mathcal{D}}$ denotes weak convergence of random variables and $\sigma(z_i, z_j)$ is as defined in the statement of Theorem 1.

Proof. We shall prove the lemma only for the case $\kappa = 4$. Clearly,

$$
\begin{aligned}
&\sum_{j=1}^{r} a_j \xi_n(z_j) \\
&= b_n((N_n h_n^d)^{-1} - |R_n|^{-1}) \sum_{j=1}^{r} a_j \int_{R_n} (I(Z(\mathbf{s}) \le z_j) - P(Z(\mathbf{s}) \le z_j))\, d\mathbf{s} \\
&\quad + b_n (N_n h_n^d)^{-1} \sum_{j=1}^{r} a_j \sum_{\mathbf{i} \in J_n} \Big[\int_{\Gamma(1,\mathbf{i})} (I(Z(\mathbf{s_i}) \le z_j) - I(Z(\mathbf{s}) \le z_j))\, d\mathbf{s} \\
&\qquad - \int_{\Gamma(1,\mathbf{i})} (P(Z(\mathbf{s_i}) \le z_j) - P(Z(\mathbf{s}) \le z_j))\, d\mathbf{s} \Big] \\
&\equiv I_{1n} + I_{2n}. \qquad (13)
\end{aligned}
$$

Note that by Lemma 1.8.1 of Ivanov and Leonenko (1989), Condition 1 and Assumption 6, I_{1n} tends to zero in mean square. Hence, it remains to prove the weak convergence of I_{2n}. For this, we employ the "blocking" method of Bernstein (1944). Let $\{\lambda_{1n}\}$ and $\{\lambda_{2n}\}$ denote two sequences of positive numbers such that $\lambda_{1n}/h_n, \lambda_{2n}/h_n \in \mathbb{Z}^+$ and, as $n \to \infty$,

$$1/\lambda_{1n} + 1/\lambda_{2n} + \lambda_{2n}/\lambda_{1n} + \lambda_{1n}/\lambda_n \to 0.$$

Let $\lambda_{3n} \equiv \lambda_{1n} + \lambda_{2n}$ and let $\Delta_n(\mathbf{i};\mathbf{0}) \equiv (\mathbf{i} + \Delta_0)\lambda_{3n}$; $\mathbf{i} \in \mathbb{Z}^d$. We further partition each cube $\Delta_n(\mathbf{i};\mathbf{0})$ into "big" blocks of (long) side length λ_{1n} and "little" parallellopipeds with at least one (short) side of length λ_{2n}. Let $I(1,i) \equiv [s_{1i}, s_{1i} + \lambda_{1n})$, $I(2,i) \equiv [s_{1i} + \lambda_{1n}, s_{1i} + \lambda_{3n})$, $\boldsymbol{\epsilon} \in \boldsymbol{\theta} \equiv \{1,2\}^d$ and $\boldsymbol{\epsilon}_0 \equiv (1,\ldots,1)'$. Then, $\Delta_n(\mathbf{i};\boldsymbol{\epsilon}) \equiv I(\epsilon_1, i_1) \times \cdots \times I(\epsilon_d, i_d)$; $\mathbf{i} \in \mathbb{Z}$, $\boldsymbol{\epsilon} \in \boldsymbol{\theta}$, defines the blocks, yielding the "big" block for $\boldsymbol{\epsilon} = \boldsymbol{\epsilon}_0$ and the "little" parallellopipeds for all $\boldsymbol{\epsilon} \ne \boldsymbol{\epsilon}_0$. Note that the volume of a block of "type $\boldsymbol{\epsilon}$" is $|\Delta(\mathbf{i};\boldsymbol{\epsilon})| = \lambda_{1n}^q \lambda_{2n}^{d-q}$, where q is the total number of "long" sides. Hence, $|\Delta(\mathbf{i};\boldsymbol{\epsilon})| = o(|\Delta(\mathbf{i};\boldsymbol{\epsilon}_0)|)$, as $n \to \infty$, for all $\boldsymbol{\epsilon} \ne \boldsymbol{\epsilon}_0$.

Next, we express I_{2n} in terms of the "big" and "little" blocks. Let $J_{4n} \equiv \{\mathbf{i} \in \mathbb{Z}^d : \Delta_n(\mathbf{i};\mathbf{0}) \subset R_n\}$ be the index set of nonboundary blocks and let $J_{5n} \equiv \{\mathbf{i} \in \mathbb{Z}^d : \Delta_n(\mathbf{i};\mathbf{0}) \cap R_n \ne \varnothing, \Delta_n(\mathbf{i};\mathbf{0}) \cap R_n^c \ne \varnothing\}$. Also, for $\mathbf{i} \in J_n$, let

$$Y_{\mathbf{i}} \equiv b_n (N_n h_n^d)^{-1} \sum_{j=1}^{r} a_j \int_{\Gamma(1,\mathbf{i})} (I(Z(\mathbf{s_i}) \le z_j) - I(Z(\mathbf{s}) \le z_j))\, d\mathbf{s}.$$

For $\mathbf{i} \in J_{4n}$ (i.e., over the nonboundary blocks) and $\boldsymbol{\epsilon} \in \boldsymbol{\theta}$, define the variables,

$$Y(\mathbf{i};\boldsymbol{\epsilon}) \equiv \sum_{\mathbf{j}:\Gamma(\mathbf{j}) \subset \Delta_n(\mathbf{i};\boldsymbol{\epsilon})} (Y_{\mathbf{j}} - EY_{\mathbf{j}}).$$

For $\mathbf{i} \in J_{5n}$ (i.e., over the boundary blocks), define

$$Y(\mathbf{i};\mathbf{0}) \equiv \sum_{\mathbf{j}:\Gamma(\mathbf{j}) \cap R_n \cap \Delta_n(\mathbf{i};\mathbf{0}) \ne \varnothing} (Y_{\mathbf{j}} - EY_{\mathbf{j}}).$$

Then, it is easy to check that

$$\begin{aligned}
I_{2n} &= \sum_{\boldsymbol{\epsilon}\in\boldsymbol{\theta}} \sum_{\mathbf{i}\in J_{4n}} Y(\mathbf{i};\boldsymbol{\epsilon}) + \sum_{\mathbf{i}\in J_{5n}} Y(\mathbf{i};\mathbf{0}) \\
&= \sum_{\mathbf{i}\in J_{4n}} Y(\mathbf{i};\boldsymbol{\epsilon}_0) + \sum_{\boldsymbol{\epsilon}\neq\boldsymbol{\epsilon}_0} \sum_{\mathbf{i}\in J_{4n}} Y(\mathbf{i};\boldsymbol{\epsilon}) + \sum_{\mathbf{i}\in J_{5n}} Y(\mathbf{i};\mathbf{0}) \\
&\equiv I_{21n} + I_{22n} + I_{23n}.
\end{aligned} \tag{14}$$

Note that $K_2 \equiv |J_{5n}| \leq C(d)(\lambda_n/\lambda_{3n})^{d-1}$, and that $|\Delta_0(\mathbf{i};\mathbf{0})| = \lambda_{3n}^d$. Hence, by Assumption 1 and Lemma 4(a),

$$\begin{aligned}
E(I_{23n})^2 &= E\Big(\sum_{\mathbf{i}\in J_{5n}} Y(\mathbf{i};\mathbf{0})\Big)^2 \\
&\leq \sum_{k=0}^{K_2-1} |J_{5n}| k^{d-1} \rho\big((k-1)^+\lambda_{3n};\lambda_{3n}^d\big) \max_{\mathbf{i}\in J_{5n}} E\big(Y(\mathbf{i};\mathbf{0})\big)^2 \\
&\leq C(d,\rho,G)(\lambda_{3n}/\lambda_n)(\lambda_{3n}^{d\theta-\zeta}).
\end{aligned} \tag{15}$$

By similar arguments, noting that $K_3 \equiv |J_{4n}| = O(\lambda_n/\lambda_{3n})^d$, $|\Delta(\mathbf{i},\boldsymbol{\epsilon})| \leq \lambda_{1n}^{d-1}\lambda_{2n}$ for all $\boldsymbol{\epsilon} \neq \boldsymbol{\epsilon}_0$, and that the distance between $\Delta(\mathbf{i};\boldsymbol{\epsilon})$ and $\Delta(\mathbf{j};\boldsymbol{\epsilon})$ is $(|\mathbf{i}-\mathbf{j}|-1)^+\lambda_{3n} + \lambda_{2n}$, we obtain,

$$\begin{aligned}
E(I_{22n})^2 &= E\Big(\sum_{\boldsymbol{\epsilon}\neq\boldsymbol{\epsilon}_0} \sum_{\mathbf{i}\in J_{4n}} Y(\mathbf{i};\boldsymbol{\epsilon})\Big)^2 \\
&\leq 2^{2^d-2} \sum_{\boldsymbol{\epsilon}\neq\boldsymbol{\epsilon}_0} E\Big(\sum_{\mathbf{i}\in J_{4n}} Y(\mathbf{i};\boldsymbol{\epsilon})\Big)^2 \\
&\leq C(d)|J_{4n}|\Big(1+\sum_{k=1}^{K_3} k^{d-1}\rho\big((k-1)\lambda_{3n}+\lambda_{2n};\lambda_{1n}^{d-1}\lambda_{2n}\big)\Big) \\
&\qquad\qquad \times \max_{\mathbf{i}\in J_{4n}} E\big(Y(\mathbf{i};\boldsymbol{\epsilon})\big)^2 \\
&\leq C(d)(\lambda_n/\lambda_{3n})^{d\theta}(\lambda_{3n}^{d\theta-\zeta})(\lambda_{2n}/\lambda_n)^{\theta}(\lambda_{2n}/\lambda_{3n}).
\end{aligned} \tag{16}$$

Hence, choosing $\lambda_{1n} \sim \lambda_n/\log\lambda_n$ and $\lambda_{2n} \sim \lambda_n^{d\theta/\zeta}\log\lambda_n$, from (13)–(16), we obtain, as $n \to \infty$,

$$E\Big(\sum_{j=1}^{r} a_j\xi_n(z_j) - \sum_{\mathbf{i}\in J_{4n}} Y(\mathbf{i};\boldsymbol{\epsilon}_0)\Big)^2 \to 0. \tag{17}$$

Next, note that for the given choices of $\lambda_{1n}, \lambda_{2n}$,

$$\left|E\exp\left(it\sum_{\mathbf{i}\in J_{4n}} Y(\mathbf{i};\epsilon_0)\right) - \prod_{\mathbf{i}\in J_{4n}} E\exp(itY(\mathbf{i};\epsilon_0))\right|$$
$$\leq C(d)|J_{4n}|\rho(\lambda_{2n};|R_n|)$$
$$\leq C(d)(\lambda_n/\lambda_{3n})^{d+d\theta}\lambda_{3n}^{d\theta}\lambda_{2n}^{-\zeta}. \quad (18)$$

Let $\{X(\mathbf{i}) : \mathbf{i} \in J_{4n}, n \geq 1\}$ denote a triangular array of independent random variables with $X(\mathbf{i})$ having the same distribution as $Y(\mathbf{i};\epsilon_0)$. Then, using Lemmas 1–3, one can verify that $\{X(\mathbf{i}) : \mathbf{i} \in J_{4n}, n \geq 1\}$ satisfies Lyapounov's Condition (cf. Billingsley, 1968, p. 44). Hence, Lemma 5 now follows from (17), (18), and the central limit theorem for sums of independent variables. □

Proof of Theorem 1. In view of Assumption 2 and Lemma 3, by standard arguments, it is enough to show that the time-scaled process $\tilde{\xi}_n(t) = [\xi_n(F^{-1}(t)) - E\xi_n(F^{-1}(t))]$; $t \in [0,1]$, converges in distribution to $W(F^{-1}(t))$; $t \in [0,1]$, as random elements in $D[0,1]$, the space of all right continuous functions on $[0,1]$ with left-hand limits, equipped with the Skorohod metric. By Lemmas 3 and 5, we have the weak convergence of the finite-dimensional distributions of $\tilde{\xi}_n(t)$ to $W(F^{-1}(t))$. Hence, it suffices to show the tightness of the sequence $\{\tilde{\xi}_n(t) : n \geq 1\}$ and the almost sure continuity of the sample paths of $W(F^{-1}(t))$. By Theorem 15.5 and the proof of Theorem 22.1 of Billingsley (1968), both results hold if we can show that for all $\epsilon > 0, \eta > 0$, there exists a $0 < \delta < 1$ such that for sufficiently large n,

$$P(\sup\{|\tilde{\xi}_n(t) - \tilde{\xi}_n(s)| : s \leq t \leq (s+\delta)\wedge 1\} \leq \epsilon) \leq \eta\delta, \quad (19)$$

for any given $0 \leq s \leq 1$. Fix $0 < \epsilon, \eta < 1$, and $s \in [0,1]$. Let $p(n,\epsilon,\eta) \equiv C(\epsilon,\eta)(\lambda_n 6dh_n^{\kappa})^{-1/\gamma}$. Then, by the fact that $(\lambda_n^d h_n^{\kappa})^{-1} \leq (\lambda_n^d h_n^{(2\gamma+1)\kappa})^{-1}$ and Assumption 6, there exists $n_1 \equiv n_1(\epsilon,\eta) \geq 1$ such that for all $n \geq n_1$,

$$\begin{aligned} b_n p &= \left(\lambda_n^{d(\gamma-2)} h_n^{-\kappa(2+\gamma)}\right)^{1/2\gamma} \\ &= \left((\lambda_n^d h_n^{(2\gamma+1)\kappa})(\lambda_n^d h_n^{\kappa})^{1-\gamma}\right)^{-1/2\gamma} \\ &\leq \left(\lambda_n^d h_n^{(2\gamma+1)\kappa}\right)^{-(2-\gamma)/2\gamma} \\ &< \epsilon/4. \end{aligned} \quad (20)$$

Note that by Assumption 2, there exists a constant K such that for any $0 \leq t_1 < t_2 \leq 1$,

$$E\Big(F_n(F^{-1}(t_2)) - F_n(F^{-1}(t_1))\Big) \leq K(t_2 - t_1), \quad (21)$$

$$E\Big(F_\infty(F^{-1}(t_2)) - F_\infty(F^{-1}(t_1))\Big) \leq K(t_2 - t_1). \quad (22)$$

Next, using (21), (22), and the monotonicity of distribution functions, for any $0 \leq t_1 \leq t \leq t_2 \leq 1$, we obtain $\tilde{\xi}_n(t) \leq \tilde{\xi}_n(t_2) + b_n(\tilde{F}_\infty(t_2) - \tilde{F}_\infty(t_1)) +$

$2K|t_2 - t_1|$ and $\tilde{\xi}_n(t) \geq \tilde{\xi}_n(t_1) - b_n(\tilde{F_\infty}(t_2) - \tilde{F_\infty}(t_1)) - 2K|t_2 - t_1|$, where $\tilde{F_\infty}(u) \equiv F_\infty(F^{-1}(u)) - EF_\infty(F^{-1}(u))$; $u \in (0,1)$. Let m be a (large) positive integer to be chosen later. Then, setting $\delta = mp$ and using (20) and the above inequalities, for sufficiently large n, we have

$$\begin{aligned} &P(\sup\{|\tilde{\xi}_n(t) - \tilde{\xi}_n(s)| : s \leq t \leq (s+\delta) \wedge 1\} > \epsilon) \\ &\quad \leq P(\max\{|\tilde{\xi}_n([s+ip] \wedge 1) - \tilde{\xi}_n(s)| : 1 \leq i \leq m\} > \epsilon/6) \\ &\qquad + P(\max\{b_n|\tilde{F_\infty}([s+ip] \wedge 1) \\ &\qquad\qquad - \tilde{F_\infty}([s+(i-1)p] \wedge 1)| : 1 \leq i \leq m\} > \epsilon/4). \end{aligned} \tag{23}$$

Next, using the simple inequality that for a (bounded) random variable X, $\mathrm{Var}(X) \leq EX^2$, and using Lemma 1 and Lemma 4(b), one can show that for any $0 \leq t_1 < t_2 \leq 1$,

$$\begin{aligned} &E(\tilde{\xi}_n(t_2) - \tilde{\xi}_n(t_1))^4 \\ &\qquad \leq C(d,\kappa) h_n^{-2\kappa} \lambda_n^{-2d} (\lambda_n^{2d} h_n^{2\kappa} |t_2 - t_1|^{2\gamma} + \lambda_n^d h_n^\kappa |t_2 - t_1|^\gamma) \\ &\qquad \leq C(d,\kappa) |t_2 - t_1|^{2\gamma}. \end{aligned} \tag{24}$$

Also, by similar arguments, for any $0 \leq t_1 < t_2 \leq 1$,

$$\begin{aligned} &E\Big(b_n(\tilde{F_\infty}(t_2) - \tilde{F_\infty}(t_1))\Big)^4 \\ &\qquad\qquad \leq C(d,\kappa,K) h_n^{-2\kappa} \lambda_n^{-2d} (\lambda_n^{2d} |t_2 - t_1|^2 + \lambda_n^d |t_2 - t_1|). \end{aligned} \tag{25}$$

Now, inequality (24) and Theorem 12.2 of Billingsley (1968) yield,

$$\begin{aligned} &P(\max\{|\tilde{\xi}_n(s+ip) - \tilde{\xi}_n(s)| : 1 \leq i \leq m\} > \epsilon/6) \\ &\qquad\qquad \leq C(d,\kappa)\epsilon^{-4}(mp)^{2\gamma}, \end{aligned} \tag{26}$$

and, by (20) and (25),

$$\begin{aligned} &P(\max\{b_n|\tilde{F}(s+ip) - \tilde{F}(s+(i-1)p)| : 1 \leq i \leq m\} > \epsilon/2) \\ &\qquad \leq \sum_{i=1}^{m} P(b_n|\tilde{F}(s+ip) - \tilde{F}(s+ip-p) - p| > \epsilon/4) \\ &\qquad \leq C(d,\kappa)\epsilon^{-4} m \lambda_n^{-2d} h_n^{-2\kappa} (\lambda_n^{2d} p^2 + \lambda_n^d p) \\ &\qquad \leq C(d,\kappa)\epsilon^{-4}(mp)(\lambda_n^d h_n^{(2\gamma+1)\kappa})^{-1/\gamma} \\ &\qquad \leq (\eta\delta)/2. \end{aligned} \tag{27}$$

Finally, let $\delta \equiv \delta(\eta, \epsilon) > 0$ be such that $C(d,\kappa)\epsilon^{-4}\delta^{2\gamma-1} < \eta/2$ (we can do so since $\gamma > \frac{1}{2}$) and δ/p is an integer. Then, we choose the integer m through $m \equiv \delta/p$, which goes to infinity with n. Thus, (19) follows from (23), (26) and (27). This completes the proof of Theorem 1. □

Acknowledgments: This research was partially supported by the U.S. Environmental Protection Agency under Assistance Agreement no. CR 822919-01-0 and the Office of Naval Research under grant no. N00014-93-1-0001 and no. N00014-99-1-0214. We would like to thank Professor Marc Moore for some helpful discussions and for a number of constructive suggestions.

5 References

Anderson, T.W. (1971). *The Statistical Analysis of Time Series.* New York: Wiley.

Bernstein, S.N. (1944). Extension of the central limit theorem of probability theory to sums of dependent random variables. *Uspekhi Matematicheskikh Nauk 10*, 65–114.

Billingsley, P. (1968). *Convergence of Probability Measures* (2nd ed.). Wiley Series in Probability and Statistics: Probability and Statistics. New York: Wiley.

Cressie, N.A.C. (1993). *Statistics for Spatial Data* (revised ed.). Wiley Series in Probability and Mathematical Statistics: Applied Probability and Statistics. New York: Wiley.

Grenander, U. (1954). On estimation of regression coefficients in the case of an autocorrelated disturbance. *The Annals of Mathematical Statistics 25*, 252–272.

Hall, P. and P. Patil (1994). Properties of nonparametric estimators of autocovariance function for stationary random fields. *Probability Theory and Related Fields 99*, 399–424.

Härdle, W. and P.D. Tuan (1986). Some theory on M-smoothing of time series. *Journal of Time Series Analysis 7*, 191–204.

Ivanov, A.V. and N.N. Leonenko (1989). *Statistical Analysis of Random Fields.* Mathematics and its Applications (Soviet Series). Dordrecht: Kluwer Academic Publishers.

Lahiri, S.N. (1996). On inconsistency of estimators under infill asymptotics for spatial data. *Sankhyā. The Indian Journal of Statistics. Series A 58*, 403–417.

Lahiri, S.N. (1999). Asymptotic distribution of the empirical spatial cumulative distribution function predictor and prediction bands based on a subsampling method. *Probability Theory and Related Fields 114*, 55–84.

Lahiri, S.N., M.S. Kaiser, N. Cressie, and N.J. Hsu (1999). Prediction of spatial cumulative distribution functions using subsampling (with discussions). *Journal of the American Statistical Association 94*, 86–110.

Majure, J.J., D. Cook, N. Cressie, M.S. Kaiser, S.N. Lahiri, and J. Symanzik (1995). Spatial cdf estimation and visualization with applications to forest health monitoring. *Computing Science and Statistics 27*, 93–101.

Overton, W.S. (1989). Effects of measurements and other extraneous errors on estimated distribution functions in the National Surface Water Surveys. Technical Report 129, Department of Statistics, Oregon State University.

Sherman, M. and E. Carlstein (1994). Nonparametric estimation of the moments of a general statistic computed from spatial data. *Journal of the American Statistical Association 89*, 496–500.

2

Robustness Problems in the Analysis of Spatial Data

Marc G. Genton

ABSTRACT Kriging is a widely used method of spatial prediction, particularly in earth and environmental sciences. It is based on a function which describes the spatial dependence, the so called variogram. Estimation and fitting of the variogram, as well as variogram model selection, are crucial stages of spatial prediction, because the variogram determines the kriging weights. These three steps must be carried out carefully, otherwise kriging can produce noninformative maps. The classical variogram estimator proposed by Matheron is not robust against outliers in the data, nor is it enough to make simple modifications such as the ones proposed by Cressie and Hawkins in order to achieve robustness. The use of a variogram estimator based on a highly robust estimator of scale is proposed. The robustness properties of these three variogram estimators are analyzed by means of the influence function and the classical breakdown point. The latter is extended to a spatial breakdown point, which depends on the construction of the most unfavorable configurations of perturbation. The effect of linear trend in the data and location outliers on variogram estimation is also discussed. Variogram model selection is addressed via nonparametric estimation of the derivative of the variogram. Variogram estimates at different spatial lags are correlated, because the same observation is used for different lags. The correlation structure of variogram estimators has been analyzed for Gaussian data, and then extended to elliptically contoured distributions. Its use for variogram fitting by generalized least squares is presented. Results show that our techniques improve the estimation and the fit significantly. Two new SPLUS functions for highly robust variogram estimation and variogram fitting by generalized least squares, as well as a MATLAB code for variogram model selection via nonparametric derivative estimation, are available on the Web at `http://www-math.mit.edu/~genton/`.

1 Introduction

In statistics, the concept of robustness is usually defined as the lack of sensitivity of a particular procedure to departures from the model assumptions. For example, a proportion of 10–15% of contaminated observations, called outliers, can sometimes be found in real data sets (Hampel, 1973, Huber, 1977). These outlying values may be due to gross errors, measure-

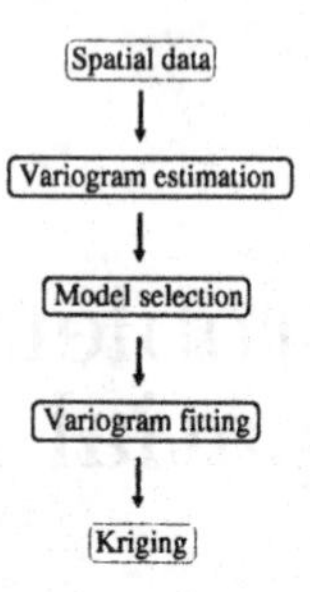

Figure 1. Illustration of the steps from the data to the spatial prediction called kriging. We want to robustify the variogram estimation, model selection, and variogram fitting steps.

ment mistakes, faulty recordings, and can seriously affect the results of a statistical analysis. Therefore, it is of prime interest to provide robust, i.e., reliable, estimators and methods. However, robust procedures are typically more difficult to develop in the spatial statistics context than in the classical one, because different types of outliers may occur. For instance, outliers can replace, or be added to, some observations of the underlying stochastic process. Furthermore, the configuration of spatial locations, where the contaminations occur, becomes important: isolated and patchy outliers can have different effects, and both have been observed in practice. In fact, the main problem is that estimators which take account of the spatial structure of the observations, are not invariant under permutation of the data, as in the case of estimators for independent and identically distributed (i.i.d.) observations. The analysis of spatial data often follows the steps summarized in Figure 1. From the spatial data, one usually models the spatial dependence structure between the observations, i.e. the so called variogram. The first step consists in estimating the variogram at various lag distances. Next, a valid variogram model has to be selected, and is fitted to the variogram estimates in the first step. Finally, this variogram is used in the spatial prediction procedure called kriging (Cressie, 1993). As one can see, the modeling of the variogram is an important stage of spatial prediction, because it determines the kriging procedure. Therefore, it is important to have a variogram estimator which remains close to the true underlying variogram, even if outliers are present in the data. Otherwise, kriging can produce noninformative maps. Of course, one might argue that any reasonable exploratory data analysis would identify and remove outliers in the data. However, this approach often contains a subjective aspect (Genton and Furrer, 1998a) that we would like to avoid, or at least to minimize.

In this paper, we bring together several ideas (Genton, 1998a,b,c, 1999, Gorsich and Genton, 1999) about the robustification of the three steps involved in the modeling of the variogram. First, in the next section, we use a highly robust estimator of the variogram and derive some of its properties.

Second, in Section 3, the issue of variogram model selection is addressed via nonparametric estimation of the derivative of the variogram. Finally, in Section 4, we use the correlation structure of variogram estimates to fit a variogram model by generalized least squares.

2 Highly Robust Variogram Estimation

Variogram estimation is a crucial stage of spatial prediction, because it determines the kriging weights. The most widely used variogram estimator is certainly the one proposed by Matheron (1962), although it is highly non-robust to outliers in the data (Cressie, 1993, Genton, 1998a,c). One single outlier can destroy this estimator completely. The main reasons for this popularity are its simple appealing formulation and unbiasedness property. If $\{Z(\mathbf{x}) : \mathbf{x} \in D \subset \mathbb{R}^d\}$, $d \geq 1$, is a spatial stochastic process, ergodic and intrinsically stationary, then Matheron's classical variogram estimator, based on the method-of-moments, is

$$2\widehat{\gamma}(\mathbf{h}) = \frac{1}{N_{\mathbf{h}}} \sum_{N(\mathbf{h})} \left(Z(\mathbf{x}_i) - Z(\mathbf{x}_j)\right)^2, \quad \mathbf{h} \in \mathbb{R}^d, \tag{1}$$

where $Z(\mathbf{x}_1), \ldots, Z(\mathbf{x}_n)$ is a realization of the process, $N(\mathbf{h}) = \{(\mathbf{x}_i, \mathbf{x}_j) : \mathbf{x}_i - \mathbf{x}_j = \mathbf{h}\}$, and $N_{\mathbf{h}}$ is the cardinality of $N(\mathbf{h})$. It is not enough to make simple modifications to formula (1), such as the ones proposed by Cressie and Hawkins (1980), in order to achieve robustness. In this section, we advocate the use of a highly robust variogram estimator (Genton, 1998a)

$$2\widehat{\gamma}(\mathbf{h}) = (Q_{N_{\mathbf{h}}})^2, \quad \mathbf{h} \in \mathbb{R}^d, \tag{2}$$

which takes account of all the available information in the data. It is based on the sample $V_1(\mathbf{h}), \ldots, V_{N_{\mathbf{h}}}(\mathbf{h})$ from the process of differences $V(\mathbf{h}) = Z(\mathbf{x}+\mathbf{h}) - Z(\mathbf{x})$ and the robust scale estimator $Q_{N_{\mathbf{h}}}$, proposed by Rousseeuw and Croux (1992, 1993)

$$Q_{N_{\mathbf{h}}} = 2.2191 \, \{|V_i(\mathbf{h}) - V_j(\mathbf{h})|; i < j\}_{(k)}, \tag{3}$$

where the factor 2.2191 is for consistency at the Gaussian distribution, $k = \binom{[N_{\mathbf{h}}/2]+1}{2}$, and $[N_{\mathbf{h}}/2]$ denotes the integer part of $N_{\mathbf{h}}/2$. This means that we sort the set of all absolute differences $|V_i(\mathbf{h}) - V_j(\mathbf{h})|$ for $i < j$ and then compute its kth order statistic (approximately the $\frac{1}{4}$ quantile for large $N_{\mathbf{h}}$). This value is multiplied by the factor 2.2191, thus yielding $Q_{N_{\mathbf{h}}}$. Note that this estimator computes the kth order statistic of the $\binom{N_{\mathbf{h}}}{2}$ interpoint distances. At first sight, the estimator $Q_{N_{\mathbf{h}}}$ appears to need $O(N_{\mathbf{h}}^2)$ computation time, which would be a disadvantage. However, it can be computed using no more than $O(N_{\mathbf{h}} \log N_{\mathbf{h}})$ time and $O(N_{\mathbf{h}})$ storage, by means of the fast algorithm described in Croux and Rousseeuw (1992). An SPLUS function for the highly robust variogram estimator, denoted `variogram.qn`, is available on the Web at `http://www-math.mit.edu/~genton/`.

The variogram estimator (2) possesses several interesting properties of robustness. For instance, its influence function (Hampel et al., 1986), which describes the effect on the estimator of an infinitesimal contamination, is bounded. This means that the worst influence that a small amount of contamination can have on the value of the estimator is finite, in opposition to Matheron's classical variogram estimator and Cressie and Hawkins' proposal. Another important robustness property is the breakdown point ε^* of a variogram estimator, which indicates how many data points need to be replaced by arbitrary values to make the estimator explode (tend to infinity) or implode (tend to zero). The highly robust variogram estimator has an $\varepsilon^* = 50\%$ breakdown point on the differences $V(\mathbf{h})$, which is the highest possible value. On the contrary, Matheron's classical variogram estimator and Cressie and Hawkins' estimator both have only an $\varepsilon^* = 0\%$ breakdown point, which is the lowest possible value. More details about the use and properties of this estimator, including some simulation studies, are presented in Genton (1998a,c).

The breakdown point discussed in the previous paragraph is based on the process of differences $V(\mathbf{h})$. However, in spatial statistics, one is also interested in the breakdown point related to the initial process Z. In this case, the effect of the perturbation of a point located on the boundary of the spatial domain D, or inside of it, can be quite different and depends notably on the lag vector $\mathbf{h}$. Therefore, a concept of spatial breakdown point of variogram estimators is introduced by Genton (1998c). Denote by I_m a subset of size m of $\{1, \ldots, n\}$, and let $\mathcal{Z} = \{Z(\mathbf{x}_1), \ldots, Z(\mathbf{x}_n)\}$. The spatial sample breakdown point of a variogram estimator $2\widehat{\gamma}(\mathbf{h}) = (S_{N_\mathbf{h}})^2$, based on a scale estimator $S_{N_\mathbf{h}}$, is defined by

$$\varepsilon_n^{Sp}(2\widehat{\gamma}(\mathbf{h}), \mathcal{Z}) = \max\left\{ \frac{m}{n} \;\middle|\; \sup_{I_m} \sup_{\mathcal{Z}(I_m)} S_n(\mathcal{Z}(I_m)) < \infty \text{ and } \inf_{I_m} \inf_{\mathcal{Z}(I_m)} S_n(\mathcal{Z}(I_m)) > 0 \right\}, \quad (4)$$

where $\mathcal{Z}(I_m)$ is the sample of size n, obtained by replacing m observations of $\mathcal{Z}$, indexed by I_m, by arbitrary values. This definition takes into account the configuration, i.e. the spatial location, of the perturbation, and seeks its worst spatial configuration for a fixed amount of perturbation and fixed lag vector $\mathbf{h}$. The spatial breakdown point of Matheron's classical variogram estimator, as well as the one of Cressie and Hawkins, is still zero. Figure 2 shows the spatial breakdown point of the highly robust variogram estimator, represented by the black curve, for each lag distance h and a sample of size $n = 100$. Upper and lower bounds on the spatial breakdown point are computed in Genton (1998c), and represented by light grey curves in Figure 2. The interpretation of this figure is as follows. For a fixed h, if the percentage of perturbed observations is below the black curve, the esti-

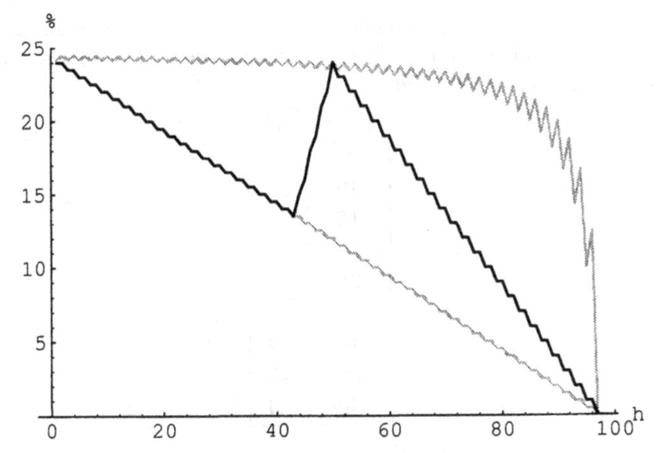

Figure 2. The spatial breakdown point (in black) as a function of the lag distance h, for the highly robust variogram estimator. The upper and lower bounds are drawn in light grey. Reprinted from *Mathematical Geology 30*(7), 853–871, with permission of the International Association for Mathematical Geology.

mator is never destroyed. If the percentage is above the black curve, there exists at least one configuration which destroys the estimator. This implies that highly robust variogram estimators are more resistant at small lags h or around $h = n/2$, than at large lags h or before $h = n/2$, according to Figure 2. The spatial breakdown point is a theoretical tool, indicating the worst-case behavior of the variogram estimator for each lag h. It allows to judge the resistance of the variogram estimates, and consequently their respective reliability. This is a local concept. However, in practice, one is generally confronted with a fixed configuration of perturbed data, which does not change with the lag h. Applied geostatisticians are usually concerned about the global effects (i.e., at all lags h) of a given configuration of perturbations on the estimation of the variogram. For that reason, Genton (1998c) carried out some simulations on the global effects of perturbations, and supported them by theoretical results. Table 1 presents the simulation (over 1000 replications) of the average percentage $\bar{p}$ of perturbed differences when m/n percent of observations of a regular grid of size $n = 10 \times 10 = 100$ are perturbed. For example, if $m/n = 20\%$ of observations Z of the grid are perturbed, then, on average, $\bar{p} = 36\%$ of differences V will be perturbed. Therefore, if a highly robust variogram estimator is used, with 50% breakdown point (on differences V), then it will have a global resistance to roughly 30% of outliers in the initial observations Z. Moreover, it turns out that this result is the same for every lag h. We note that if m/n is small, $\bar{p}$ equals approximatively $2m/n$, whereas it is slightly smaller if m/n is large. This decrease is due to differences taken between two perturbed observations. Simulations carried out on irregular grids showed similar behavior.

Table 1. Simulation (over 1000 replications) of the average percentage $\bar{p}$ of perturbed differences when m/n percent of observations of a regular grid of size $n = 10 \times 10 = 100$ are perturbed.

m/n	$\bar{p}$
5	10
10	19
15	28
20	36
25	44
30	51
40	64
50	75
60	84

Cressie (1993) investigated the effect of a linear trend on the estimation of the variogram. He considered a spatial stochastic process $Z(x)$ in $\mathbb{R}^1$, defined by $Z(x) = S(x) + \epsilon(x - (n+1)/2)$, $x = 1, \ldots, n$, where $S(x)$ is a zero-mean, unit variance, second-order stationary process, and ϵ is the degree of contamination. He showed that for Matheron's classical variogram estimator (1):

$$2\widehat{\gamma}_Z(h) \cong 2\widehat{\gamma}_S(h) + \epsilon^2 h^2, \tag{5}$$

in probability, where $2\widehat{\gamma}_Z(h)$ and $2\widehat{\gamma}_S(h)$ are the variograms of Z and S respectively. Thus, the effect of a linear trend contamination term of magnitude ϵ is that of an upward shift of magnitude $\epsilon^2 h^2$. This effect does not happen when using Q_{N_h}. Effectively, for the highly robust variogram estimator (2):

$$2\widehat{\gamma}_Z(h) = 2\widehat{\gamma}_S(h). \tag{6}$$

The highly robust variogram estimator is not affected by the linear trend contamination, because it is based on differences of the process $V(h)$.

Another interesting issue pointed out by Peter Guttorp during the workshop is the robustness of variogram estimators towards misspecification of the location of the data, i.e. towards location outliers. For instance, consider the very simple situation where the observations $Z(x_1), \ldots, Z(x_n)$ have a unidimensional and regular support. Assume that one location is misspecified, resulting in the exchange of the data at two locations x_i and x_j, i.e. new values $Z^*(x_i)$ and $Z^*(x_j)$. What is the effect on the estimation of the variogram? First, note that the set $N(h)$ will be modified for most lags h, depending on the spatial locations of x_i and x_j with regard to the border of the domain D. The resulting set $N^*(h)$ may be quite different from $N(h)$. Therefore, the final effect on variogram estimations depends on the sensitivity of the variogram estimator to changes in the set $N(h)$. Because Matheron's classical variogram estimator has no robustness properties, it is more sensitive than the highly robust variogram estimator. As an illustrative example, consider the five locations given in Figure 3, with

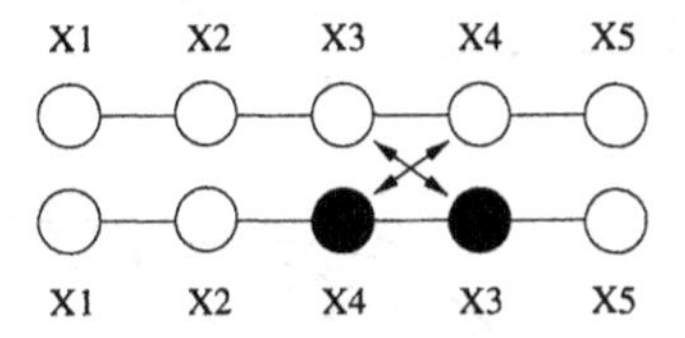

Figure 3. Illustration of one location outlier at x_3, resulting in the exchange of the two data at locations x_3 and x_4.

a location outlier at x_3. The locations x_3 and x_4 have been exchanged, resulting in the following sets $N(h)$ and $N^*(h)$:

$$\begin{aligned}
N(1) &= \{(x_1,x_2),(x_2,x_3),(x_3,x_4),(x_4,x_5)\}\\
N(2) &= \{(x_1,x_3),(x_2,x_4),(x_3,x_5)\}\\
N(3) &= \{(x_1,x_4),(x_2,x_5)\}\\
N(4) &= \{(x_1,x_5)\}
\end{aligned}$$

$$\begin{aligned}
N^*(1) &= \{(x_1,x_2),(x_2,x_4),(x_3,x_4),(x_3,x_5)\}\\
N^*(2) &= \{(x_1,x_4),(x_2,x_3),(x_4,x_5)\}\\
N^*(3) &= \{(x_1,x_3),(x_2,x_5)\}\\
N^*(4) &= \{(x_1,x_5)\}
\end{aligned}$$

Although $N(4)$ and $N^*(4)$ are the same, only half of the elements of $N(1)$ (respectively $N(3)$) are the same as $N^*(1)$ (respectively $N^*(3)$). The worst case is for $h = 2$, where $N(2) \cap N^*(2) = \varnothing$. As a consequence, the estimation of the variogram at $h = 2$ could be seriously biased, depending on the robustness of the variogram estimator. Of course, these effects may be even worse depending on the number of location outliers, the intensity of the perturbation of locations, the dimension $d > 1$ of the spatial domain D, and the irregularity of the locations. Further research is needed to characterize the effects of such situations.

3 Variogram Model Selection

The variogram estimates obtained in the previous section cannot be used directly for kriging because they are not necessarily valid, i.e. conditionally negative definite. Therefore, a valid parametric variogram model must be chosen and fitted to the variogram estimates. The choice of the variogram model is important, since it affects the kriging procedure. Unfortunately, no selection technique can be found in the literature, apart from a priori knowledge about the underlying process, and the user's subjectivity. Figure 4 presents an example where the choice among a spherical, exponential, or Gaussian variogram is unclear, although the true underlying variogram of the data is exponential. In order to reduce the subjectivity in the choice

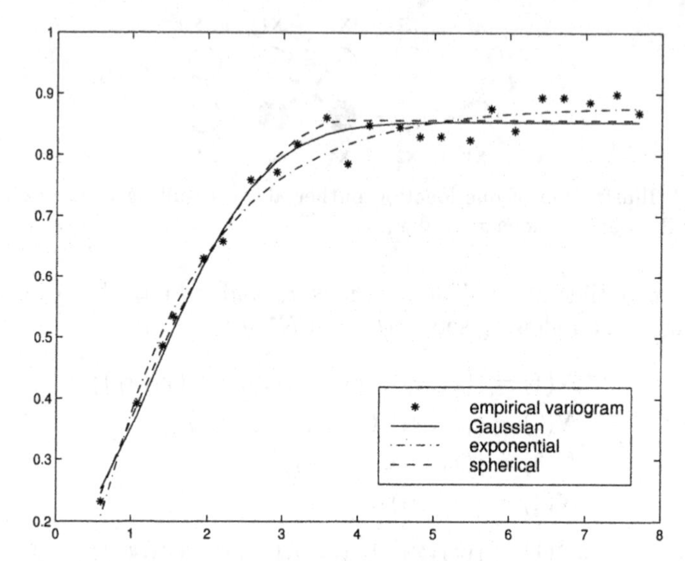

Figure 4. Example of the subjectivity involved in the selection of a variogram model: spherical, exponential, or Gaussian models are fitted equally well, although the true underlying variogram of the data is exponential.

of a variogram model, Gorsich and Genton (1999) have suggested to estimate the derivative of the variogram in a nonparametric way as a help for the selection. Effectively, although many variogram models appear very similar, their derivatives with respect to the lags are not, as is shown in Figure 5.

In order to estimate the derivative without assuming a prior model, a nonparametric variogram estimator which guarantees its conditional negative definiteness is needed (Shapiro and Botha, 1991, Cherry et al., 1996, Cherry, 1997). It is based on the spectral representation of positive definite functions (Bochner, 1955), and consists in finding by nonnegative least squares, positive jumps $p_1, \ldots, p_n$ corresponding to the nodes $t_1, \ldots, t_n$ in

$$2\widehat{\gamma}(h) = 2\sum_{j=1}^{n} p_j\bigl(1 - \Omega_r(ht_j)\bigr), \tag{7}$$

where $\Omega_r(x) = (2/x)^{(r-2)/2}\Gamma(r/2)J_{(r-2)/2}(x)$. Here Γ is the gamma function, and J_v is the Bessel function of the first kind of order v. The parameter r controls the smoothness of the fit, and $r \geq d$ is required in order to maintain positive definiteness, where d is the dimension of the spatial domain D. Note that the nugget effect, range, and sill are not well defined in the nonparametric fit. The derivative of the variogram can now be estimated by differentiating the estimator (7) or by using finite differences. It turns out that the first approach does not work well because the basis Ω_r is causing aliasing problems. The second approach is more appropri-

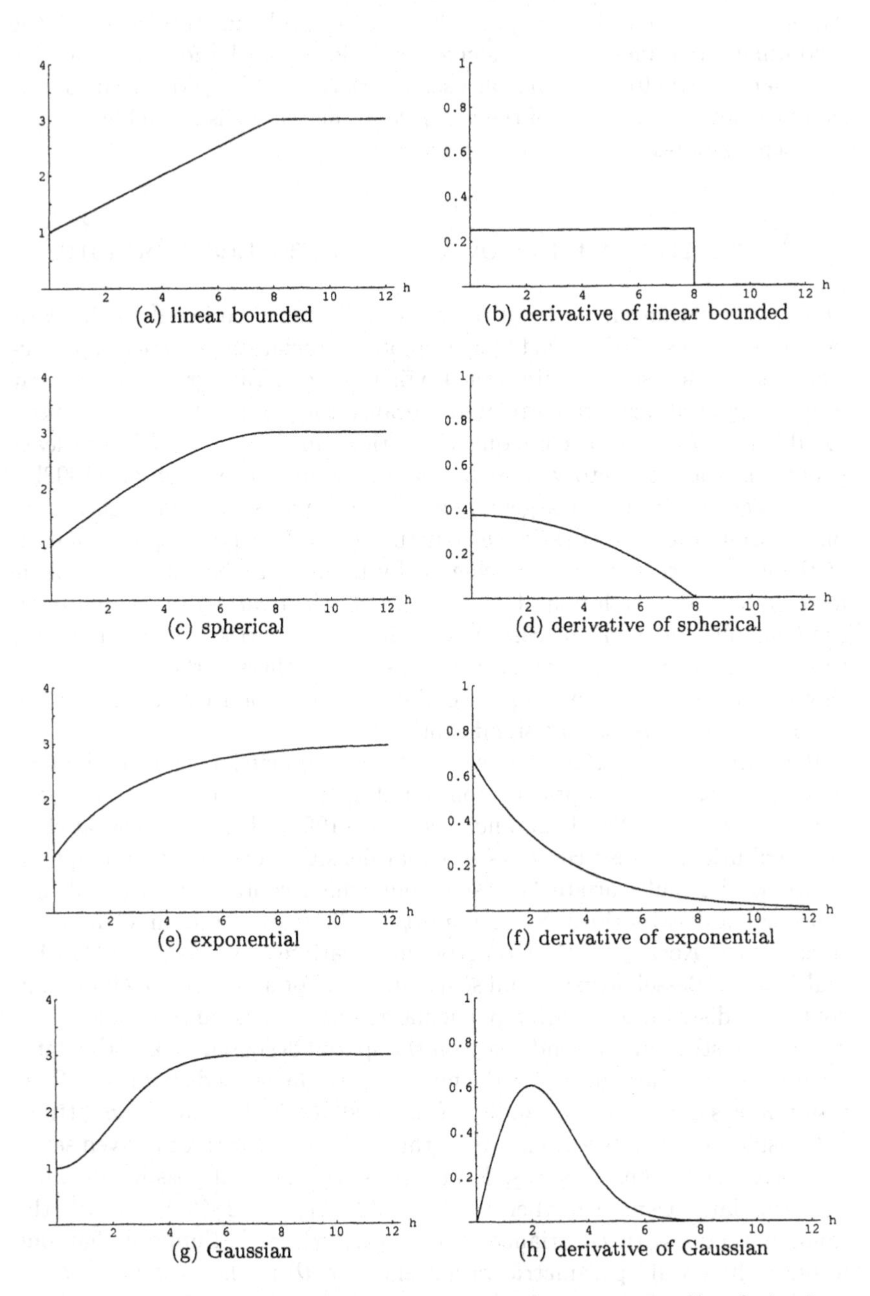

(a) linear bounded (b) derivative of linear bounded

(c) spherical (d) derivative of spherical

(e) exponential (f) derivative of exponential

(g) Gaussian (h) derivative of Gaussian

Figure 5. Some typical variogram models with their corresponding derivatives.

ate and can be used to select a variogram model based on its derivative. More details about this method, the aliasing problem, the choice of the smoothness parameter r, and some simulations, can be found in Gorsich and Genton (1999). A graphical user interface (GUI) in MATLAB for the nonparametric estimation of the variogram's derivative is available to users at `http://www-math.mit.edu/~gorsich/`.

4 Variogram Fitting by Generalized Least Squares

Once a variogram model has been selected, it must be fitted to the variogram estimates. Variogram fitting is another crucial stage of spatial prediction, because it also determines the kriging weights. Variogram estimates at different spatial lags are correlated, because the same observation is used for different lags. As a consequence, variogram fitting by ordinary least squares is not satisfactory. This problem is addressed by Genton (1998b), who suggests the use of a generalized least squares method with an explicit formula for the covariance structure (GLSE). A good approximation of the covariance structure is achieved by taking into account the explicit formula for the correlation of Matheron's classical variogram estimator in the Gaussian independent case. Simulations were carried out with several types of underlying variograms, as well as with outliers in the data. Results showed that the GLSE technique, combined with a robust estimator of the variogram, improves the fit significantly.

Recently, Genton (1999) has extended the explicit formula for the correlation structure to elliptically contoured distributions (Fang et al., 1989, Fang and Zhang, 1990, Fang and Anderson, 1990). This is a general class of distributions whose contours of equal density have the same elliptical shape as the multivariate Gaussian, but which contains long-tailed and short-tailed distributions. Some important elliptically contoured distributions are the Kotz type, Pearson type, multivariate t, multivariate Cauchy, multivariate Bessel, logistic, and scale mixture. For a subclass of elliptically contoured distributions with a particular family of covariance matrices, the correlation structure depends only on the spatial design matrix of the data, i.e. it is exactly the same as for the multivariate Gaussian distribution. This result allows to extend the validity of the GLSE method of variogram fitting.

Consider an omnidirectional variogram estimator $2\widehat{\gamma}(h)$ for a given set of lags $h_1, \ldots, h_k$, where $1 \leq k \leq K$ and K is the maximal possible distance between data. Denote further by $2\widehat{\gamma} = (2\widehat{\gamma}(h_1), \ldots, 2\widehat{\gamma}(h_k))^T \in \mathbb{R}^k$ the random vector with covariance matrix $\mathrm{Var}(2\widehat{\gamma}) = \Omega$. Suppose that one wants to fit a valid parametric variogram $2\gamma(h, \boldsymbol{\theta})$ to the estimated points $2\widehat{\gamma}$. The method of generalized least squares consists in determining the estimator $\widehat{\boldsymbol{\theta}}$ which minimizes

$$G(\boldsymbol{\theta}) = (2\widehat{\gamma} - 2\gamma(\boldsymbol{\theta}))^T \Omega^{-1} (2\widehat{\gamma} - 2\gamma(\boldsymbol{\theta})), \tag{8}$$

where $2\gamma(\boldsymbol{\theta}) = (2\gamma(h_1,\boldsymbol{\theta}),\ldots,2\gamma(h_k,\boldsymbol{\theta}))^T \in \mathbb{R}^k$ is the vector of the valid parametric variogram, and $\boldsymbol{\theta} \in \mathbb{R}^p$ is the parameter to be estimated. Note that $2\gamma(h,\boldsymbol{\theta})$ is generally a nonlinear function of the parameter $\boldsymbol{\theta}$. Journel and Huijbregts (1978) suggest to use only lag vectors h_i such that $N_{h_i} > 30$ and $0 < i \leq K/2$. This empirical rule is often met in practice. The GLSE algorithm is the following:

(1) Determine the matrix $\Omega = \Omega(\boldsymbol{\theta})$ with element Ω_{ij} given by

$$\mathrm{Corr}(2\widehat{\gamma}(h_i), 2\widehat{\gamma}(h_j))\gamma(h_i,\boldsymbol{\theta})\gamma(h_j,\boldsymbol{\theta})/\sqrt{N_{h_i}N_{h_j}}.$$

(2) Choose $\boldsymbol{\theta}^{(0)}$ and let $l = 0$.

(3) Compute the matrix $\Omega(\boldsymbol{\theta}^{(l)})$ and determine $\boldsymbol{\theta}^{(l+1)}$ which minimizes

$$\left(2\widehat{\gamma} - 2\gamma(\boldsymbol{\theta})\right)^T \Omega(\boldsymbol{\theta}^{(l)})^{-1}\left(2\widehat{\gamma} - 2\gamma(\boldsymbol{\theta})\right).$$

(4) Repeat (3) until convergence to obtain $\widehat{\boldsymbol{\theta}}$.

In step (1), an element of the matrix Ω is given by

$$\begin{aligned}\Omega_{ij} &= \mathrm{Cov}\left(2\widehat{\gamma}(h_i), 2\widehat{\gamma}(h_j)\right)\\ &= \mathrm{Corr}\left(2\widehat{\gamma}(h_i), 2\widehat{\gamma}(h_j)\right)\sqrt{\mathrm{Var}\left(2\widehat{\gamma}(h_i)\right)\mathrm{Var}\left(2\widehat{\gamma}(h_j)\right)}.\end{aligned} \tag{9}$$

The correlation $\mathrm{Corr}\left(2\widehat{\gamma}(h_i), 2\widehat{\gamma}(h_j)\right)$ can be approximated by the one in the independent case. An explicit formula can be found in Genton (1998b), which depends only on the lags h_i and h_j, as well as on the sample size n. The variances in equation (9) are replaced by their asymptotic expressions (Genton, 1998b), yielding the formula given in step (1). In step (2), the initial choice $\boldsymbol{\theta}^{(0)}$ can be carried out randomly, or with the result of a fit by ordinary least squares (OLS) or by weighted least squares (WLS). An SPLUS function for variogram fitting by generalized least squares (GLSE), denoted `glse.fitting`, is available on the Web at `http://www-math.mit.edu/~genton/`.

We present the estimation and fitting of a variogram for a simulated data set, having a spherical underlying variogram given by:

$$\gamma(h,a,b,c) = \begin{cases} 0 & \text{if } h = 0,\\ a + b\left(\frac{3}{2}\left(\frac{h}{c}\right) - \frac{1}{2}\left(\frac{h}{c}\right)^3\right) & \text{if } 0 < h \leq c,\\ a + b & \text{if } h > c,\end{cases}$$

with parameter $\boldsymbol{\theta} = (a,b,c)^T = (1,2,15)^T$. Figures 6 and 7 present the effects of outliers on estimation with Matheron's classical variogram estimator or the highly robust one, and fitting by weighted least squares

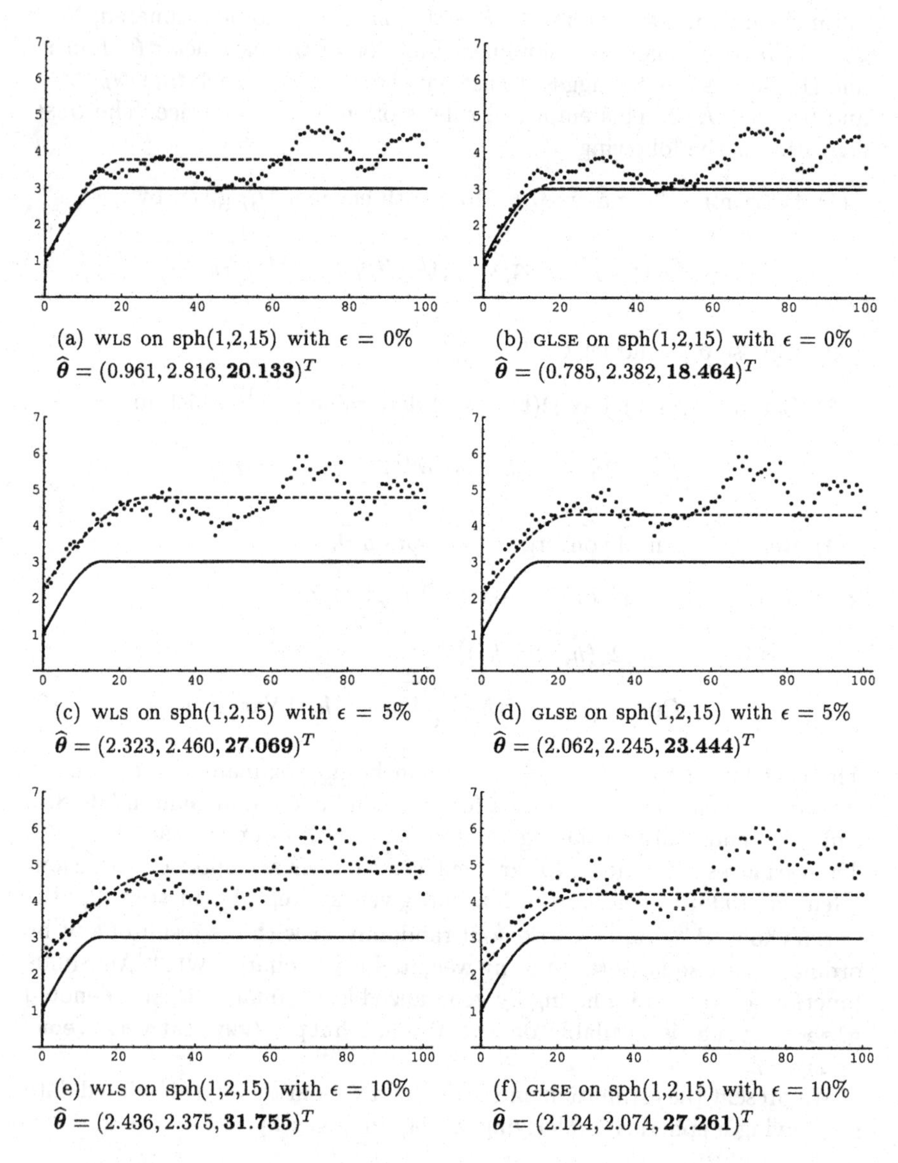

(a) WLS on sph(1,2,15) with $\epsilon = 0\%$
$\widehat{\boldsymbol{\theta}} = (0.961, 2.816, \mathbf{20.133})^T$

(b) GLSE on sph(1,2,15) with $\epsilon = 0\%$
$\widehat{\boldsymbol{\theta}} = (0.785, 2.382, \mathbf{18.464})^T$

(c) WLS on sph(1,2,15) with $\epsilon = 5\%$
$\widehat{\boldsymbol{\theta}} = (2.323, 2.460, \mathbf{27.069})^T$

(d) GLSE on sph(1,2,15) with $\epsilon = 5\%$
$\widehat{\boldsymbol{\theta}} = (2.062, 2.245, \mathbf{23.444})^T$

(e) WLS on sph(1,2,15) with $\epsilon = 10\%$
$\widehat{\boldsymbol{\theta}} = (2.436, 2.375, \mathbf{31.755})^T$

(f) GLSE on sph(1,2,15) with $\epsilon = 10\%$
$\widehat{\boldsymbol{\theta}} = (2.124, 2.074, \mathbf{27.261})^T$

Figure 6. An example of estimation with Matheron's classical variogram estimator of the perturbed spherical variogram and fit with WLS or GLSE (the underlying variogram is the solid line and the fitted variogram is the dashed line). Reprinted from *Mathematical Geology 30*(4), 323–345, with permission of the International Association for Mathematical Geology.

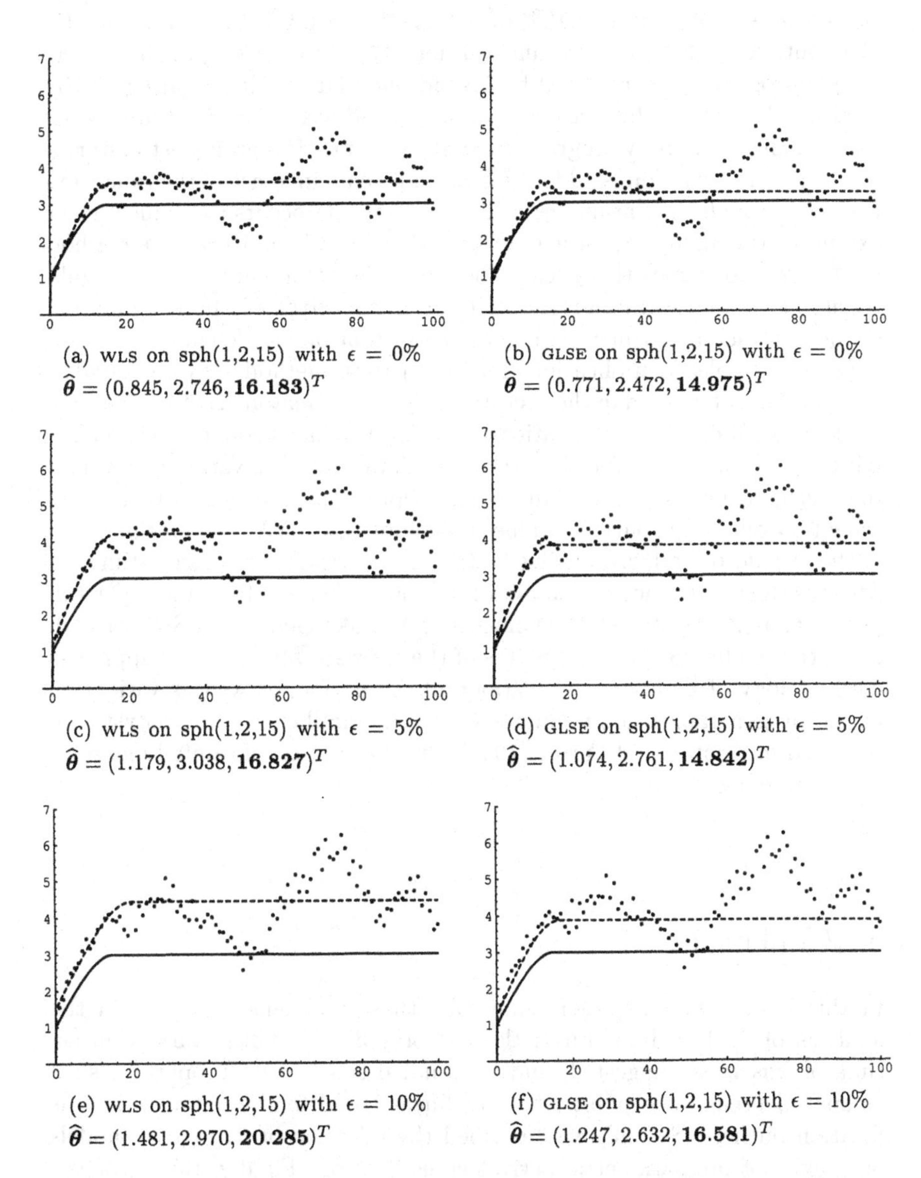

(a) WLS on sph(1,2,15) with $\epsilon = 0\%$
$\widehat{\boldsymbol{\theta}} = (0.845, 2.746, \mathbf{16.183})^T$

(b) GLSE on sph(1,2,15) with $\epsilon = 0\%$
$\widehat{\boldsymbol{\theta}} = (0.771, 2.472, \mathbf{14.975})^T$

(c) WLS on sph(1,2,15) with $\epsilon = 5\%$
$\widehat{\boldsymbol{\theta}} = (1.179, 3.038, \mathbf{16.827})^T$

(d) GLSE on sph(1,2,15) with $\epsilon = 5\%$
$\widehat{\boldsymbol{\theta}} = (1.074, 2.761, \mathbf{14.842})^T$

(e) WLS on sph(1,2,15) with $\epsilon = 10\%$
$\widehat{\boldsymbol{\theta}} = (1.481, 2.970, \mathbf{20.285})^T$

(f) GLSE on sph(1,2,15) with $\epsilon = 10\%$
$\widehat{\boldsymbol{\theta}} = (1.247, 2.632, \mathbf{16.581})^T$

Figure 7. An example of estimation with the highly robust variogram estimator of a perturbed spherical variogram and fit with WLS or GLSE (the underlying variogram is the solid line and the fitted variogram is the dashed line). Reprinted from *Mathematical Geology 30*(4), 323–345, with permission of the International Association for Mathematical Geology.

(Cressie, 1985) WLS or generalized least squares GLSE. The data is perturbed by $\epsilon = 5\%$ and $\epsilon = 10\%$ of observations from a Gaussian $N(0, 25)$ distribution, with mean zero and variance 25. On each graph, the underlying variogram is represented by a solid line and the fitted variogram by a dashed line. The effect of perturbations is noticeable by a greater vertical variability of the variogram estimations. For Matheron's estimator, a horizontal deformation is added, which leads to an increase of the range, expressed through the parameter c. This phenomena occurs to a much lesser extent for the highly robust variogram estimator. When fitting, the method GLSE tends to reduce this effect. Therefore, the combination of the highly robust variogram estimator and of GLSE fitting gives the best estimation of the parameter c, which is the most important one for kriging.

Some examples of application of the robustness methodology discussed in this paper can be found in the literature. Eyer and Genton (1999) present an application of highly robust variogram estimation in astronomy, where outlying values can sometimes be present in data sets. The variogram is used to determine a pseudo-period in the pulsation of variable stars. Simulations show that one single outlying value can completely mask the determination of the pseudo-period when using Matheron's classical variogram estimator, whereas the highly robust estimator remains unaffected. Furrer and Genton (1999) analyze a data set of sediments from Lake Geneva, in Switzerland, using the module S+SPATIALSTATS of the software SPLUS. They apply the methodology of highly robust variogram estimation, as well as variogram fitting by generalized least squares (GLSE). A similar type of analysis can be found in Genton and Furrer (1998b), in the context of rainfall measurements in Switzerland.

5 Conclusion

In this paper, we addressed some robustness problems occurring in the analysis of spatial data. First, the use of a highly robust variogram estimator has been suggested, and we studied some of its properties, such as spatial breakdown point, effect of linear trend in the data, as well as location outliers. Second, we described the selection of variogram models by means of nonparametric derivative estimation. Finally, the fit of the variogram model by generalized least squares has been discussed. Two new SPLUS functions for highly robust variogram estimation and variogram fitting by generalized least squares are made available on the Web at `http://www-math.mit.edu/~genton/`, as well as a MATLAB code for variogram model selection via nonparametric derivative estimation. Note that the kriging step has not been robustifyied yet, and further research is needed in this direction.

6 References

Bochner, S. (1955). *Harmonic Analysis and the Theory of Probability.* Berkeley: University of California Press.

Cherry, S. (1997). Non-parametric estimation of the sill in geostatistics. *Environmetrics 8*, 13–27.

Cherry, S., J. Banfield, and W.F. Quimby (1996). An evaluation of a non-parametric method of estimating semi-variogram of isotropic spatial processes. *Journal of Applied Statistics 23*, 435–449.

Cressie, N.A.C. (1985). Fitting variogram models by weighted least squares. *Mathematical Geology 17*, 563–586.

Cressie, N.A.C. (1993). *Statistics for Spatial Data* (revised ed.). Wiley Series in Probability and Mathematical Statistics: Applied Probability and Statistics. New York: Wiley.

Cressie, N.A.C. and D. M. Hawkins (1980). Robust estimation of the variogram. I. *Mathematical Geology 12*, 115–125.

Croux, C. and P. J. Rousseeuw (1992). Time-efficient algorithms for two highly robust estimators of scale. In Y. Dodge and J. Whittaker (Eds.), *Computational Statistics*, Volume 1, pp. 411–428. Heidelberg: Physica-Verlag.

Eyer, L. and M.G. Genton (1999). Characterization of variable stars by robust wave variograms: an application to Hipparcos mission. *Astronomy and Astrophysics, Supplement Series 136*, 421–428.

Fang, K.T. and T.W. Anderson (Eds.) (1990). *Statistical Inference in Elliptically Contoured and Related Distributions.* New York: Allerton Press.

Fang, K.T., S. Kotz, and K.W. Ng (1989). *Symmetric Multivariate and Related Distributions*, Volume 36 of *Monographs on Statistics and Applied Probability.* London: Chapman and Hall.

Fang, K.T. and Y.T. Zhang (1990). *Generalized Multivariate Analysis.* Berlin: Springer.

Furrer, R. and M.G. Genton (1999). Robust spatial data analysis of Lake Geneva sediments with S+SpatialStats. *Systems Research and Information Systems 8*, 257–272. special issue on Spatial Data Analysis and Modeling.

Genton, M.G. (1998a). Highly robust variogram estimation. *Mathematical Geology 30*, 213–221.

Genton, M.G. (1998b). Variogram fitting by generalized least squares using an explicit formula for the covariance structure. *Mathematical Geology 30*, 323–345.

Genton, M.G. (1998c). Spatial breakdown point of variogram estimators. *Mathematical Geology 30*, 853–971.

Genton, M.G. (1999). The correlation structure of Matheron's classical variogram estimator under elliptically contoured distributions. *Mathematical Geology 32*, 127–137.

Genton, M.G. and R. Furrer (1998a). Analysis of rainfall data by simple good sense: is spatial statistics worth the trouble? *Journal of Geographic Information and Decision Analysis 2*, 11–17.

Genton, M.G. and R. Furrer (1998b). Analysis of rainfall data by robust spatial statistics using S+SPATIALSTATS. *Journal of Geographic Information and Decision Analysis 2*, 126–136.

Gorsich, D.J. and M.G. Genton (1999). Variogram model selection via nonparametric derivative estimations. *Mathematical Geology 32*, 249–270.

Hampel, F.R. (1973). Robust estimation, a condensed partial survey. *Zeitschrift für Wahrscheinlichkeitstheorie und Verwandte Gebiete 27*, 87–104.

Hampel, F.R., E.M. Ronchetti, P.J. Rousseeuw, and W.A. Stahel (1986). *Robust Statistics, the Approach Based on Influence Functions.* Wiley Series in Probability and Mathematical Statistics: Probability and Mathematical Statistics. New York: Wiley.

Huber, P.J. (1977). *Robust Statistical Procedures*, Volume 27 of *CBMS-NSF Regional Conference Series in Applied Mathematics.* Philadelphia: Society for Industrial and Applied Mathematics.

Journel, A.G. and Ch.J. Huijbregts (1978). *Mining Geostatistic.* London: Academic Press.

Matheron, G. (1962). *Traité de géostatistique appliquée. I*, Volume 14 of *Mémoires du Bureau de Recherches Géologiques et Minières.* Paris: Éditions Technip.

Rousseeuw, P.J. and C. Croux (1992). Explicit scale estimators with high breakdown point. In Y. Dodge (Ed.), *L_1-Statistical Analyses and Related Methods*, pp. 77–92. Amsterdam: North-Holland.

Rousseeuw, P.J. and C. Croux (1993). Alternatives to the median absolute deviation. *Journal of the American Statistical Association 88*, 1273–1283.

Shapiro, A. and J.D. Botha (1991). Variogram fitting with a general class of conditionally nonnegative definite functions. *Computational Statistics & Data Analysis 11*, 87–96.

3

Standards Adoption Dynamics and Test for Nonspatial Coordination

C. Hardouin
X. Guyon

ABSTRACT When n individuals choose a standard according to "local rules" depending on their position on a set $S = \{1, 2, \ldots, n\}$, there is a spatial correlation occuring between these choices. On the contrary, other situations lead to spatial independency. We study here some models with or without spatial coordination and present tests for absence of spatial coordination.

1 Introduction

We first describe the model considered. Suppose n consumers are placed on a finite regular set of sites $S = \{1, 2, \ldots, n\}$ and let $\mathcal{G}$ be a symmetric graph on S: $\langle i, j \rangle$ means that two different sites i and j are neighbours with respect to $\mathcal{G}$. We denote by ∂i the neighbours of i :

$$\partial i = \{j \in S : \langle i, j \rangle\}.$$

For instance, S can be the square (or the torus) $\{1, 2, \ldots, N\}^2$ of $\mathbf{Z}^2$, $n = N^2$, equipped with the four nearest neighbours relation graph.

Each individual i chooses a standard $y_i \in E$, according to a rule to be defined. For simplicity, we consider only binary set of standards $E = \{0, 1\}$, but the results can be generalized to a qualitative finite set (see Section 4.4). Let us denote $y = (y_i, i = 1, \ldots, n)$ the configuration of the n choices, $y^i = (y_j, j \neq i)$ the configuration except at i, y_A the configuration of y on a set A and y^A the configuration of y outside A.

In the two following sections, we describe several rules for standards adoption. First we present models without *spatial coordination* (Section 2), and then look for models where the adoption rule depends on the *local context* of the previous neighbours; in that case, we define spatial coordination.

We begin in Section 3.1 with the case of a one step sequential choice: let σ be a permutation of $[1, 2, \ldots n]$ (we also call it a sweep of S); we first choose the value $y_{\sigma(1)}$ at the site $\sigma(1)$, then $y_{\sigma(2)}$ at the site $\sigma(2)$, and so on up to the site $\sigma(n)$. The distribution of y_i at the site $i = \sigma(k)$ (kth step) depends on the previous choices made by the neighbours of i. In spite of the simplicity of these rules, their analytic features, like the nearest neighbours correlation, cannot be explicited because of the complexity related to the geometry of the permutation. On the other hand, this simplicity allows an empirical approach by simulations (Bouzitat and Hardouin, 2000).

Repeating the scans, and noting $Y(k)$ the configuration on all the set S after the kth scan ($k \geq 1$) (this is a step by step learning), the model becomes easy to specify. In fact, $Y(1), Y(2), \ldots$ is an ergodic Markov chain for which the stationary distribution can be characterized. When the scans of S are sequential (or asynchronous, see Section 3.2), we get the *Gibbs sampler dynamics* (Geman and Geman, 1984, Geman, 1990, Guyon, 1995, Robert, 1996); the stationary distribution is then the distribution defining the rules of choice. If the scans are synchronous (see Section 3.3), the n values being simultaneously modified, the chain is still ergodic, but it is more difficult to determine explicitly the stationary distribution, which differs from the previous one.

Contrary to the models presented in Section 2, models of Section 3 are characterized by spatial correlation. We give, in Section 4, tests for absence of spatial coordination. Two different situations occur. If the spatial model is parametric, depending on a parameter $\theta \in \mathbb{R}^p$, in such a way that the nonspatial coordination is characterized by the nullity of some componants of θ, then we suggest parametric tests derived from the usual estimation procedures (maximum likelihood or pseudo maximum likelihood for instance) and their asymptotic properties (see e.g. Besag, 1974, Besag and Moran, 1975, Cliff and Ord, 1981, or Guyon, 1995, §5.3). If the model is nonparametric, we propose other methods which do not need an a priori spatial knowledge; these methods are based on the estimation of spatial correlation indicators and their asymptotic distribution (like for instance the Moran's indicator, see Cliff and Ord, 1981). We give tests based on the empirical estimation of the nearest neighbours correlation and its asymptotic distribution under the null hypothesis. We finally compare these tests by computing their power under two alternatives for Y; an unconditional Ising model, and an Ising model conditional to the sum of the Y_i's value.

2 Models without Spatial Coordination

We define the model without spatial coordination as follows.

1. We choose a distribution π for a variable $X = (X_i, i = 1, \ldots, n) \in \{0,1\}^n$. This distribution π depends on $(2^n - 1)$ parameters, the probabilities $\pi_A^X = P(X_A = 1_A, X^A = 0^A)$, where $A \in \mathcal{P}_n$, the set of all subsets of $I = [1, \ldots, n]$, under the constraint $\sum_A \pi_A^X = 1$.

2. Then we choose a permutation σ of $[1, 2, \ldots, n]$ which assigns x_i to

the site $\sigma(i)$. So, we get $y_i = x_{\sigma^{-1}(i)}$ (or $y_{\sigma(i)} = x_i$). The permutation σ is chosen according to a distribution μ on Ξ_n, the set of the permutations of $[1, 2, \ldots, n]$. The distribution of Y is then parametrized by (π, μ).

Let us set $S_X = \sum_{i=1}^n X_i = S_Y$, $p = E(S_X)/n$ and define, for each $0 \leq k \leq n$, the probabilities

$$\pi(k) = P(S_X = k) = \sum_{\substack{A \in \mathcal{P}_n \\ |A| = k}} \pi_A^X \quad \text{and} \quad p_k = \binom{n}{k}^{-1} \pi(k)$$

where $|A|$ denotes the cardinal number of A. We have the following property.

Proposition 1. *We assume that σ is uniformly chosen on Ξ_n.*

(1) *For any subset B of $I = [1, \ldots, n]$, we have $\pi_B^Y = p_{|B|}$. In particular, the variables $Y_1, Y_2, \ldots, Y_n$ are exchangeable: for any $\delta \in \Xi_n$, all the variables $Y_{(\delta)} = (Y_{\delta(1)}, Y_{\delta(2)}, \ldots, Y_{\delta(n)})$ have the same distribution.*

(2) *The distribution of the variables Y_i is Bernoulli $\mathcal{B}(p)$, with covariance for $i \neq j$,*

$$\mathrm{Cov}(Y_i, Y_j) = \frac{\mathrm{Var}\, S_X}{n(n-1)} + \frac{(E(S_X))^2}{n^2(n-1)} - \frac{E(S_X)}{n(n-1)}.$$

This covariance is of order $O(1/n)$ as soon as $\mathrm{Var}\, S_X = O(n)$.

(3) *Conditionally to $S_X = b$, $0 \leq b \leq n$, the distribution of Y is uniform on the subsets of S of size b, and verifies: $E(Y_i) = b/n = \tilde{p}$, and, for $i \neq j$, $\mathrm{Cov}(Y_i, Y_j) = -\tilde{p}(1-\tilde{p})/(n-1)$.*

Proof. (1) Let B be a subset of cardinal number b. The realization $(Y_B = 1_B, Y^B = 0^B)$ comes from the realizations $(X_A = 1_A, X^A = 0^A)$, where A is of cardinal number b. Once A is set, there are $b!\,(n-b)!$ permutations from A to B. This gives the announced form for π_B^Y. As these probabilities depend only on b, the variables $Y_1, Y_2, \ldots, Y_n$ are exchangeable.

(2) In particular, the distribution is Bernoulli with parameter p given by

$$P(Y_i = 1) = \sum_{y^i} P(Y_i = 1, y^i) = \sum_{k \geq 1} \frac{\binom{n-1}{k-1}}{\binom{n}{k}} \pi(k) = \frac{1}{n} E(S_X) = p.$$

The following computation leads to the covariance function:

$$P(Y_i = Y_j = 1) = \sum_{k \geq 2} \frac{\binom{n-2}{k-2}}{\binom{n}{k}} \pi(k) = \sum_{k \geq 2} \frac{k(k-1)}{n(n-1)} \pi(k).$$

(3) follows from (1) and (2). □

It is worth pointing out that conditionally to $S_X = b$ (among the n individuals, b have chosen the standard 1), the variables Y_i's are identically distributed, and hardly correlated if n is large enough, the joined distribution being uniform on the subsets of size b. Only the information $S_X = S_Y = b$ is important, the permutation σ and the initial distribution of X do not occur any more in the distribution of Y.

Example 1 ($X_i \sim B(p_i)$ independent, $i = 1, n$). Let $\bar{p}$ be the average of the p_i's, and $V(p) = \overline{p^2} - \bar{p}^2$ be their variance. The variables Y_i's have a Bernoulli distribution with parameter $\bar{p}$, covariance $-(V(p))/(n-1)$, whose absolute value is bounded by $1/(4(n-1))$. They are independent if and only if the p_i are all equal. In which case, the Y_i's as well as the X_i's form a Bernoulli sample.

Example 2 (Polya's urn, without spatial interaction (Feller, 1957, p. 120)). Let $b > 0$ and $r > 0$ be the initial numbers of white and red balls in an urn. We draw balls with replacement and at each drawing, we add $c \geq 0$ balls of the same color than the one which has just been drawn. With the notation $X_i = 1$ if a white ball is drawn at the ith drawing, the Polya's urn model without spatial interaction is characterized by the sequence of distributions of X_i satisfying

$$P(X_i = 1 \mid X_1 = x_1, \ldots, X_{i-1} = x_{i-1}) = p_i$$

where $p_i = (b + c\sum_{j<i} x_j)/(b + r + (i-1)c)$, $i = 1, n$. Thus, the conditional distribution of X_i is Bernoulli with parameter p_i evolving with i. Whenever the permutation σ is at random, the conditional distribution $(Y \mid S_X)$ is independent of the p_i's.

3 Models with Spatial Coordination

We consider now the situation where the choice of a standard at a given site is made according to the past choices made in the neighbourhood of this site. Typically, spatial dependency appears; it will be positive in case of cooperation, and negative if there is competition.

If we consider one scan only, it is not possible to obtain analytically the main characteristics of these models (see Section 3.1). On the contrary, if we repeat indefinitely the scanning (or if there is a large number of scans), these characteristics can be made explicit (see Section 3.2).

3.1 One Step Sequential Choices

The model. Once the position σ of the n individuals is set, the model is the following:

1. At the site $\sigma(1)$, we choose $X_1 = Y_{\sigma(1)} \sim \mathcal{B}(p_1)$. Let us suppose that the following choices have occured: $x_1, x_2, \ldots, x_{i-1}$, $i \leq n$.

2. At the site $\sigma(i)$, we choose Y according to a contextual rule

$$X_i = Y_{\sigma\theta(i)} \sim \mathcal{B}\big(p_i(x^{i-}, \sigma)\big)$$

where $p_i(x^{i-}, \sigma) = p_i(x_j : j < i \text{ and } \langle\sigma(i), \sigma(j)\rangle)$.

Example 3 (Progressing Ising model). We consider the d-dimensional torus $S = T^d = \{1, 2, \ldots, N\}^d$, with $n = N^d$ sites, equipped with the $2d$-nearest neighbours relation (with the usual linking conditions at the boundary). The state space is $E = \{0, 1\}$. We define $n_i(x) = \sum x_j$, where the sum is on the $j < i$ such that $\langle\sigma(i), \sigma(j)\rangle$. In our model, $n_i(x) \in \{0, 1, \ldots, 2d\}$ typifies the effect of the neighbourhood of $\sigma(i)$ at the time i, the distribution of X_i being given by

$$p_i(x_i \mid x^{i-}) = \frac{\exp x_i\{\alpha + \beta n_i(x)\}}{1 + \exp\{\alpha + \beta n_i(x)\}}, \quad x_i \in \{0, 1\}.$$

where α, β, are two real parameters. The parameter β controls spatial coordination: $\beta > 0$ ensures cooperation while $\beta < 0$ expresses competition.

Example 4 (Majority assignment model). In the previous model we have $\alpha + \beta n_i(x) = (\alpha + d\beta) + \beta(n_i(x) - d)$. Thus, if we impose the constraint $\alpha + d\beta = 0$ and if we let $\beta \to +\infty$, we find the rule of *majority assignment*: we set $x_i = 1$ (resp. $x_i = 0$) if $n_i(x) > d$ (resp. $n_i(x) < d$), choosing on $\{0, 1\}$ at random if $n_i(x) = d$.

We can still write down the joined distribution of X

$$\pi(x) = \frac{1}{n!} \sum_{\sigma \in \Xi_n} \prod_{i=1,n} p_i(x_i \mid x^{i-}, \sigma),$$

but it is reliant on the geometric complexity due to the permutations σ. This explains why it is impossible to express analytically the characteristics of π, like the nearest neighbours correlation. However, these can be obtained by simulations (Bouzitat and Hardouin, 2000).

3.2 Repeating Assignments: the Gibbs Sampler Dynamics

After the first scan of S, we repeat the procedure. A main difference with what happens at the first scan is that now, from the second scan, all the neighbours of any site are already occupied. Let us note k the number of the scan, $y = Y(k-1)$ and $z = Y(k)$ the configuration before and after the kth scan σ. Then, $Y = \big(Y(k)\big)_{k \geq 2}$ is a *Markov chain* on $\{0, 1\}^S$ with transition probabilities

$$P_\sigma(y, z) = \prod_{i=1,n} p_i(z_i \mid z_1, z_2, \ldots, z_{i-1}, y_{i+1}, \ldots, y_n).$$

The sequential feature is expressed as follows; the system changes step by step the values at each site, conditionally to the new assignments z and to the past ones y which remain to be modified. The new scans can be seen as new learning steps.

Let us note σ_k the scan at the kth step and $P_k = P_{\sigma_k}$; Y is an inhomogeneous chain. The distribution of $Y(k)$ is

$$Y(k) \sim \mu P_1 P_2 P_3 \cdots P_k.$$

where μ stands for the initial distribution of $Y(0)$. If the scanning order is the same at each scan, $\sigma_k \equiv \sigma$ for all $k \geq 1$, the chain is then homogeneous with transition probabilities $P = P_\sigma$, and $Y(k) \sim \mu P^k$.

The following ergodicity property holds.

Proposition 2. *Let us assume that* $\delta = \inf\{p_i(y_i \mid y^i),\, y \in \{0,1\}^S\} > 0$.

(1) *The homogeneous case: the chain* Y *is ergodic with limiting distribution* π*, the unique invariant distribution of* P *being such that* $\pi P = \pi$.

(2) *The inhomogeneous case: we assume that there exists a common invariant distribution* π *for all permutation* $\sigma(\forall\sigma, \pi P_\sigma = \pi)$. *Then the inhomogeneous chain* Y *is ergodic converging towards* π.

Proof. (1) We have $\inf_{y,z} P(y,z) \geq \delta^n > 0$; the transition probability is regular. The state space being finite, Y is ergodic (Kemeny and Snell, 1960).

(2) The chain is inhomogeneous and we use the result of Isaacson and Madsen (1976); see also Guyon (1995, Theorem 6.1.1). Their sufficient conditions ensuring ergodicity are satisfied here because π is invariant for all P_σ, and because the contraction coefficients verify

$$\forall\sigma : c(P_\sigma) \leq (1 - 2\delta)^n < 1. \qquad \square$$

Remark 1. (i) The results still hold in the case of scans σ_k with possible repetitions, but covering all the set S, and of bounded size.

(ii) We can control the rate of convergence with respect to δ. Let $\|\cdot\|$ be the norm in variation; then we have

$$\|\mu P_1 P_2 P_3 \cdots P_k - \pi\| \leq 2[1 - (2\delta)^n]^k.$$

(iii) The required hypothesis in the inhomogeneous case is automatically verified for the Gibbs sampler; we describe it below.

Example 5 (The Gibbs sampler (Geman and Geman, 1984, Geman, 1990, Guyon, 1995, Robert, 1996)). Let π be some distribution over E^S fulfilling the *positivity condition*

(P) $$\forall y \in E^S, \pi(y) > 0.$$

We denote by $\pi_i(y_i \mid y^i)$ the conditional distributions at each site $i \in S$. The algorithm defined with $p_i(x_i \mid x^i) = \pi_i(x_i \mid x^i)$ is called a *Gibbs sampler of the distribution* π. As the positivity condition on π implies the positivity of the distributions π_i, we have $\delta > 0$. Moreover, π being invariant for each π_i, it is again invariant for each P_σ. Whether the scans are homogeneous or not, the chain Y is ergodic with limiting distribution π. We have here a sampling algorithm of π.

Example 6 (Markov field sampler). We consider the two-dimensional torus $S = T^2$ equipped with the 4 nearest neighbours relation, and we assume the values of Y are in $\{0, 1\}$. Suppose π is given by

$$\pi(y) = Z^{-1} \exp\left\{\alpha \sum_{i \in S} y_i + \beta \sum_{\langle i,j \rangle} y_i y_j\right\} \tag{1}$$

where Z is the appropriate normalizing constant. The two standards are equiprobable if $\alpha + 2\beta = 0$. Indeed, $\alpha \sum_{i \in S} y_i + \beta \sum_{\langle i,j \rangle} y_i y_j = \sum_i y_i\{(\alpha + 2\beta) + \beta(v_i - 2)\}$ where $v_i = \sum_{j \in \partial i} y_j$ is the sum on the nearest neighbours of i. If $\alpha + 2\beta > 0$, 1 is more probable than 0. The conditional distribution at the site i is

$$\pi_i(y_i \mid y^i) = \frac{\exp y_i(\alpha + \beta v_i)}{1 + \exp(\alpha + \beta v_i)} \tag{2}$$

depending on the neighbourhood configuration *via* v_i.

We give in Section 4.3 (Figures 2 and 4) the plots of the correlation at distance 1, $\beta \to \rho(\beta)$, the correlations being estimated by simulation (see Pickard, 1987, for exact computation).

3.3 Synchronous Choices

We assume now that in the transition from $Y(k-1) = y$ to $Y(k) = z$, the n choices are synchronous (or simultaneous). We write the new transition

$$Q(y, z) = \prod_{i=1,n} p_i(z_i \mid y^i).$$

Q remains ergodic, but its (unique) stationary distribution μ is very different from the one provided by a sequential sweeping. Usually, we cannot give explicit analytic form for μ. However, if $p_i = \pi_i$, the π_i being the conditional distributions of the nearest neighbours Ising model, it is possible to write μ explicitly; we precise it in the following example.

Example 7 (Synchronous sampling relative to an Ising model (Trouvé, 1988)). Let π be the distribution for the Ising model (1) on the two-dimensional torus. We consider the measure

$$\mu(y) = \Gamma^{-1} \prod_{i=1,n} \{1 + \exp(\alpha + \beta v_i)\}$$

where Γ is the appropriate normalizing constant and v_i the sum on the nearest neighbours values. We have $\mu Q = \mu$. It is the limit distribution (called *virtual*) of the synchronous sampling relative to π. Let us note that we have

$$\mu(y) = \Gamma^{-1} \exp\left\{\sum_{i \in S} \Phi(y_{\partial i})\right\}$$

where the so called potentials are $\Phi(y_{\partial i}) = \log\{1 + \exp(\alpha + \beta v_i)\}$. The *cliques* defining the *Markovian graph* of μ (see Besag, 1974, Guyon, 1995) are the squares $\partial i = \{i_1, i_2, i_3, i_4\}$ centered in $i \in S$, and with the diagonal sides of length $\sqrt{2}$. We note two significant differences between π and μ.

- $\langle i, j \rangle_\mu \Leftrightarrow \|i - j\|_2 = \sqrt{2}$; the neighbour sites for μ are the nearest sites on the diagonals (for π, $\langle i, j \rangle_\pi \Leftrightarrow \|i - j\|_1 = 1$).
- Let us set $S_+ = \{i = (i_1, i_2) \in S, i_1 + i_2 \text{ even}\}$, S_- the complementary set, y_+ (resp. y_-) the configuration y on S_+ (resp. S_-). If the side N of the torus is even, we have:

$$\mu(y) = \mu_+(y_+)\mu_-(y_-)$$

with $\mu_+ = \mu_-$ and $\mu_+(y_+) = \Gamma^{-1/2} \prod_{i \in S_-} \{1 + \exp(\alpha + \beta v_i)\}$. Thus, for μ, the configurations on S_+ and S_- are independent.

4 Tests for Nonspatial Coordination: the Case of a Regular Lattice with ν Nearest Neighbours

Let $S = \{1, 2, \ldots, n\}$ be a regular lattice equipped with the ν nearest neighbours relation. In all the following we consider the nonspatial coordination framework, as described in section 2, and we suppose the variables X_i are independent. We consider the models:

(B) The *i.i.d.* Bernoulli model with probability π. As the X_i are i.i.d. with distribution $\mathcal{B}(\pi)$, the variables Y_i are also i.i.d. $\mathcal{B}(\pi)$.

(HG) The "Hypergeometric" model conditional to $b = \sum_i X_i = \sum_i Y_i$.

A natural statistic to test the hypothesis (H_0) of nonspatial coordination is based on C_n, the estimator of the nearest neighbours covariance c. The statistic C_n itself is expressed with M_n:

$$M_n = \frac{2}{n\nu} \sum_{\langle i,j \rangle} Y_i Y_j.$$

Concerning the Y_i, there is nonspatial coordination if $Y_i = X_{\sigma^{-1}(i)}$ is "independent" of $Y_j = X_{\sigma^{-1}(j)}$; in the context of (B), this corresponds to the nonspatial correlation, i.e., $c = \mathrm{Cov}(Y_i, Y_j) = 0$; but in the context of (HG), we do not have $c \neq 0$. We give the explicit expression of the statistic C_n in the three contexts.

- (B) and π *known.*

$$C_n = M_n - \pi^2.$$

- (B) and π *unknown.* When π is unknown, π^2 is estimated without bias by $n(\overline{Y}_n^2 - 1/n\overline{Y}_n)/(n-1)$. We then use the following statistic in place of C_n

$$\Gamma_n = M_n - \frac{n}{n-1}\left(\overline{Y}_n^2 - \frac{1}{n}\overline{Y}_n\right).$$

- the case (HG)

$$C_n = M_n - \frac{b(b-1)}{n(n-1)}.$$

There are $n\nu/2$ pairs $\{i,j\}$ of neighbour sites, and all the above statistics are centered under (H_0). It remains to determine their distribution. First, we calculate their variances; then, we study the asymptotic Gaussian behaviour.

4.1 Identification of C_n's Variance

Proposition 3. (a) *For the Bernoulli model* (B),

$$\mathrm{Var}_\mathrm{B}(C_n) = \sigma_\mathrm{B}^2 = \frac{2\pi^2(1-\pi)}{\nu n}\{1 + (2\nu - 1)\pi\}.$$

(b) *For the conditional hypergeometric model* (HG),

$$\mathrm{Var}_\mathrm{HG}(C_n) = A + B,$$

with

$$A = \frac{2b(b-1)}{\nu n^2(n-1)}\left\{1 + 2(\nu-1)\frac{b-2}{n-2} - (2\nu-1)\frac{(b-2)(b-3)}{(n-2)(n-3)}\right\},$$

$$B = \frac{b(b-1)(n-b)[6(n+b-1) - 4nb]}{n^2(n-1)^2(n-2)(n-3)}.$$

If $n \to \infty$ and $b/n \to \pi$, the asymptotic variance is equivalent to

$$\mathrm{Var}_\mathrm{HG}(C_n) \sim \sigma_\mathrm{HG}^2 = \frac{2\pi^2(1-\pi)^2}{\nu n}.$$

In particular, the variance of C_n is larger under the unconditional model (B) *than under the conditional model* (HG).

Proof. The proof is based on the computation of the expectation of C_n^2. Its terms, their number, and their respective expectations under (B) and (HG) are:

- $Y_i^2 Y_j^2 = Y_i Y_j$, $\langle i,j \rangle$, there are $\nu n/2$, of respective expectations π^2 (B), and $b(b-1)/n(n-1)$ (HG).
- $Y_i^2 Y_j Y_k = Y_i Y_j Y_k$, $\langle i,j \rangle$, $\langle i,k \rangle$ and $j \neq k$, there are $n\nu(\nu-1)$, of respective expectations π^3, and $b(b-1)(b-2)/n(n-1)(n-2)$.
- $Y_i Y_j Y_k Y_l$, $\langle i,j \rangle$, $\langle k,l \rangle$ and $\{i,j\} \cap \{k,l\} = \varnothing$, there are $(\nu n/2) \times (\nu n/2 - 2\nu + 1)$, of respective expectations π^4 and $b(b-1)(b-2) \times (b-3)/[n(n-1)(n-2)(n-3)]$.

Writing

$$\mathrm{Var}_{\mathrm{B}}(C_n) = E(C_n^2) - \pi^4, \quad \mathrm{Var}_{\mathrm{HG}}(C_n) = E(C_n^2) - \{b(b-1)/n(n-1)\}^2,$$

a straightforward computation gives the announced results. □

4.2 Asymptotic Normality and Test for Nonspatial Correlation

The Bernoulli Model (B)

We want to establish the asymptotic normality of C_n and Γ_n. To simplify the presentation, we assume that S is either the cube $\{1, 2, \ldots, N\}^d$ of $\mathbb{Z}^d$, or the d-dimensional torus of the same size, both of them being equipped with the $\nu = 2d$ nearest neighbours relation.

Proposition 4. *Test of absence of spatial coordination under* (B).

(1) *If π is known,*

$$\sqrt{\frac{\nu n}{2}} C_n \xrightarrow{d} \mathcal{N}(0, \pi^2(1-\pi)\{1 + (2\nu - 1)\pi\}).$$

(2) *If π is unknown,*

$$\sqrt{\frac{\nu n}{2}} \Gamma_n \xrightarrow{d} \mathcal{N}(0, \pi^2(1-\pi)^2).$$

Proof. We set $U_n = (C_n, \overline{Y}_n)$.

First step. Let us show that

$$\mathrm{Var}(U_n) = 2\frac{\pi^2(1-\pi)}{n} \begin{pmatrix} \frac{1+(2\nu-1)\pi}{\nu} & 1 \\ 1 & \frac{1}{2\pi} \end{pmatrix}$$

The C_n 's variance has been already determined and the variance of $\overline{Y}_n$ is equal to $\pi(1-\pi)/n$. It remains to calculate the covariance between C_n and $\overline{Y}_n$. For this purpose, it is sufficient to identify the terms of the following product and evaluate their number and their expectations:

$$\Big\{\sum_{i\in S} Y_i\Big\}\Big\{\sum_{\langle j,k\rangle} Y_jY_k\Big\}.$$

- $Y_iY_jY_k$, i, j and k nonequal, there are $\nu n(n-2)/2$, each of expectation π^3.
- $Y_i^2Y_j = Y_iY_j$. there are νn, each of expectation π^2.

This implies the announced result.

Second step. U_n is asymptotically Gaussian. We first look at the case where S is a cube of $\mathbf{Z}^d$. Let us note ∂i^+ the d positively oriented directions among the $2d$ directions, $\Delta_i = Y_i\{\sum_{j\in\partial i^+} Y_j\}$ and $Z_i = (Y_i, \Delta_i)$. The variables Z_i are bounded and identically distributed, $M_n = 1/nd\sum_i \Delta_i$ and $\overline{Y}_n = 1/n\sum_i Y_i$.

Moreover, if $\|i-j\|_1 > 1$, Z_i and Z_j are independent. We may then apply a central limit theorem for weakly dependent and bounded fields (Bolthausen, 1982, Guyon, 1995, Theorem 3.3.1).

Suppose now S is the torus; we write $\sum_{i\in S}\Delta_i = \sum_{i\in S^-}\Delta_i + R_n$, where $S^-$is the cube $\{1,2,\ldots,N-1\}^d$ and R_n is the sum of the N^{d-1} border terms. We apply the previous result on S^- and note that $\big(R_n - E(R_n)\big)/\sqrt{n}$ $\to 0$ in probability to obtain the result. Indeed, let us note $R_{n,j}$ the terms of R_n in the direction j, so $R_n = \sum_{j=1}^d R_{n,j}$, $\operatorname{Var} R_n \le d^2 \operatorname{Var} R_{n,1} \le d^2N^{d-1}C$ where C is the variance of the product of two neighbour terms and then $\operatorname{Var} R_n/\sqrt{n} = 0(N^{d-1}/N^d)$.

Third step. Asymptotic normality of Γ_n. The normality and the identification of the asymptotic variance follow from the following property; if $\sqrt{n}(Z_n-\mu) \xrightarrow{d} \mathcal{N}_p(0,\Sigma)$, and if $g:\mathbb{R}^p\to\mathbb{R}^q$ is a continuously differentiable mapping on a neighbourhood of μ, then $\sqrt{n}\big(g(Z_n)-g(\mu)\big) \xrightarrow{d} \mathcal{N}_q(0, J\Sigma^tJ)$ where J is the Jacobian matrix of g at μ. Let $\Gamma'_n = g(U_n) = C_n - \bar{X}_n^2$. We see that $\Gamma_n - \Gamma'_n = o_P(1)$ in probability and also $E(U_n) = (0,\pi)$, $\Sigma = n\operatorname{Var}(U_n)$ which has been calculated at the first step, g is regular everywhere, $J = (1,-2\pi)$, and we verify that the asymptotic variance of $\sqrt{n}\Gamma'_n$ is the one announced. This completes the proof. □

Remark 2. When π is unknown (estimated by $\overline{Y}_n$), the statistic Γ_n has the same variance than the one obtained under the model (HG) conditional to $S_Y = n\overline{Y}_n$,

$$\operatorname{Var}_{\mathrm{HG}}(C_n) = \operatorname{Var}_{\mathrm{B}}(\Gamma_n) < \operatorname{Var}_{\mathrm{B}}(C_n).$$

The Conditional Model (HG)

We assume the following asymptotic framework:

$$\text{(S)} \qquad \frac{b}{n} = \frac{1}{n} S_Y \xrightarrow[n\to\infty]{} \pi,$$

i.e. there is stabilization of the standards' proportion. Conditionally to S_Y, and for $i \neq j$,

$$\mathrm{Cov}(Y_i, Y_j) = -\frac{\overline{Y}_n(1-\overline{Y}_n)}{n-1}.$$

The vector $(M_n, \overline{Y}_n)$ is not asymptotically Gaussian because the conditional expectation is quadratic in $\overline{Y}_n$,

$$E(M_n \mid \overline{Y}_n) = \frac{\overline{Y}_n(n\overline{Y}_n - 1)}{(n-1)}.$$

We obtain the conditional normality of $(M_n \mid \overline{Y}_n)$ using a result of J. Bretagnolle (Prum, 1986, p. 168, Theorems 2 and 3). This allows us to present the following result which leads to a test for the absence for spatial coordination under (HG).

Proposition 5. *Under the stabilization condition* (S),

$$\sqrt{\frac{\nu n}{2}}\left(M_n - \frac{\overline{Y}_n(n\overline{Y}_n - 1)}{(n-1)}\right) \xrightarrow{d} \mathcal{N}(0, \pi^2(1-\pi)^2).$$

In fact, as $\Gamma_n - C_n = O_P(1/n)$, we find another way to obtain the result given for (B) when π is unknown.

4.3 Other Tests and an Empirical Study of Power

We can consider other statistics to test nonspatial coordination, based for instance on another assessment of spatial correlation. Among two tests we will prefer the one with the higher power. We compare here three tests under two kinds of alternative hypothesis.

For simplicity, we consider a square two-dimensional torus with $n = N^2$ sites. Let Y be a realization of our binary process. As previously studied, the nonspatial coordination can be tested using the estimation M_n of the four nearest neighbours correlation

$$M_n = \frac{1}{2n} \sum_{\langle i,j \rangle} Y_i Y_j.$$

We can also use the statistic M_n^* of the 8 nearest neighbours correlation (4 adjacent neighbours plus 4 diagonal neighbours)

$$M_n^* = \frac{1}{4n} \sum_{\langle\langle i,j \rangle\rangle} Y_i Y_j.$$

Proposition 4 gives the variance for these two statistics. In fact, looking at the proof of this proposition, we see that the number of neighbours is relevant, but not the distance between them. Under the model (B) (resp. (HG), for $n \to \infty$ and $b/n \to \pi$), we obtain:

$$\mathrm{Var_B}(M_n) = \frac{\pi^2}{2n}(1-\pi)(1+7\pi) \quad \text{and} \quad \mathrm{Var_{HG}}(M_n) \simeq \frac{\pi^2}{2n}(1-\pi)^2$$
$$\mathrm{Var_B}(M_n^*) = \frac{\pi^2}{4n}(1-\pi)(1+15\pi) \quad \text{and} \quad \mathrm{Var_{HG}}(M_n^*) \simeq \frac{\pi^2}{4n}(1-\pi)^2.$$

The asymptotic normality of M_n^* is obtained in a similar way as for M_n.

Power of the Tests

We consider two kinds of alternative hypothesis for Y; for each, we evaluate by simulation the power of the two tests M_n and M_n^*, built at the 5% level.

Alternative 1 (Y is an Ising model (versus $(\mathbf{H}_0)$=(B))). Y is the Ising model (1) with $\alpha + 2\beta = 0$. This implies that the marginals are uniform on $\{0,1\}$. The energy of the distribution π is

$$U(y) = -2\beta \sum_i y_i + \beta \sum_{\langle i,j \rangle} y_i y_j, \tag{3}$$

and the conditional distribution at the site j is given by

$$(H_\beta) \quad : \quad \pi(y_j \mid \beta, v_j) = \frac{\exp\{y_j(-2\beta + \beta v_j)\}}{1 + \exp\{-2\beta + \beta v_j\}}.$$

The null hypothesis (H_0) corresponds to $\beta = 0$ in (H_β).

We cannot give explicitly the correlation at distance 1, $\beta \to \rho(\beta)$, but we can find an estimation *via* a Monte Carlo procedure. We simulate toric fields (H_β) of size $N \times N$ using the Gibbs sampler. We initialize the field by drawing values 0 and 1 independently and at random. The scans are periodic, we run along the lines from left to right, one after the other, downward. We consider asynchronous sweeps, as described in Section 3.2. We scan 100 times.

We consider two sizes for the field, $N = 6$ and $N = 10$. In each case, the parameter β varies from 0 to 1.3 by step of 0.1. At the end of each simulation we calculate the test statistics and the correlation at distance one, and this whenever π is known or unknown (where $\pi = P(Y_i = 1)$).

We consider the following test statistics:

Model (B), π known

$$C4 = \frac{\sqrt{2n}(M_n - \pi^2)}{\sqrt{\pi^2(1-\pi)(1+7\pi)}}, \quad C8 = \frac{\sqrt{4n}(M_n^* - \pi^2)}{\sqrt{\pi^2(1-\pi)(1+15\pi)}}$$

Model (B), π unknown

$$G4 = \frac{\sqrt{2n}(M_n - \widehat{\pi^2})}{\widehat{\pi}(1-\widehat{\pi})} \quad G8 = \frac{\sqrt{4n}(M_n^* - \widehat{\pi^2})}{\widehat{\pi}(1-\widehat{\pi})}$$

where $\widehat{\pi^2} = n(\overline{Y}_n^2 - \overline{Y}_n/n)/(n-1)$ and $\widehat{\pi} = \overline{Y}_n$. In the framework (B) and π unknown, we have also studied the test statistic based on the two nearest horizontal neighbours

$$G2 = \frac{\sqrt{n}(M_n^{**} - \widehat{\pi^2})}{\widehat{\pi}(1-\widehat{\pi})}, \text{ with } M_n^{**} = \frac{1}{n} \sum_{\substack{s,s',t \\ |s-s'|=1}} Y_{s,t} Y_{s',t}.$$

We compute the power values on the basis of 500 simulations.

We display in Figure 1 the empirical power functions for these five statistics and for $N = 10$; we note that:

(i) the tests based on the 4 nearest neighbours are more powerful than those using the 2 or 8 nearest neighbours. This is observed whether π is known or not.

(ii) the tests based on statistics computed with an estimate of π are more powerful than those using the true value of π.

We offer an heuristic explanation regarding the first observation. When we consider two neighbours, we estimate the correlation at distance 1 without bias, but with less observations than with the 4 nearest neighbours statistic. With regard to the 8 nearest neighbours statistic, we have more observations, but the diagonal correlation being smaller than the one at distance 1, the statistic is negatively biased, and thus less powerful.

We have empirically estimated the correlation at distance 1. Figure 2 shows the results for $N = 10$, π being known and estimated, and for $N = 6$ with π estimated. We see that the correlation increases with N for a fix β.

Alternative 2 (Conditional Ising model (versus (H_0): (HG))). Y is the Ising model with energy given by (3) but conditional to $S_Y = n/2$. We use the Metropolis dynamics of spin exchange, the two choices being in parity in the initial configuration (see Guyon, 1995, 6.2.1). The kth repetition provides $Y(k) = y$. Next we choose at random two sites s and t; if the values y_s and y_t are different, we exchange them and we note z the new realization ($z_s = y_t$ at s and $z_t = y_s$ at t). We end as usual the Metropolis procedure on the basis of the quantity $\pi(z)/\pi(y) = \exp\{U(z) - U(y)\}$ noting that $U(z) - U(y)$ is local in $\{s, t\}$. We repeat the procedure $50 \times N \times N$ times.

We consider the same sizes $N = 6$ and $N = 10$, and we let β vary from 0 to 2.6 by step of 0.1. We look at the two following test statistics:

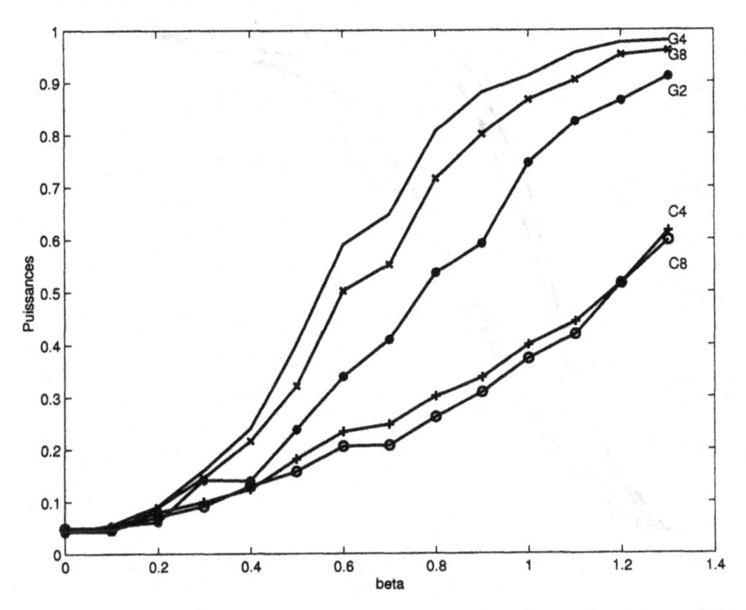

Figure 1. Model (B). Powers of the 5 statistics, based on the 2 n.n. (G2,$*-*$), the 4 n.n. ((C4,$+-+$) and (G4,—)), and the 8 n.n. ((C8,$\circ-\circ$) and (G8,$\times-\times$)); the level is 5%; the torus' dimension is 10×10.

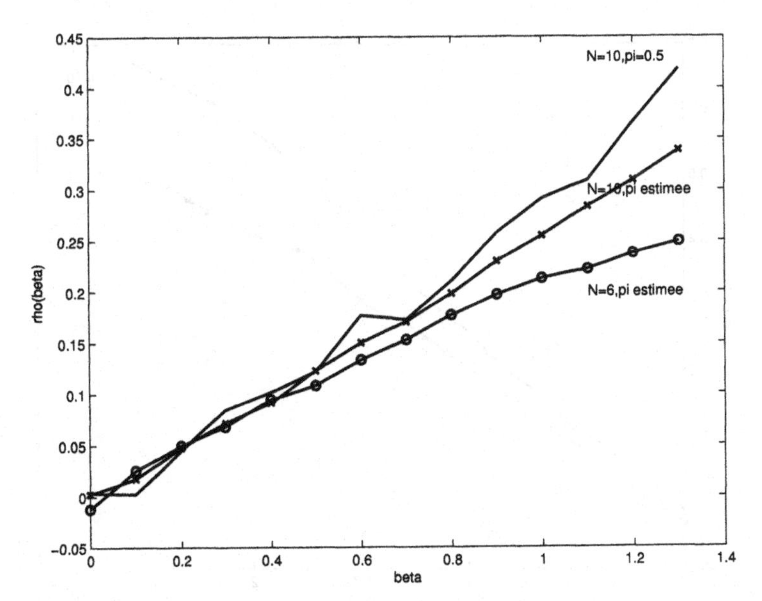

Figure 2. Model (B). Estimated correlation $\beta \rightarrow \rho(\beta)$ at distance 1. (a) $N = 6$ and π estimated ($\circ-\circ$); (b) $N = 10$ and π estimated ($\times-\times$), (c) $N = 10$ and $\pi = 0.5$ (—).

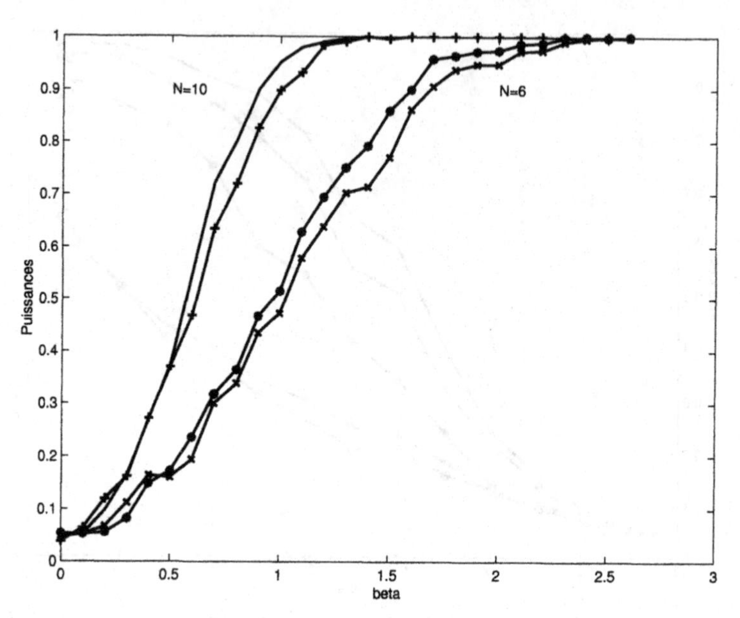

Figure 3. Model (HG). Powers of the tests based on the statistics using the 4 n.n. and the 8 n.n. For $N = 10$, (a) C4: (—), (b) C8: (+–+) . For $N = 6$, (c) C4: ($*$–$*$), and (d) C8: (×–×).

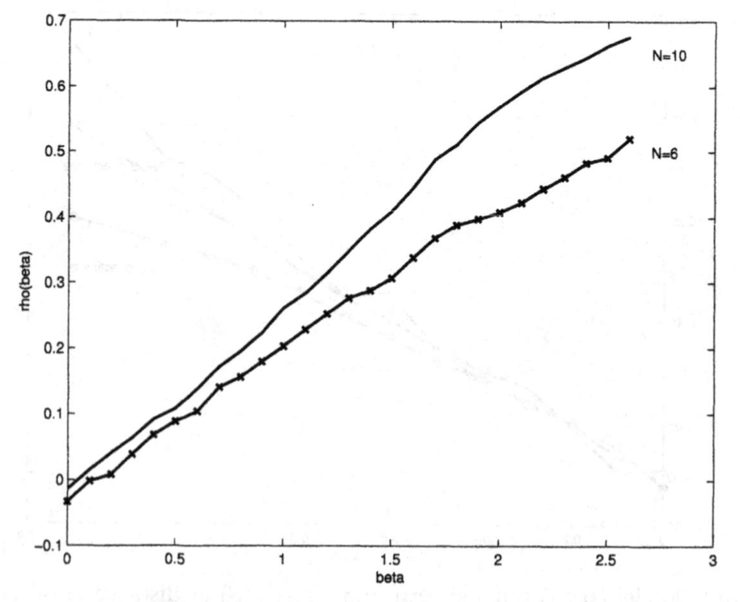

Figure 4. Model (HG). Correlation $\beta \to \rho(\beta)$ at distance 1. (a) $N = 6$ (×–×) ; (b) $N = 10$ (—).

Model (HG)

$$C4 = \frac{\sqrt{2n}(M_n - (n-2)/4(n-1))}{\sqrt{A_4 + B_4}}$$

$$C8 = \frac{\sqrt{4n}(M_n^* - (n-2)/4(n-1))}{\sqrt{A_8 + B_8}}$$

where A_4, B_4, (resp. A_8, B_8) are the constants given in Proposition 3 with $\nu = 4$ (resp. $\nu = 8$) and $b = n/2$.

The power functions are estimated (Figure 3) on the basis of 500 simulations and for the 5% level. Figure 3 shows, as for the alternative 1, that the tests based on the 4 nearest neighbours are more powerful than those using the 8 nearest neighbours. Also, the power increases rapidly with N.

Finally, Figure 4 shows the empirical correlations at distance 1 for both $N = 6$ and $N = 10$. The correlation increases with N for a given $\beta > 0$.

4.4 The Case of More Than Two Standards

We assume that there are $r + 1 \geq 2$ standards $\{a_0, a_1, \dots, a_r\}$. At a site i, the realization is encoded by the multidimensional variable $Y_i \in \{0,1\}^r$ of $\mathbb{R}^r$

$$Y_i(l) = \begin{cases} 1 & \text{if } Y_i = a_l \\ 0 & \text{otherwise,} \end{cases} \quad l = 1, \dots, r.$$

So, $Y_i = 0$ is the coding of a_0, and $\sum_{l=1,r} Y_i(l) = 1$. When there are two standards, $r = 1$ and we have the notations used for the binary model $Y_i \in \{0,1\}$.

Let us consider for instance the Bernoulli model; for each $l = 1, \dots, r$, the variables $Y_i(l)$ follow a Bernoulli distribution $B(\pi_l)$, they are independent in i, but linked in l for a same i. Let us note $\pi = (\pi_1, \dots, \pi_r)^t$. We use the following statistic in order to test nonspatial coordination

$$C_n = \left\{ \frac{2}{n\nu} \sum_{\langle i,j \rangle} Y_i {Y_j}^t \right\} - \pi\pi^t.$$

Using similar tools as those used previously, we obtain the asymptotic behaviour of C_n under the hypothesis of independence; $\sqrt{n}C_n$ is asymptotically Gaussian $\mathcal{N}_r(0, \Sigma)$, Σ being calculated *via* the multinomial distribution properties.

5 References

Besag, J. (1974). Spatial interaction and the statistical analysis of lattice systems (with discussion). *Journal of the Royal Statistical Society. Series B. Methodological 36*, 192–236.

Besag, J.E. and P.A.P. Moran (1975). On the estimation and testing of spatial interaction for Gaussian lattice processes. *Biometrika 62*, 555–562.

Bolthausen, E. (1982). On the central limit theorem for stationary mixing random fields. *The Annals of Probability 10*, 1047–1050.

Bouzitat, C. and C. Hardouin (2000). Adoption de standards avec coordination spatiale : une approche empirique. (in preparation).

Cliff, A.D. and J.K. Ord (1981). *Spatial autocorrelation* (2nd ed.). London: Pion.

Feller, W. (1957). *An Introduction to Probability Theory and its Applications* (2nd ed.), Volume 1. New York: Wiley.

Geman, D. (1990). Random fields and inverse problem in imaging. In *École d'été de Probabilités de Saint-Flour XVIII—1988*, Volume 1427 of *Lectures Notes in Mathematics*, pp. 113–193. Berlin: Springer.

Geman, S. and D. Geman (1984). Stochastic relaxation, Gibbs distributions, and the Bayesian restoration of images. *IEEE Transactions on Pattern Analysis and Machine Intelligence 6*, 721–741.

Guyon, X. (1995). *Random Fields on a Network: Modeling, Statistics and Applications.* Probability and its Applications. New York: Springer.

Isaacson, D.L. and R.Q. Madsen (1976). *Markov Chain. Theory and Applications.* Wiley Series in Probability and Mathematical Statistic. Wiley.

Kemeny, J.G. and J. L. Snell (1960). *Finite Markov Chain.* The University Series in Undergraduate Mathematics. Van Nostrand.

Pickard, D.K. (1987). Inference for Markov fields: the simplest non-trivial case. *Journal of the American Statistical Association 82*, 90–96.

Prum, B. (1986). *Processus sur un réseau et mesure de Gibbs. Applications.* Techniques Stochastiques. Paris: Masson.

Robert, C. (1996). *Méthodes de simulation en statistiques.* Economica.

Trouvé, A. (1988). Problèmes de convergence et d'ergodicité pour les algorithmes de recuit parallélisés. *Comptes Rendus des Séances de l'Académie des Sciences. Série I. Mathématique 307*, 161–164.

4

Random Fields on $\mathbf{Z}^d$, Limit Theorems and Irregular Sets

G. Perera

To the memory of Gonzalo Pérez Iribarren

ABSTRACT Consider a stationary random field $X = (X_n)_{n \in \mathbf{Z}^d}$ and an additive functional of the form

$$Z_N(A; X) = \sum_{n \in A_N} f_N(X_n),$$

where $A_N = A \cap [-N, N]^d$, A is an infinite subset of $\mathbf{Z}^d$ and f_N is a real function, for instance $f_N(x) = \big(x - E(X_0)\big)/(2N+1)^{d/2}$ (normalized averages) or $f_N(x) = \mathbf{1}_{\{x > u_N\}}$ for u_N tending to infinity with N (high-level exceedances). We present results showing that if X is a weakly dependent random field, and A satisfies certain geometrical conditions, limit theorems for Z_n can be obtained. On the other hand, if A does not fulfill these conditions, then limit theorems can fail to hold, even for m-dependent X. We also apply these results to derive limit theorems for additive functionals of some nonstationary models.

1 Introduction

In this paper we present a survey of recent results concerning the asymptotic behaviour of additive functionals of random fields over irregular sets.

Problem 1: Testing Homogeneity of a Random Field

Let us first consider a very simple problem: testing the homogeneity of the mean of a random field. Suppose we have a function defined over $\mathbf{Z}^d$ that may assume two unknown values μ, τ. This binary function $\theta = (\theta_n)_{n \in \mathbf{N}}$ could be, for instance, the concentration of a pollutant at point n, where $\theta_n = \mu (\theta_n = \tau)$ means "the concentration at point n is normal" (is not normal). Assume this binary function is perturbed by a random noise that can be modeled by a centered random field $\varepsilon = (\varepsilon_n)_{n \in \mathbf{Z}^d}$, and that the observed variables are

$$X_n = \theta_n + \varepsilon_n.$$

Let us assume that ε is stationary and weakly dependent (say mixing, or associated, as defined later). We suspect that a given set $A \subset \mathbf{Z}^d$ could correspond to the "risk points," and we would like to confirm this suspicion by hypothesis testing. More precisely, we want to perform the following test

$$H_0 : \theta^{-1}(\mu) = \{n \in \mathbf{Z}^d : \theta_n = \mu\} = \mathbf{Z}^d \quad \text{versus} \quad H_1 : \theta^{-1}(\mu) = A^c.$$

We construct a decision rule based on the observation of X over the window $[-N, N]^d$ for large values of N. Denote

$$A_N = A \cap [-N, N]^d, v_N(A) = \frac{\text{card}(A_N)}{(2N+1)^d},$$

and assume that A satisfies

$$\lim_N v_N(A) = v(A) \in (0, 1),$$

that is A and A^c take both a significant proportion of the window.

Let us remark that if θ is itself a random field, A is a random set that may be extremely complex from a geometrical point of view, its "shape" may be very "irregular."

The I.I.D. Case

Let us give a naive way to solve this problem when X is i.i.d. with bounded second moments. The procedure is not the most efficient for this specific problem, but can be easily extended to more complex situations and will illustrate the motivation of the topics we discuss later on.

Let us introduce the following notation,

$$S_N(B; X) = \frac{1}{(2N+1)^{d/2}} \sum_{n \in B_N} (X_n - E(X_n)), \quad B \subset \mathbf{Z}^d.$$

Suppose that we introduce an enumeration of $A = \{a_m\}_{m \in \mathbf{N}}$ such that, if we define

$$C = \{a_{2m}\}_{m \in \mathbf{N}}, \quad D = \{a_{2m+1}\}_{m \in \mathbf{N}},$$

C and D split A in two disjoint subsets of "equal size," in the sense that $v(C) = v(D) = v(A)/2$. If σ^2 stands for the common variance of the coordinates of X, an application of the Central Limit Theorem (CLT, for short) to $(S_N(C), S_N(D), S_N(A^c))$ shows that

$$T_N = \frac{1}{1 - v_N(A)} \frac{\left((1 - v_N(A))S_N(A) - v_N(A)S_N(A^c)\right)^2}{\left(S_N(C) - S_N(D)\right)^2} \overset{w}{\underset{N}{\Longrightarrow}} \mathbf{F}(1, 1),$$

where $\mathbf{F}(a, b)$ stands for the Fischer distribution with a, b degrees of freedom.

Observe now that, if H_0 holds then T_N is asymptotically equivalent to

$$E_N = \frac{1}{1 - v_N(A)} \frac{(\sum_{n \in A_N} X_n - \sum_{n \in A_N^c} X_n)^2}{(\sum_{n \in C_N} X_n - \sum_{n \in D_N} X_n)^2},$$

and the CLT plus some calculus shows that if H_1 holds, there is a bias difference between both statistics that is asymptotically equivalent to

$$b_N = \frac{(\tau - \mu)^2}{\sigma^2} (1 - v(A)) (v(A)) \frac{(2N+1)^d}{J_N},$$

where J_N is a random variable that converges in law to a $\chi^2(1)$ (chi-square distribution with one degree of freedom).

Therefore, if we reject H_0 when

$$E_N \geq \mathbf{F}_{1-\alpha}(1,1),$$

where $F_{1-\alpha}(1,1)$ stands for the $1-\alpha$ percentile of the $\mathbf{F}(1,1)$ distribution, we have an asymptotic test with level of significance α and which is consistent against fixed alternatives (its power tends to one as the sample size increases to infinity if the alternative H_1 is fixed).

The Weakly Dependent Case: CLT and Irregular Sets

Let us see what happens if we drop the hypothesis of independence, and we assume that X is weakly dependent. We may try to mimic the previous procedure. After a brief inspection of what we have done, we can see that the same procedure will work without major changes *if a CLT holds* for $(S_N(C), S_N(D), S_N(A^c))$. First, we examine if a CLT is available for $S_N(A)$. Let us start with some basic definitions about weak-dependence.

Mixing coefficients

If $T \subset \mathbf{Z}^d$ we denote by $\Sigma^X(T)$ the σ-algebra generated by $\{X_n : n \in T\}$. The α-mixing coefficients of X are defined by

$$\alpha^X(r) = \sup\{|P(A \cap B) - P(A)P(B)| : A \in \Sigma^X(T), B \in \Sigma^X(S), \\ S, T \subset \mathbf{Z}^d, d(S,T) \geq r\}, \quad r \in \mathbf{N}, \quad (1)$$

and X is said to be *mixing* if $\lim_r \alpha^X(r) = 0$. Weaker mixing assumptions can also be defined, by assuming additional restrictions on the size or shape of the sets S, T considered in (1).

In particular, if for some m, $\alpha^X(r) = 0 \; \forall r > m$, we have that $\Sigma^X(T)$, $\Sigma^X(S)$ are independent if $d(S,T) > m$ and X is said to be *m-dependent.*

A vast literature on mixing conditions and limit theorems for random fields is available, see for example Bradley (1986), Doukhan (1995), Guyon (1995) for an overview, and Bernstein (1944), Bradley (1993), Bolthausen (1982), Davydov (1973), Dedecker (1998), Dobrushin (1968), Kolmogorov and Rozanov (1960), Rio (1993), Rosenblatt (1956) for an extensive account.

Association

We say that a random vector $(X_1, \ldots, X_n)$ is *positively associated* (or that it satisfies the FKG inequality) if, for any coordinate-wise nondecreasing functions $g, h : \mathbf{R}^n \to \mathbf{R}$ such that $E(g(X_1, \ldots, X_n)^2) < \infty$, $E(h(X_1, \ldots, X_n)^2) < \infty$, we have

$$\text{Cov}\big(g(X_1, \ldots, X_n), h(X_1, \ldots, X_n)\big) \geq 0.$$

If for any nonempty subset B of $\{1, \ldots, n\}$ and any pair of coordinate-wise nondecreasing functions, $g : \mathbf{R}^B \to \mathbf{R}$, $h : \mathbf{R}^{B^c} \to \mathbf{R}$ such that $E(g(X_t : t \in B)^2) < \infty$, $E(h(X_t : t \in B^c)^2) < \infty$, we have

$$\text{Cov}\big(g(X_t : t \in B), h(X_t : t \in B^c)\big) \leq 0,$$

we say that $(X_1, \ldots, X_n)$ is *negatively associated.*

A random field is said to be positively (negatively) associated if all its finite-dimensional projections are positively (negatively) associated. We will say that a random field is *associated* if it is either positively associated or negatively associated. Association is a weak-dependence assumption alternative to mixing: many limit theorems have been obtained for associated random fields, and it is easy to provide examples of mixing, nonassociated random fields or associated, nonmixing random fields (see Andrews, 1984, Pitt, 1982). For further reading concerning association and limit theorems, see Esary et al. (1967), Birkel (1988), Louhichi (1998), Newman (1980, 1984), Roussas (1994), Yu (1993).

A more comprehensive definition of "weak dependence", that includes mixing and association (as well as other dependence structures) as particular examples, have been given recently by Doukhan and Louhichi (1997).

Coming back to our problem, we know from Perera (1994, 1997b), that, *even for m-dependent processes, the viability of a CLT depends on the geometry of A*. More precisely, a set $A \in \mathbf{Z}^d$ is said to be *asymptotically measurable* (AMS, for short) if for any $r \in \mathbf{Z}^d$, the following limit exists

$$\lim_N \frac{\text{card}\{A_N \cap (A_N - r)\}}{(2N+1)^d} = F(r; A). \tag{2}$$

Then, if A is AMS, for any weakly dependent X the CLT holds for $S_N(A; X)$. But, if A is nonasymptotically measurable, there exists an m-dependent Gaussian field X such that CLT fails to hold for $S_N(A; X)$. This means that there are some "shapes" of sets that we will not be able to recognize by the "naive" statistical approach presented before. In Section 2 we discuss in detail this result, and we provide several examples where condition (2) holds and where it does not. We can anticipate some examples of asymptotically measurable sets: sets with "thin" border, or with a periodic structure, or level sets of stationary random fields. Further, if A is AMS, our problem of hypothesis testing can be solved in a very simple way for

X weakly dependent, using the techniques of Section 2 (see Perera, 1994). Hence, the first step in this paper will be the study of the class of asymptotically measurable sets and its relations with Central Limit Theorems for weakly dependent random fields.

The second step, developed in Section 3, concerns the application of Section 2 to obtain CLT for random fields that can be written in the following way:

$$X_n = \varphi(\xi_n, Y_n), \tag{3}$$

where ξ is a stationary and weakly dependent random process, but Y is a process that may be nonstationary and even nonweakly dependent, ξ being independent of Y, and both ξ, Y are nonobservable. As a simple example assume that each point n of $\mathbf{Z}^d$ is occupied by an individual that is susceptible to a certain epidemic disease. Let Y_n be a variable describing its clinical condition (for instance, its number of parasites if the disease is parasitic); typically the process Y depends on many geographical and sociological factors, hence it is not stationary, and its dependence structure may be complex. Furthermore Y is not directly observed: our data (X) correspond to the result of a serological analysis, which can be assumed to depend on Y and on a "nice" stationary process ξ, independent of Y, that corresponds to the "pure error" of the analytic methodology. We would like to obtain the asymptotic distribution of

$$S_N = \frac{1}{(2N+1)^d} \sum_{n \in [-N,N]^d} (X_n - E(X_n)).$$

This was done in Perera (1997a) using the results of Section 3. To illustrate the relation with AMS sets, let us think that Y takes just two values, say 1 or 0. Then S_N can be rewritten in the following way

$$S_N = \frac{1}{(2N+1)^{d/2}} \Big\{ \sum_{n \in A_N} \big(\varphi(\xi_n, 1) - E(\varphi(\xi_n, 1))\big) + \sum_{n \in A_N^c} \big(\varphi(\xi_n, 0) - E(\varphi(\xi_n, 0))\big) \Big\},$$

where

$$A(\omega) = Y(\omega)^{-1}(1) = \{n \in \mathbf{Z}^d : Y_n(\omega) = 1\}.$$

Under mild hypotheses, this set A turns to be asymptotically measurable with probability 1; conditioning with respect to Y, and using the independence of ξ and Y we just have to study the asymptotics of averages of two weakly dependent and stationary processes ($\varphi(\xi, 1)$ and $\varphi(\xi, 0)$) over asymptotically measurable sets, what can be reduced to the problem studied in Section 2.

Problem 2: Asymptotics for High-Level Exceedances

In Section 4 we follow a similar program, but for high-level exceedances instead of averages. In many meteorological or hydrological problems, relevant features are related to exceedances of high levels by some time series. In particular, current standards for ozone regulation involve the exceedances of high levels. In this case it is clear that the time series of actual ozone level depends on some nonstationary meteorological variables, like temperature or wind speed. Therefore, a reasonable model for the ozone level at time t (say X_t) should be again of the form

$$X_t = \varphi(\xi_t, Y_t), \tag{4}$$

where ξ_t is "pure noise," corresponding to local fluctuations of measurements systems, Y_t is a vector of "explicative" variables at time t and φ is some suitable regression function. Indeed, as we will see later, Y may contain not only actual values, but also previous values (for instance, temperatures of the last q days). We may also consider t as a d-dimensional parameter, corresponding to space and time. Once again, we may assume that ξ is a "nice" process in the sense that it is stationary and weakly dependent, but Y may not be stationary; for instance in the case of temperature, besides seasonal effects that affect the time scale, spatial variations due to differences between urban and rural areas make the assumption of stationarity nonreasonable. Furthermore, even if Y may satisfy some ergodic properties (as the Law of Large Numbers) it is not reasonable to expect mixing, association or any particular weak dependence structure. Here, for the sake of simplicity, we will write all the results in the context of time series (with $\mathbf{N}$ as parameter space), but results and techniques are still valid when the parameter space is $\mathbf{Z}^d$.

The I.I.D. Case

We will observe the exceedances of $X_1, \ldots, X_n$ of a level u_n that grows to infinity with n in a suitable way. When $X = (X_t : t \in \mathbf{N})$ is i.i.d., simple arguments show that the point process

$$N_n(B) = \sum_{t/n \in B} \mathbf{1}_{\{X_t > u_n\}}, \quad B \in \mathcal{B}, \tag{5}$$

where $\mathcal{B}$ stands for the Borel σ-algebra of $[0,1]$, converges to a Poisson process of intensity

$$\lambda = \lim_n nP(\{X_0 > u_n\}).$$

The Weakly Dependent Case: CPLT

Recall that a point process N is a Compound Poisson Process with intensity measure ν (denoted by CP(ν)), where ν is a positive finite measure on $\mathbf{N}$, if:

- For any $h \in \mathbf{N}$, if $B_1, \ldots, B_h$ are disjoint Borel sets, then $N(B_1), \ldots,$ $N(B_h)$ are independent.
- For any Borel set B, the Laplace transform of $N(B)$ is:

$$L(B; s) = \exp\left(m(B) \sum_{j=1}^{+\infty} \nu_j (\exp(-sj) - 1) \right),$$

where m denotes Lebesgue measure, and $\nu_j = \nu(\{j\}) \forall j \in \mathbf{N}$.

If X is stationary and weakly dependent, clustering of exceedances may occur and one obtains a CP limit. In the sequel, if X is a random process such that (for some sequence u_n) the point process in (5) converges to a CP process, we shall say that X satisfies a Compound Poisson Limit Theorem (CPLT, for short). CPLT for stationary processes satisfying some mixing conditions are known (see Cohen, 1989, Dziubdziela, 1988, Ferreira, 1993, Hsing et al., 1988, Leadbetter and Nandagopalan, 1989, Leadbetter and Hsing, 1990, Leadbetter, 1991, 1995), for Markov Chains (see Hsiau, 1997). Some results are also available for X weakly dependent but nonstationary; see Alpuim et al. (1995), Dziubdziela (1995), Hüsler (1993), Hudson et al. (1989). For a nice summary of many related result see Falk et al. (1994). The authoritative text of Leadbetter et al. (1983) is a basic reference for exceedances, extremes and related topics, so is Leadbetter and Rootzén (1988). For continuous-time results see Volkonskii and Rozanov (1959, 1961), Wschebor (1985) and the very nice monograph of Berman (1992). In some cases, rates of convergence can also be obtained, by means of Stein-Chein method; see Barbour et al. (1992), and Brown and Xia (1995). For the application of point process exceedances to modeling ozone data, see Smith and Shively (1994).

However, models like (4) when Y is neither stationary nor weakly dependent, do not satisfy the hypotheses required for those results. It is proved in Bellanger and Perera (1997) that, under a mild condition that we present below, the model (4) satisfies a CPLT. This result generalizes the preceding ones; it is not assumed that X has a particular weak-dependence structure (like mixing, association, Markov, etc.), hence previous results do not apply to (4). For instance, we may have X defined by (4) where ξ is mixing but Y is merely ergodic, but the assumptions of Bellanger and Perera (1997) still hold. Also, we will see that without additional effort, the limit distribution of N_n when Y (hence X) presents long-range dependence can be obtained: in that case the limit distribution is no longer Compound Poisson but a mixture of several Compound Poisson distributions.

As mentioned before, the technique used here to deal with high-level exceedances is similar to that used for averages. We will not give all the details here, but try to emphasize analogies and differences with respect to the case of the CLT. It is interesting to remark that the geometrical conditions required on A are stronger than those for the CLT. Asymptotical measurability is not enough to get a CPLT for any stationary and weakly dependent process. A strictly stronger condition, called asymptotical ponderability, must be imposed. A set $A \in \mathbf{N}$ is called *asymptotically ponderable* (AP, for short) if for any $k \in \mathbf{N}$ and any $\vec{r} = (r_1, \ldots, r_k) \in \mathbf{N}^k$ the following limit exists

$$\lim_N \frac{\text{card}\{\bigcap_{i=1}^k (A_N - r_i)\}}{N} = F(\vec{r}; A). \tag{6}$$

If we compare this condition to (2) we see that (despite the minor fact of the substitution of $\mathbf{Z}^d$ by $\mathbf{N}$ as parameter space, what changes the size of the observation window from $(2N+1)^d$ to N) in (2) we only require the existence of the limit of (6) for $k = 2$. It is then clear that asymptotical ponderability implies asymptotical measurability; an example of an asymptotically measurable set that is not asymptotically ponderable is presented in Section 4.

We pay a particular attention to the geometrical (or arithmetical) properties of the sets introduced, giving examples of the different type of classes of sets.

2 CLT for Weakly Dependent Stationary Random Fields over Irregular Sets

2.1 *For Some Irregular Sets, CLT Fails*

Consider a stationary random field $X = (X_n)_{n \in \mathbf{Z}^d}$ and assume that $E(X_0^2) < \infty$, $E(X_0) = 0$. Let us denote $r^X(k) = E(X_0 X_k)\ \forall k \in \mathbf{Z}^d$.

Given $A \subset \mathbf{Z}^d$ recall that $S_N(A; X) = 1/(2N+1)^{d/2} \sum_{n \in A_N} X_n$. Its variance is, by an elementary computation,

$$E(S_N(A;X)^2) = \sum_{k \in \mathbf{Z}^d} r^X(k) \frac{\text{card}\{A_N \cap (A_N + k)\}}{(2N+1)^d}. \tag{7}$$

Assume that the set A is non-AMS; this means that there exist $r_0 \in \mathbf{Z}^d$ such that the limit (2) fails to exist; it is clear that we must have $r_0 \neq 0$. We will apply now (7) to a particular random field. Pick $\rho \in (0, \frac{1}{2})$ and consider $\xi = (\xi_n)_{n \in \mathbf{Z}^d}$ i.i.d. with common law $N(0,1)$. Define $a = (1 + \sqrt{1 - 4\rho^2})/2$ and set

$$X_n = \sqrt{a}\xi_n + \sqrt{1-a}\xi_{n+r_0}.$$

It is clear that X is a Gaussian, stationary, centered and $\|r_0\|$-dependent random field (we always consider the sup-norm). On the other hand, we have that

$$r^X(0) = 1, \quad r^X(r_0) = r^X(-r_0) = \sqrt{a(1-a)} = \rho > 0, \quad r^X(k) = 0$$

in any other case. Going back to (7) this implies that for this X

$$E(S_N(A;X)^2) = \frac{\text{card}(A_N)}{(2N+1)^d} + 2\frac{\text{card}\{A_N \cap (A_N - r_0)\}}{(2N+1)^d}\rho. \tag{8}$$

On the right-hand side of this expression the first term converges to $v(A)$ while the second is a sequence on $[0,1]$ that has no limit. We conclude that there exist two subsequences such that $E(S_N(A;X)^2)$ have different limits; since $S_N(A;X)$ is a centered Gaussian variable, we deduce that $S_N(A;X)$ admit two subsequences with different weak limits, what means that $S_N(A;X)$ does not converge in law.

In conclusion, *if A is non-AMS, there exists a stationary, Gaussian, m-dependent, centered process X such that $S_N(A;X)$ has no weak limit.*

2.2 An Example of Non-AMS

We give an example of a set A that fails to be AMS. Define, for $n \in \mathbf{N}$, $I(n) = [100^{2^{n-1}}, 100^{2^n})$ and $A(n,0) = I(n) \cap (5\mathbf{N})$, $A(n,1) = I(n) \cap [(10\mathbf{N}) \cup (10\mathbf{N}+1)]$, and set $A = \bigcup_{i=1}^{\infty}(A(2i,0) \cup A(2i+1,1))$. What we are doing here is alternating two different patterns (multiples of 5 and multiples of 10 plus its consecutives) over a sequence of intervals that grows very fast. After straightforward computation, we verify that $F(1;A) = \lim_N \text{card}\{A_N \cap (A_N - 1)\}/N$ does not exist, and hence, A is non-AMS. More precisely, take the subsequence $n_k = 100^{2^{2k}}$; we can easily note that the pattern "multiples of 5" is predominant on A_N, and hence the limit of $\text{card}\{A_N \cap (A_N+1)\}/N$ is smaller than the same limit for the subsequence $100^{2^{2k+1}}$, where the pattern $10\mathbf{Z} \cap (10\mathbf{Z}+1)$ is the most relevant. In Section 4 we give a more elaborated example using a similar construction.

Finally, let $x \in [0,1]$ and $x = \sum_{n=1}^{\infty} x(n)/2^n$ its binary expansion. If we define the set $\Phi(x) = \bigcup_{i=1}^{\infty}\big(A\big(2i, x(2i)\big) \cup A\big(2i+1, x(2i+1)\big)\big)$, then for a nondyadic x, $\Phi(x)$ fails to be an AMS. Here we have a collection of non-AMS sets, and this collection has the power of the continuum.

2.3 If A is an AMS, Then CLT Holds

If A is an AMS, then the CLT holds for stationary and weakly dependent random fields. We present our result in the more general context of asymptotically measurable collections.

Definition 1. Let $\mathcal{A} = (A^1, \ldots, A^k)$ be a collection of subsets of $\mathbf{Z}^d$, $k \in \mathbf{N}$. We will say that $\mathcal{A}$ is an asymptotically measurable collection (AMC, for short) if for any $i = 1, \ldots, k$, $\lim_N \text{card}\{(A^i)_N\}/(2N+1)^d = v(A^i) > 0$ and for any pair A^i, A^j with $i \neq j$ and any $r \in \mathbf{Z}^d$ we have that the following limit exists

$$\lim_N \frac{\text{card}\{A_N^i \cap (A_N^j - r)\}}{(2N+1)^d} = F(r; A^i, A^j).$$

Remark 1. It is clear that A is an AMS iff (A, A^c) is an AMC. On the other hand, AMC is an hereditary property: if a collection is an AMC, so is any subcollection.

We denote by $N_r(0, \Sigma)$ a r-dimensional centered Gaussian law with covariance matrix Σ.

Let $X = (X^1, \dots, X^k)$ be an R^k-valued centered stationary random field; for a collection of sets $A^1, \dots, A^k$ denote

$$M_N(A^1, \dots, A^k; X) = (S_N(A^1, X^1), S_N(A^2, X^2), \dots, S_N(A^k, X^k)).$$

We assume that X satisfies the following conditions.

(C1) $\sum_{n \in \mathbf{Z}^d} |E\{X_0^i X_n^j\}| < \infty$, $\sum_{n \in \mathbf{Z}^d} E\{X_0^i X_n^i\} > 0$, $i, j = 1, \dots, k$.

(C2) There exists a sequence $b(J)$ such that $\lim_{J \to \infty} b(J) = 0$ and for any $A^1, \dots, A^k \subset \mathbf{Z}^d$ we have

$$E\{(M_N(A^1, \dots, A^k; X - X^J))^2\} \le b(J) \frac{\sum_{i=1}^k |A_N^i|}{(2N+1)^d},$$

where X^J is the truncation by J of the random field X, i.e.,

$$X_n^J = X_n \mathbf{1}_{\{\|X_n\| \le J\}} - E\{X_n \mathbf{1}_{\{\|X_n\| \le J\}}\},$$

and $\|\cdot\|$ stands for the Euclidean norm.

(C3) For each $J > 0$ there is a number $c(X, J)$ depending only on X and J such that for all $N \ge 1$, $A \subset [-N, N]^d$, $i = 1, \dots, k$ we have

$$E\{(S_N(A; X^i))^4\} \le c(X, J) \left(\frac{|A_N|}{(2N+1)^d} \right)^2.$$

(C4) There exists a sequence $d(J)$ such that $\lim_{J \to \infty} d(J) = 0$, and a real function g such that for any pair A, $B \subset \mathbf{Z}^d$ that satisfy $d(A, B) \ge J$ we have

$$|\operatorname{Cov}\{\exp[iS_N(A; \langle t, X \rangle)], \exp[iS_N(B; \langle t, X \rangle)]\}| \le d(J) g(t), \quad t \in \mathbf{R}^k.$$

Then, we have the following result.

Proposition 1. *For a random field* X *satisfying* (C1)–(C4), *if* $A^1, \dots, A^k$ *is an AMC, then* $M_N(A^1, \dots, A^r; X^1, \dots, X^r) \underset{N}{\overset{w}{\Longrightarrow}} N_r(0, \Sigma)$, *where*

$$\Sigma(i,j) = \sum_{n \in \mathbf{Z}^d} F(n; A^i, A^j) E\{X_0^i, X_n^j\}, \quad i,j = 1, \ldots, k.$$

This result has been obtained by means of Bernshtein method (see Perera, 1997b,a).

The assumptions made can be verified in different contexts, like mixing or association. For instance, if $E\{\|X_0\|^4\} < \infty$, $\sum_{m=1}^{\infty} m^{d-1} \alpha^X(m) < \infty$ and $m^2 \alpha^X(m)$ is bounded, then it is easy to check, using covariance inequalities for mixing random fields (see Bradley, 1986, Doukhan, 1995) that conditions (C1) to (C4) hold. We refer again to Doukhan and Louhichi (1997) to find different weak dependence assumptions that ensure that Proposition 1 applies.

2.4 Examples of AMC

If d is the sup-norm distance on $\mathbf{Z}^d$, and $A \subset \mathbf{Z}^d$, define

$$\partial A = \{n \in A : d(n, A^c) = 1\}.$$

We have (see Perera, 1997b)

$$\operatorname{card}\{A_N \cap (A_N^c - r)\} \leq d(\|r\| + 1) \operatorname{card}\{(\partial A)_N\}, \quad \forall r \in \mathbf{Z}^d.$$

This implies that if $v(\partial A) = 0$ then A is an AMS with $F(r; A) = v(A) \forall r \neq 0$, and that $M_N(A, A^c; X)$ converges weakly to a couple of independent Gaussian variables.

Let now $Y = (Y_n)_{n \in \mathbf{Z}^d}$ be a random field and $B^1, \ldots, B^k$ given Borel sets; define

$$A^j(\omega) = \{n \in \mathbf{Z}^d : Y_n(\omega) \in B^j\}.$$

Note that

$$\frac{\operatorname{card}\big(A_N^i \cap (A_N^j - r)\big)}{(2N+1)^d} = \frac{1}{(2N+1)^d} \sum_{n \in A_N} \mathbf{1}_{\{Y_n \in B^i, Y_{n+r} \in B^j\}}. \tag{9}$$

Observe that if Y is stationary, then by the Ergodic Theorem (see Guyon, 1995, p. 108), the limit of (9) exists for any r with probability one (hence A is almost surely an AMS) and its limit equals $\mu^r(B^i \times B^j)$, where μ^r is a (random) probability measure with respect to the Borel sets of $\mathbf{R}^2$. This give examples of our next definition: we will say that a process Y is *asymptotically measurable* (AM, for short) if the limit of (9) exists for any r with probability one and defines a probability measure on the Borel sets of $\mathbf{R}^2$. If in addition, for any r the limit measure is nonrandom, we will say that Y is *regular*. Stationary random fields are AM, ergodic random fields are regular; we give later further examples.

To end this section, let us just show a trick based on the previous results on level sets. In Problem 1, when dealing with i.i.d. data, we have split

A in two parts of the "same size" C and D, and then used a CLT for $M_N(C,D,A^c;X,X,X)$. The splitting was done with no particular care, because we considered an *iid* situation, and therefore, the shape of sets was not relevant, only their size. But if we try to do the same for X weakly dependent (hence A is an AMS), it is clear that we will need to split A carefully, because (C,D,A^c) must be an AMC. It seems to be hard to describe a deterministic algorithm for this splitting, but it is very simple to do it by randomization. Take $U=(U_n)_{n\in\mathbf{Z}^d}$ *iid* with common Ber$(\frac{1}{2})$ law and define

$$C(\omega)=\{n\in A:U_n(\omega)=1\},\quad D(\omega)=A-C(\omega);\forall\omega.$$

Since U is regular, (C,D,A^c) is an AMC with probability one.

3 CLT for $X_n=\varphi(\xi_n,Y_n)$

We deal here with models (3), where ξ and Y are independent, Y is AM and ξ is a φ-random field defined as follows:

Definition 2. Let $\varphi:\mathbf{R}^2\to\mathbf{R}$ be a function and $\xi=(\xi_n)_{n\in\mathbf{Z}^d}$ a stationary random field that satisfy

(H1) φ is continuous and for each $(x,y)\in\mathbf{R}^2$, $\partial\varphi/\partial y(x,y)$ exists and is a continuous function of the second argument.

(H2) For each $y\in\mathbf{R}$, $E\{\varphi(\xi_0,y)\}=0$.

(H3) For each pair $y,z\in\mathbf{R}$, setting $\Gamma(n;y,z)=E\{\varphi(\xi_0,y)\varphi(\xi_n,z)\}$, $n\in\mathbf{Z}^d$, we have

$$\sum_{n\in\mathbf{Z}^d}|\Gamma(n;y,z)|<\infty,$$

and if we define $\Psi:\mathbf{R}^2\to\mathbf{R}$, $\psi:\mathbf{R}^2\to\mathbf{R}$, by

$$\Psi(y,z)=\sum_{n\in\mathbf{Z}^d}|\Gamma(n;y,z)|,\quad \psi(y,z)=\sum_{n\in\mathbf{Z}^d}\Gamma(n;y,z),$$

then Ψ is bounded and ψ is continuous in $\mathbf{R}^2$.

(H4) For each $k\in\mathbf{N}$, for each AMC $A^1,\ldots,A^k$ and any $(y_1,\ldots,y_k)\in\mathbf{R}^k$ we have

$$M_N(A^1,\ldots,A^k;\varphi(\xi,y_1),\ldots,\varphi(\xi,y_k))\underset{N}{\overset{w}{\Longrightarrow}}N_r(0,\Sigma),$$

where

$$\Sigma(i,j)=\sum_{n\in\mathbf{Z}^d}\Gamma(n;y_i,y_j)F(n;A^i,A^j).$$

(H5) For each pair $w, z \in \mathbf{R}$ we have

$$\eta(w,z) = \sum_{n \in \mathbf{Z}^d} \left| E\left\{ \frac{\partial \varphi}{\partial y}(\xi_0, w) \frac{\partial \varphi}{\partial y}(\xi_n, z) \right\} \right| < \infty,$$

and η is bounded over compact subsets of $\mathbf{R}^2$.

Then ξ is said to be a φ-random field.

Example 1. It is not difficult to check conditions (H1)–(H5) under weak dependence structures. For instance, consider a C^1-function, $\varphi : \mathbf{R}^2 \to \mathbf{R}$, odd with respect to the first variable and assume

$$H_K(x) = \sup\left\{ \left| \frac{\partial \varphi}{\partial y}(x,y) \right| : |y| \leq K \right\} < \infty, \quad \forall K \in \mathbf{N},$$
$$h(x) = \sup\{|\varphi(x,y)| : y \in \mathbf{R}\} < \infty, \qquad \forall x \in \mathbf{R}.$$

Consider further a stationary random field ξ such that the law of ξ_0 is symmetric, $E\{\xi_0^4\} < \infty$, $E\{H_K(\xi_0)^2\} < \infty$ for any $K > 0$, $E\{h(\xi_0)^2\} < \infty$, $\sum_{m=1}^{\infty} m^{d-1} \alpha^{\xi}(m) < \infty$ and $m^2 \alpha^{\xi}(m)$ is bounded. Then it is easy to check that ξ is a φ-random field.

3.1 Examples of AM Random Fields

We have seen that if Y is a stationary random field, then it is AM. Observe that, given $h \in \mathbf{N}$, the limit of (9) does not depend on $(Y_t : \|t\| \leq h)$ (a finite set of coordinates does not affect the limit of averages) and hence, we deduce that the limit measure μ^r is measurable with respect to the σ-algebra

$$\sigma_\infty^Y = \bigcap_{h=1}^{\infty} \sigma(Y_t : \|t\| \geq h).$$

Therefore, if σ_∞^Y is trivial, Y is regular.

We say that a centered random field $Z = \{Z_n : n \in \mathbf{Z}^d\}$ with bounded second moment, satisfies a Rosenthal inequality of order $q > 2$ if there exists a constant $\mathbf{C}(q)$ such that for any finite $F \subset \mathbf{Z}^d$ we have

$$E\left(\left| \sum_{m \in F} Z_m \right|^q \right) \leq \mathbf{C}(q) \left[\sum_{m \in F} E(|Z_m|^q) + \left(\sum_{m \in F} E(Z_m^2) \right)^{q/2} \right].$$

If X is nonstationary but satisfies a Rosenthal inequality of order $q > 2$ and for any $r \in \mathbf{N}$ there exists a probability measure in $\mathbf{R}^2$, μ^r, such that for any Borel sets B, C we have:

$$\lim_n \frac{1}{n} \sum_{m=1}^{n} E(\mathbf{1}_{\{Y_m \in B, Y_{m+r} \in C\}}) = \mu^r(B \times C),$$

then a simple Borel–Cantelli argument proves that Y is regular.

For instance, if the field Y satisfies the mixing condition $\rho^X(1) < 1$, then, for any Borel sets B, C and for any $r \in \mathbf{Z}^d$, the Strong Law of Large Numbers applies to

$$Z_m = \mathbf{1}_{\{Y_m \in B\}} \mathbf{1}_{\{Y_{m-r} \in C\}} - P(Y_m \in B, Y_{m-r} \in C),$$

(by Bryc and Smolenski, 1993, Theorem1).

Rosenthal inequalities can be obtained under different mixing assumptions (see Bryc and Smolenski, 1993, Rio, 1993), association (see Birkel, 1988), and for linear or Markov random fields; (see Doukhan and Louhichi, 1997) for a synthetic overview of different contexts were Rosenthal inequalities apply.

Example 2. Now we give a very simple example of the preceding situation in $\mathbf{N}$. Consider $U = (U_t)_{t \in \mathbf{N}} iid$ variables with common distribution μ absolutely continuous with respect to Lebesgue measure. Let V be a random variable taking a finite number of values ($V \in \{1, \ldots, S\}$) and independent of U and let $(a_t)_{t \in \mathbf{N}}$ be a sequence of real numbers satisfying $\lim_n a_{kn+h} = a(h)$, $h = 0, \ldots, k-1$ (for example, a periodic sequence). Define $Y_t = U_t + a_t V$. An elementary computation shows that Y is AM, and that for any $r \in \mathbf{N}$,

$$\mu^r(B \times C)(\omega) = \frac{1}{k} \sum_{h=0}^{k-1} \mu(B^i - a(h)V(\omega)) \mu(B^j - a(m(h,r))V(\omega)) \quad \forall r,$$

where $0 \leq m(h,r) \leq k-1$ satisfies $h + r = m(h,r) \pmod{k}$. If $a(h) = 0$, $h = 0, \ldots, k-1$, then Y is regular.

3.2 The CLT

We are interested in the limit of

$$S_N(X) = \frac{1}{(2N+1)^{d/2}} \sum_{n \in [-N,N]^d} X_n.$$

Definition 3. Consider a random field $X = \{X_n : n \in \mathbf{Z}^d\}$. We say that X is *I-decomposable* if there exist two random fields, $\xi = \{\xi_n : n \in \mathbf{Z}^d\}$ and $Y = \{Y_n : n \in \mathbf{Z}^d\}$ and a function $\varphi : \mathbf{R}^2 \to \mathbf{R}$ such that ξ is a φ-random field, Y is AM, ξ is independent of Y and X follows the regression model

$$X_n = \varphi(\xi_n, Y_n) \quad \text{for each } n \in \mathbf{Z}^d.$$

The essential idea of the CLT is presented in the following propositions, and corresponds to the simplified case where Y takes a finite number of values.

Proposition 2. *Assume that X is I-decomposable and that the coordinates of Y take values in a finite set $F = \{y_1, \ldots, y_r\}$. Let us denote by P^Y the distribution of the process Y, defined on $F^{\mathbf{Z}^d}$ and, for any $n \in \mathbf{Z}^d$, let $\mu^n(\cdot)(\omega)$ be the probability measure on $\mathbf{R}^2$ defined by the limit of (9). Then,*

(a) *there exists a $E \subset F^{\mathbf{Z}^d}$ such that $P^Y(E) = 1$ and for each $\mathbf{y} \in E$ we have*

$$S_N(X)/Y = \mathbf{y} \underset{N}{\overset{w}{\Longrightarrow}} N(0, \sigma^2(\mathbf{y})),$$

where

$$\sigma^2(\mathbf{y}) = \sum_{i,j=1}^{r} \sum_{n \in \mathbf{Z}^d} \Gamma(n; y_i, y_j) \mu^n(\{y_i\} \times \{y_j\})(Y^{-1}(\mathbf{y})),$$

(b) $S_N(X) \underset{N}{\overset{w}{\Longrightarrow}} \sigma(Y)N$, *with $N \sim N(0,1)$, independent of Y,*

(c) *if Y is regular, $S_N(X)$ satisfies the CLT (with a Gaussian limit).*

Proof. (b) and (c) are obtained using (a), previous definitions, the fact that

$$P(S_N(X) \le t) = \int_E P(S_N(X) \le t/Y = \mathbf{y}) dP^Y(\mathbf{y}),$$

and the dominated convergence theorem.

To prove a) let $E = Y(\Omega^*)$, where Ω^* is the intersection of the sets of probability one where convergence of (9) hold; a) follows by properties (H1) to (H4) and the definition of μ^n $\diamond$. □

Remark 2. It is clear that if μ^n is one of the limit measures of (9), and if ω, ω^* are such that $Y(\omega) = Y(\omega^*)$ then $\mu^n(\cdot)(\omega) = \mu^n(\cdot)(\omega^*)$ and therefore, there is no ambiguity in the expression $\mu^n(\cdot)(Y^{-1}(\mathbf{y}))$.

Finally, let us present the main result of this section.

Theorem 1. *Let X be I-decomposable. With the notation of the previous proposition, we have*

(a) *there exists a $E \subset F^{\mathbf{Z}^d}$ such that $P^Y(E) = 1$ and for each $\mathbf{y} \in E$ we have,*

$$(S_N(X)/Y = \mathbf{y}) \underset{N}{\overset{w}{\Longrightarrow}} N(0, \sigma^2(\mathbf{y}))$$

where

$$\sigma^2(\mathbf{y}) = \int_{\mathbf{R}^2} \sum_{n \in \mathbf{Z}^d} \Gamma(n; y, z) d\mu^n(y, z)(Y^{-1}(\mathbf{y})),$$

(b) $S_N(X) \underset{N}{\overset{w}{\Longrightarrow}} \sigma(Y)N$, *with $N \sim N(0,1)$, independent of Y,*

(c) *if Y is regular, $S_N(X)$ satisfies the CLT (with a Gaussian limit).*

Idea of the proof. The case where Y takes only a finite number of values is already done; (H3) and (H5) allow us, by standard approximation arguments, to reduce the problem (by truncation and discretization of Y) to the previous case. For a detailed proof, see Perera (1997a) □

Remark 3. (a) It is easy to check that a similar result holds in the multidimensional case (i.e., φ is $\mathbf{R}^d$-valued, ξ is $\mathbf{R}^a$-valued, and Y is $\mathbf{R}^b$-valued). The proof is obtained following the previous one, without substantial changes.

(b) Assume now that φ is defined over $\mathbf{R}^{\mathbf{Z}^d} \times \mathbf{R}^{\mathbf{Z}^d}$, pick $v \in \mathbf{Z}^d$ and denote by θ the shift that v induces on $\mathbf{R}^{\mathbf{Z}^d}$, $\theta(x)_n = x_{n-v}, n \in \mathbf{Z}^d$. Consider the regression model

$$X_n = \varphi(\theta^n(\xi, Y)).$$

Then, if we assume that φ can be suitably approximated by a sequence of functions $(\varphi_k : k \in \mathbf{N})$, such that for each k, φ_k is defined over $\mathbf{R}^{(2k+1)^d}$ and the multidimensional CLT applies to

$$X_n^k = \varphi_k(\xi_n^k, Y_n^k),$$

$$\xi_n^k = (\xi_{n+m} : \|m\| \leq k), \quad Y_n^k = (Y_{n+m} : \|m\| \leq k),$$

then we deduce by standard approximation arguments that the CLT hold for X (see Billingsley, 1968, p. 183).

(c) In a forthcoming paper, we use the fact that the limit is Gaussian in the case where Y is regular, while for a nonregular Y the limit is a mixture of Gaussian laws, to test the regularity of Y.

4 The Same Program for High-Level Exceedances

4.1 *The Compound Poisson Limit Theorem (CPLT)*

We follow here the same program as in Sections 2 and 3. However, proofs in this context are more complex than in the previous one; for the sake of brevity, we omit details that can be found in Bellanger and Perera (1997). Let us also remind that in this section we will just take $\mathbf{N}$ as the parameter space. Since N is the standard notation for counting process, the variable of the asymptotics will be n from now on, and the parameter will be denoted by t.

For $A \subset \mathbf{N}$ and $n \in \mathbf{N}$ we set $A_n = A \cap [1, n]$. If $B \subset \mathbf{N}$, $\vec{r} \in \mathbf{N}^d$, $d \geq 1$, and $\mathcal{A} = (A^1, \ldots, A^d)$ is any (ordered) finite collection of subsets of $\mathbf{N}$, set, for any $n \in \mathbf{N}$:

$$T_n(\vec{r}; B) = \bigcap_{i=1}^{i=d} (B_n - r_i), \quad G_n(\vec{r}; \mathcal{A}) = \bigcap_{i=1}^{i=d} (A_n^i - r_i).$$

Definition 4. Let A be a subset of $\mathbf{N}$. A is an *asymptotically ponderable set* (APS, for short) if for any $d \geq 1$, $\vec{r} \in \mathbf{N}^d$, the following limit exists:

$$\lim_n \frac{\text{card}\big(T_n(\vec{r}; A)\big)}{n} = \tau(\vec{r}; A).$$

If $(A^1, \ldots, A^h)$ is a collection of subsets of $\mathbf{N}$, we will say that $(A^1, \ldots, A^h)$ is an *asymptotically ponderable collection* (APC, for short), if for any $d \geq 1$, $\vec{r} \in \mathbf{N}^d$, $\{i_1, \ldots, i_d\} \in \{1, \ldots, h\}$ and any sub-collection $\mathcal{A} = (A^{i_1}, \ldots, A^{i_d})$, the following limit exists:

$$\lim_n \frac{\text{card}\big(G_n(\vec{r}; \mathcal{A})\big)}{n} = \gamma(\vec{r}; \mathcal{A}).$$

Remark 4. (a) A is APS if and only if $\mathcal{A} = (A)$ is an APC.

(b) Asymptotic ponderability is an hereditary property: if a collection is an APC, so is any sub-collection.

(c) If $(A^1, \ldots, A^h)$ is APC with $\tau(0, A^j) > 0$ for all j, then $(A^1, \ldots, A^h)$ is AMC. Indeed, a set A is AMS if the convergence in Definition 4 holds for $d = 1, 2$. Therefore, the non-AMS sets of Section 2.2 are non-APS. We will give later on an example of an AMS that is non-APS.

(d) Consider $\{Y_t : t \in \mathbf{N}\}$ a stationary and ergodic random process, such that Y_0 takes values on $\{1, \ldots, k\}$ and let $A^j(\omega) = \{t \in \mathbf{N} : Y_t(\omega) = j\}$. Then, by the Ergodic Theorem, $\mathcal{A} = (A^1, \ldots, A^k)$ is, almost surely, an APC and $\gamma(\vec{r}; \mathcal{A}) = E(\prod_{i=1}^{i=m} \mathbf{1}_{\{Y_{r_i} = j\}})$, $\vec{r} \in \mathbf{N}^m$, $m \in \mathbf{N}$.

Remark 5. Let us explain the motivation of the *APS* and *APC* concepts. Assume that we are studying the asymptotic behaviour of a functional of the form

$$Z_n = \sum_{t \in A_n} f_n(X_t),$$

where X is stationary, f_n is a real function ($f_n(x) = \mathbf{1}_{\{x > u_n\}}$ in the case of high-level exceedances, $f_n(x) = x/\sqrt{n}$ in the case of averages of centered processes).

If we are trying to show that Z_n is asymptotically Gaussian, in order to identify its limit we only need to compute asymptotic moments up to order two. If Z_n is centered, we only need to deal with the second moment

$$E(Z_n^2) = \sum_{t=0}^{\infty} nE\big(f_n(X_0) f_n(X_t)\big) \frac{\text{card}\big(T_n((0,t), A)\big)}{n}.$$

Therefore, to obtain Gaussian limits we only need to control (in the mean) the arithmetic distribution of couples of points of A.

But for a non-Gaussian limit as in the case of CPLT, we need to compute higher moments of Z_n. For moments of order three, it is easy to show that

$$E(Z_n^3) = \sum_{t=0}^{n}\sum_{s=t}^{n} nE\big(f_n(X_0)f_n(X_t)f_n(X_s)\big)\frac{\text{card}\big(T_n\big((0,t,s),A\big)\big)}{n}.$$

Hence, for computation of asymptotic moments up to order three, we need the convergence in Definition 4 for $d = 1, 2, 3$; and if all the moments are involved, convergence for any d must hold.

We now present a rough version of the CPLT, analogous to Proposition 1, that holds in this context. As in Proposition 1, a collection of examples corresponding to different "weak dependence structures" are presented. We refer to Bellanger and Perera (1997) for a complete version of this result.

Proposition 3. *CPLT for the counting process of high-level exceedances over A of a stationary and weakly dependent set holds if A is an APS. If we look to the exceedances of X^1 over A^1,..., X^k over A^k, where $(X^1, \ldots, X^k)$ is stationary and weakly dependent, the CPLT holds if $(A^1, \ldots, A^k)$ is an APC.*

The application of Definition 4 to level sets of random process leads to the following definition.

Definition 5. We will say that a real-valued process Y is *ponderable* if for every $d \in \mathbf{N}$, $\vec{r} \in \mathbf{N}^d$, there exists a (random) probability measure $\mu^{\vec{r}}(\cdot)(\omega)$ defined on the Borel sets of $\mathbf{R}^d$ such that if $B_1, \ldots, B_d$ are Borel real sets then the (random) collection $\mathcal{A}\big((B_1, \ldots, B_d)\big)(\omega) = \big(A^1(\omega), \ldots, A^d(\omega)\big)$ defined by

$$A^j(\omega) = \{t \in \mathbf{N} : Y_t(\omega) \in B_j\}$$

is an APC almost surely with $\gamma\big(\vec{r}, \mathcal{A}\big((B_1, \ldots, B_d)\big)(\omega)\big) = \mu^{\vec{r}}(B_1 \times B_2 \times \cdots \times B_d)(\omega)$.

If, in addition, the measures $\mu^{\vec{r}}$ are nonrandom, we will say that Y is *regular*.

Remark 6. (a) If σ_∞^Y is trivial, Y is regular.

(b) Observe that Y ponderable means that for any $B_1, \ldots, B_d$ Borel real sets its (mean) asymptotic occupation measure is defined a.s., i.e.,

$$\mu^{\vec{r}}(B_1 \times B_2 \times \cdots \times B_d)(\omega) = \lim_n \frac{1}{n}\sum_{t=0}^{n} \mathbf{1}_{\{Y_{t+r_j}(\omega) \in B_j, j=1,\ldots,d\}} \text{ a.s.}$$

In this way, a process is regular when a deterministic mean occupation measure exists.

Definition 6. Let ξ be a real-valued random process and $\varphi : \mathbf{R}^2 \to \mathbf{R}$ a measurable function. We will say that ξ is a *φ-noise* if for every finite

$h \in \mathbf{N}$, any vector $(y_1, \ldots, y_h) \in \mathbf{R}^h$ and any APC $\mathcal{A} = (A^1, \ldots, A^h)$ the random process

$$X_t = \sum_{j=1}^{j=h} \varphi(\xi_t, y^j) \mathbf{1}_{A^j}(t) \tag{10}$$

satisfies the CPLT.

Proposition 3 gives examples of ξ where this definition holds. So, as in Sections 2 and 3, we can place this ξ in the definition of X, and the same program of Sections 1 and 2 will work.

Finally, introduce the notation :

$$\varphi(\xi(\vec{r}), \vec{y}) = (\varphi(\xi_{r_1}, y_1), \ldots, \varphi(\xi_{r_d}, y_d), \quad \vec{r} \in \mathbf{N}_L^d,\ \vec{y} \in \mathbf{R}^d,$$

if $J \subset \mathbf{R}$ and $d \geq 1$, denote

$$J_L^d = \{(j_1, \ldots, j_d) : j_i \in J, j_i < j_{i+1} i = 1, \ldots, d-1\}.$$

If $\vec{r} \in \mathbf{N}^d$, $J \subset \mathbf{N}$, for any random process V and $u > 0$, set:

$$\{V(\vec{r}) > u\} := \bigcap_{i=1}^{d} \{V_{r_i} > u\}, \quad \{V(J) > u\} := \bigcap_{t \in J} \{V_t > u\}.$$

Definition 7. Let X be a real-valued random process. We will say that X admits an *I-decomposable regression* if there exist a ponderable process Y, a measurable function φ and a φ-noise ξ independent of Y such that $X_t = \varphi(\xi_t, Y_t) \forall t \in \mathbf{N}$ and the following conditions are fulfilled

(a) $\forall K > 0$

$$\limsup_{\delta \to 0} \left(\limsup_{n} \left(\sup_{\substack{|x-z| \leq \delta \\ |x| \leq K}} nP(\{\varphi(\xi_0, x) > u_n\} \nabla \{\varphi(\xi_0, z) > u_n\}) \right) \right) = 0$$

where $A \nabla B = (A^c \cap B) \cup (A \cap B)$, and

$$\limsup_{K} \left(\limsup_{n} \left(\sup_{|x| > K} nP(\{\varphi(\xi_0, x) > u_n\}) \right) \right) < \infty.$$

(b) $\forall x \in \mathbf{R}$, $\lim_n nP(\{\varphi(\xi_0, x) > u_n\}) = \lambda(x)$

(c) $\forall d \in \mathbf{N}$, $\vec{y} \in \mathbf{R}^d$, $\vec{r} \in \mathbf{N}_L^d$,

$$\lim_{n} P(\{\varphi(\xi(\vec{r}), \vec{y}) > u_n\} / \{\varphi(\xi_0, y_0) > u_n\}) = a(\vec{r}, \vec{y})$$

and $a(\vec{r}, \cdot)$ is continuous $\forall \vec{r}$.

(d) For any $s \in \mathbf{N}$,

$$\sum_{d=s}^{\infty} |\Theta(s;d)| \sum_{\vec{r} \in \mathbf{N}_L^d} \sup_{y \in \mathbf{R}^d} a(\vec{r}, \vec{y}) < \infty,$$

$$\lim_{s \to \infty} \sum_{d=s}^{\infty} |\Theta(s;d)| \sum_{\vec{r} \in \mathbf{N}_L^d} \sup_{y \in \mathbf{R}^d} a(\vec{r}, \vec{y}) = 0.$$

Remark 7. A straightforward computation shows that condition (a) implies that the function λ defined in (b) is uniformly continuous and bounded.

Remark 8. Let us briefly explain these conditions. (b) assumes that for fixed x, the limit intensity of $\varphi(\xi_m, x)$ is well-defined. Its continuity, as well as (c), guarantees that if Y is close to Y^*, their limit Laplace transforms are also close. (a) allows to approximate Y first, by its restriction to a compact set $[-K, K]$ and, second, its discretization. (d) is the technical hypothesis required to apply the CPLT for the approximation of Y.

Remark 9. The reader may ask where does the "I" of "*I*-decomposable" comes from. It comes from "Independent": keep in mind that X is a process that we can decompose in two *independent* random components, one mainly "local" and "weakly dependent" (ξ) and other we can control "in the mean" (Y).

Example 3. Consider $U = (U_t)_{t \in \mathbf{N}}$ i.i.d. with common absolutely continuous law μ. Let V be a random variable assuming a finite number of values ($V \in \{1, \ldots, S\}$) and independent of U and let $(a_t)_{t \in \mathbf{N}}$ be a sequence of real numbers satisfying $\lim_t a_{kt+h} = a(h)$, $h = 0, 1, \ldots, k-1$. Define $Y_t = U_t + a_t V$. Then Y is ponderable and regular if $a(h) = 0$ for any h.

Consider now $\varphi(\xi, y) = \xi g(y)$ with g a real, bounded and positive function and ξ a moving average of *iid* Cauchy variables.

It is easy to check that $X_t = \varphi(\xi_t, Y_t)$ satisfies Definition 7. Indeed, the reader can find in Section 4 of Bellanger and Perera (1997), and in particular, in the Example 4 of Section 4, a precise guide to show that ξ is a φ-noise; conditions (a) to (d) of Definition 7 are obtained by elementary computations.

This simple example clearly shows what are the essential properties required for ξ, Y and φ:

- Concerning Y we need "stationarity in the mean," in the sense that the occupation measures

$$\mu_n^{\vec{r}}(C) = \frac{1}{n} \sum_{t=1}^{n} \mathbf{1}_{\{Y_t(\vec{r}) \in C\}}$$

must converge. Those asymptotic occupation measures will be non-random if Y has "short-range memory."

- Concerning ξ we need weak dependence, like the conditions presented in Section 4 of Bellanger and Perera (1997), and a good knowledge of the conditional distribution of ξ given ξ_0.
- Finally, φ must be a smooth function, more precisely, we require a smooth control in y of the probability tails of $\varphi(\xi_0, y)$.

The main result here is the following.

Theorem 2. *If X admits an I-decomposable regression, then:*

(a) *If Y is regular, X satisfies the CPLT. More precisely, if $X = \varphi(\xi, Y)$, and*

$$N_n(B) = \sum_{t/n \in B} \mathbf{1}_{\{X_t > u_n\}}, \quad B \in \mathcal{B},$$

then N_n converges in law to N, a Compound Poisson process with Laplace transform

$$L(B; x) = \exp\left(m(B) \sum_{j=1}^{\infty} \nu_j (e^{-xj} - 1) \right),$$

with

$$\nu_j = \sum_{d=j}^{\infty} (-1)^{j+d} C_j^d \int_{\mathbf{R}^d} \sum_{\vec{r} \in \mathbf{N}_L^{d-1}} a(\vec{r}, \vec{y}) \lambda(y_0) \mu^{\vec{r}}(dy) \quad \forall j \in \mathbf{N}.$$

(b) *If Y is not regular, then N_n/Y satisfies the CPLT and N_n converges weakly to a mixture of Compound Poisson processes.*

4.2 An Example of an AMS That is Non-AMP

Consider $U = (U_t)_{t \in \mathbf{N}} iid$ with common Ber($\frac{1}{2}$) law. It is easy to see that we can construct a 3-dependent stationary process $V = (V_t)_{t \in \mathbf{N}}$, independent of U, whose two-dimensional laws are identical to those of U but such that (U_1, U_2, U_3) and (V_1, V_2, V_3) have different laws. For any $n \in \mathbf{N}$ define $I(n) = [100^{2^{n-1}}, 100^{2^n})$ and set $B = \bigcup_{r=0}^{\infty} I(2r)$; finally, define

$$A(\omega) = \{n \in B \mid U_n(\omega) = 1\} \bigcup \{n \in B^{\mathrm{c}} \mid V_n(\omega) = 1\}.$$

Straightforward computations show that A is, with probability one, an AMS, with

$$F(r; A) = \frac{1}{4} \quad \forall r \geq 1.$$

On the other hand, for $\vec{r} = (0, 1, 2) \in \mathbf{N}^3$, it is easy to show that

$$\limsup_n \frac{\mathrm{card}\{T_n(\vec{r}; A)\}}{n} \geq \frac{13}{96} > \frac{1}{8} \geq \liminf_n \frac{\mathrm{card}\{T_n(\vec{r}; A)\}}{n} \quad \text{a.s.}$$

Therefore A is non-APS with probability one.

Acknowledgments: To Marc Moore and Xavier Guyon, organizers of the CRM Workshop on Inference for Spatial Processes, who highly stimulated our work. This was also possible by the constant support and help given by Mario Wschebor, José Rafael Léon, and Didier Dacunha-Castelle. To Ernesto Mordecki, for his careful reading. Finally, a very special acknowledgement: in 1985, after a very critical period of our history, mathematical research came back to life in Uruguay. This new period of activity was made possible by the extraordinary effort and human quality of a small group of people, where Gonzalo Pérez Iribarren played an extremely relevant role.

5 References

Alpuim, M.T., N.A. Catkan, and J. Hüsler (1995). Extremes and clustering of nonstationary max-AR(1) sequences. *Stochastic Processes and their Applications 56*, 171–184.

Andrews, D.W.K. (1984). Nonstrong autoregressive processes. *Journal of Applied Probability 21*, 930–934.

Barbour, A.D., L. Holst, and S. Janson (1992). *Poisson Approximation*, Volume 2 of *Oxford Studies in Probability*. New York: Oxford University Press.

Bellanger, L. and G. Perera (1997). Compound Poisson limit theorems for high-level exceedances of some non-stationary processes. preprint 46, Université de Paris-Sud.

Berman, S.M. (1992). *Sojourns and Extremes of Stochastic Processes*. The Wadsworth & Brooks/Cole Statistics/Probability Series. Monterey, CA: Wadsworth & Brooks.

Bernstein, S.N. (1944). Extension of the central limit theorem of probability theory to sums of dependent random variables. *Uspekhi Matematicheskikh Nauk 10*, 65–114.

Billingsley, P. (1968). *Convergence of Probability Measures* (2nd ed.). Wiley Series in Probability and Statistics: Probability and Statistics. New York: Wiley.

Birkel, T. (1988). Moment bounds for associated sequences. *The Annals of Probability 16*, 1084–1093.

Bolthausen, E. (1982). On the central limit theorem for stationary mixing random fields. *The Annals of Probability 10*, 1047–1050.

Bradley, R.C. (1986). Basic properties of strong mixing conditions. In E. Eberlein and M. Taqqu (Eds.), *Dependence in Probability and Statistics: A Survey of Recent Results*, Volume 11 of *Progress in Probability and Statistics*, pp. 165–192. Boston: Birkhäuser.

Bradley, R.C. (1993). Equivalent mixing conditions for random fields. *The Annals of Probability 21*, 1921–1926.

Brown, T.C. and A. Xia (1995). On Stein-Chen factors for Poisson approximation. *Statistics & Probability Letters 23*, 327–332.

Bryc, W. and W. Smolenski (1993). Moment conditions for almost sure convergence of weakly correlated random variables. *Proceedings of the American Mathematical Society 119*, 355–373.

Cohen, J. (1989). On the compound Poisson limit theorem for high level exceedances. *Journal of Applied Probability 26*, 458–465.

Davydov, Ju.A. (1973). Mixing conditions for Markov chains. *Theory of Probability and its Applications 18*, 312–328.

Dedecker, J. (1998). A central limit theorems for stationary random fields. *Probability Theory and Related Fields 110*, 397–426.

Dobrushin, R.L. (1968). The description of a random field by its conditional distribution. *Theory of Probability and its Applications 13*, 201–229.

Doukhan, P. (1995). *Mixing: Properties and Examples*, Volume 85 of *Lectures Notes in Statistics.* Springer.

Doukhan, P. and S. Louhichi (1997). Weak dependence and moment inequalities. preprint 8, Université de Paris-Sud.

Dziubdziela, W. (1988). A compound Poisson limit theorem for stationary mixing sequences. *Revue Roumaine de Mathématiques Pures et Appliquées 33*, 39–45.

Dziubdziela, W. (1995). On the limit distribution of the sums of mixing Bernoulli random variables. *Statistics & Probability Letters 23*, 179–182.

Esary, J.D., F. Proschan, and D.W. Walkup (1967). Association of random variables, with applications. *The Annals of Mathematical Statistics 38*, 1466–1476.

Falk, M., J. Hüsler, and R.D. Reiss (1994). *Laws of Small Numbers: Extremes and Rare Events* (2nd ed.), Volume 23 of *DMV Seminar.* Basel: Birkhäuser.

Ferreira, H. (1993). Joint exceedances of high levels under a local dependence condition. *Journal of Applied Probability 30*, 112–120.

Guyon, X. (1995). *Random Fields on a Network: Modeling, Statistics and Applications.* Probability and its Applications. New York: Springer.

Hsiau, S.R. (1997). Compound Poisson limit theorems for Markov chains. *Journal of Applied Probability 34*, 24–34.

Hsing, T., J. Hüsler, and M.R. Leadbetter (1988). On the exceedance point process for a stationary sequence. *Probability Theory and Related Fields 78*, 97–112.

Hüsler, J. (1993). A note on exceedances and rare events of non-stationary sequences. *Journal of Applied Probability 30*, 877–888.

Hudson, W., H.G. Tucker, and J.A. Veeh (1989). Limit distributions of sums of m-dependent Bernoulli random variables. *Probability Theory and Related Fields 82*, 9–17.

Kolmogorov, A.N. and Ju.A. Rozanov (1960). On the strong mixing conditions for stationary Gaussian sequences. *Theory of Probability and its Applications 5*, 204–207.

Leadbetter, M.R. (1991). On a basis for 'peak over threshold' modeling. *Statistics & Probability Letters 12*, 357–362.

Leadbetter, M.R. and T. Hsing (1990). Limit theorems for strongly mixing stationary random measures. *Stochastic Processes and their Applications 36*, 231–243.

Leadbetter, M.R., G. Lindgren, and H. Rootzén (1983). *Extremes and Related Properties of Random Sequences and Processes.* New York: Springer.

Leadbetter, M.R. and S. Nandagopalan (1989). On exceedance point processes for stationary sequences under mild oscillation restrictions. In J. Hüsler and R.-D. Reiss (Eds.), *Extreme Value Theory*, Volume 51 of *Lectures Notes in Statistics*, Oberwolfach, 1987, pp. 69–80.

Leadbetter, M.R. and H. Rootzén (1988). Extremal theory for stochastic processes. *The Annals of Probability 16*, 431–478.

Leadbetter, R. (1995). On high level exceedance modeling and tail inference. *Journal of Statistical Planning and Inference 45*, 247–260.

Louhichi, S. (1998). *Théorèmes limites pour des suites positivement ou faiblement dépendantes.* Ph.D. thesis, Université de Paris-Sud.

Newman, C.M. (1980). Normal fluctuations and the FKG inequalities. *Communications in Mathematical Physics 74*, 119–128.

Newman, C.M. (1984). Asymptotic independence and limit theorems for positively and negatively dependent random variables. In *Inequalities in Probability and Statistics*, Volume 5 of *IMS Lecture Notes-Monograph Series*, Lincolln, NE, 1982, pp. 127–140. Inst. Math. Statist., Hayward, CA.

Perera, G. (1994). Spatial statistics, central limit theorems for mixing random fields and the geometry of $\mathbb{Z}^d$. *Comptes Rendus des Séances de l'Académie des Sciences. Série I. Mathématique 319*, 1083–1088.

Perera, G. (1997a). Applications of central limit theorems over asymptotically measurable sets: regression models. *Comptes Rendus des Séances de l'Académie des Sciences. Série I. Mathématique 324*, 11275–1280.

Perera, G. (1997b). Geometry of $\mathbb{Z}^d$ and the central limit theorem for weakly dependent random fields. *Journal of Theoretical Probability 10*, 581–603.

Pitt, L.D. (1982). Positively correlated normal variables associated. *The Annals of Probability 10*, 496–499.

Rio, E. (1993). Covariance inequalities for strongly mixing processes. *Annales de l'Institut Henri Poincaré. Probabilités et Statistiques 29*, 587–597.

Rényi, A. (1951). On composed Poisson distribution. II. *Acta Mathematica Academiae Scientiarum Hungaricae 2*, 83–98.

Rosenblatt, M. (1956). A central limit theorem and a strong mixing condition. *Proceedings of the National Academy of Sciences USA 42*, 43–47.

Roussas, G. (1994). Asymptotic normality of random fields of positively of negatively associated processes. *Journal of Multivariate Analysis 16*, 152–173.

Smith, R.L. and T.S. Shively (1994). A point process approach to modeling trends in tropospheric ozone based on exceedances of a high threshold. Technical Report 16, National Institute of Statistical Sciences.

Volkonskii, V.A. and Ju.A. Rozanov (1959). Some limit theorems for random functions. I. *Theory of Probability and its Applications 4*, 178–197.

Volkonskii, V.A. and Ju.A. Rozanov (1961). Some limit theorems for random functions. II. *Theory of Probability and its Applications 6*, 186–198.

Wschebor, M. (1985). *Surfaces aléatoires : mesure géométrique des ensembles de niveau*, Volume 1147 of *Lectures Notes in Mathematics*. Berlin: Springer.

Yu, H. (1993). A Glivenko-Cantelli lemma and weak convergence for empirical processes of associated sequences. *Probability Theory and Related Fields 95*, 357–370.

5

Multiscale Graphical Modeling in Space: Applications to Command and Control

Hsin-Cheng Huang
Noel Cressie

ABSTRACT Recently, a class of multiscale tree-structured models was introduced in terms of scale-recursive dynamics defined on trees. The main advantage of these models is their association with a fast, recursive, Kalman-filter prediction algorithm. In this article, we propose a more general class of multiscale graphical models over acyclic directed graphs, for use in command and control problems. Moreover, we derive the generalized-Kalman-filter algorithm for graphical Markov models, which can be used to obtain the optimal predictors and prediction variances for multiscale graphical models.

1 Introduction

Almost every aspect of command and control (C2) involves dealing with information in the presence of uncertainty. Since information in a battlefield is never precise, its status is rarely known exactly. In the face of this uncertainty, commanders must make decisions, issue orders, and monitor the consequences. The uncertainty may come from noisy data or, indeed, regions of the battle space where there are no data at all. It is this latter aspect, namely lack of knowledge, that is often most important. Statistical models that account for noisy data are well accepted by the research community but the full quantification of uncertainty due to lack of knowledge (caused either by hidden processes or missing data) falls under the purview of statistical modeling.

Statistical models for C2 are relatively new and still rather embryonic; for example, Paté-Cornell and Fischbeck (1995) consider a very simple statistical model for a binary decision in C2. Given the spatial nature of the battlefield, it seems that a logical next step is to propose and fit spatial statistical models. Because timely analysis and decision-making in C2 is crucial, very fast optimal-filtering algorithms are needed. The purpose of this article is to present spatial statistical models for which first and second posterior moments can be obtained quickly and exactly.

Consider a scene of interest made up of $M \times N$ pixels

$$\{P(p,q) : p = 1, \ldots, M, q = 1, \ldots, N\}$$

centered at pixel locations

$$\{s(p,q) : p = 1, \ldots, M, q = 1, \ldots, N\}.$$

For the moment, assume the scene is observed completely at one time instant, yielding multivariate spatial data from L sensors:

$$\boldsymbol{y}_l \equiv \big(y_l\big(s(p,q)\big) : p = 1, \ldots, M, q = 1, \ldots, N\big)'; \quad l = 1, \ldots, L,$$

where the index l corresponds to lth sensor and the datum $y_l\big(s(p,q)\big)$ is observed on pixel $P(p,q)$ centered at pixel location $s(p,q)$.

One powerful way to deal with multi-sensor information is to use a hierarchical statistical model. The multi-sensor data $\{\boldsymbol{y}_l : l = 1, \ldots, L\}$ are noisy. The "true" scene is an image

$$\boldsymbol{x} \equiv \big(x\big(s(p,q)\big) : p = 1, \ldots, M, q = 1, \ldots, N\big)'.$$

Then a hierarchical statistical model is obtained upon specifying the probability densities,

$$[\boldsymbol{y}_1, \ldots, \boldsymbol{y}_L | \boldsymbol{x}] = f_\phi$$

and

$$[\boldsymbol{x}] = g_\theta.$$

The general goal is to obtain $[\boldsymbol{x} \mid \boldsymbol{y}_1, \ldots, \boldsymbol{y}_L]$, from which predictions about the true scene can be made. In this article, we assume only one sensor and we give very fast algorithms that yield $E(\boldsymbol{x} \mid \boldsymbol{y})$ and $\mathrm{var}(\boldsymbol{x} \mid \boldsymbol{y})$. If the parameters (ϕ, θ) are estimated, then our approach could be called empirical Bayesian. A fully Bayesian approach would express the uncertainty in (ϕ, θ) through a prior distribution but, while conceptually more pleasing, the very fast optimal-filtering algorithms referred to above are no longer available.

Optimal C2 decisions, in the face of uncertainty, can be handled through statistical decision theory, provided commanders can quantify their loss functions. That is, what loss

$$L(\boldsymbol{\delta}(\boldsymbol{y}), \boldsymbol{x})$$

would be incurred if the true scene is $\boldsymbol{x}$ but an estimate $\boldsymbol{\delta}(\boldsymbol{y})$ were used? Some basic results in decision theory demonstrate that the optimal filter $\boldsymbol{\delta}^0$ is the one that minimizes $E\big(L(\boldsymbol{\delta}(\boldsymbol{y}), \boldsymbol{x}) \mid \boldsymbol{y}\big)$ with respect to measurable functions $\boldsymbol{\delta}$. For example, if the squared-error loss function:

$$L(\boldsymbol{\delta}(\boldsymbol{y}), \boldsymbol{x}) = \big(\boldsymbol{\delta}(\boldsymbol{y}) - \boldsymbol{x}\big)'\big(\boldsymbol{\delta}(\boldsymbol{y}) - \boldsymbol{x}\big),$$

is used, the optimal filter is:

$$\delta^0(\boldsymbol{y}) = E(\boldsymbol{x} \mid \boldsymbol{y}).$$

This is the loss function we shall use henceforth in this article. It is an appropriate loss function for drawing maps of $\boldsymbol{x}$, although it is not all that useful for identifying nonlinear features (e.g., extremes in $\boldsymbol{x}$).

It is typically the case that the battle commander has a different "aperture" than a platoon commander. The battle commander needs more global, aggregated data, whereas the platoon commander is often making decisions based on local, more focused information. Yet the sum of the parts is equal to the whole and it can be important for the battle commander and those in a chain of command all the way to the platoon commander, to query and receive information at different scales. Our intent here is not to decide which scales belong to which level of command but rather to develop a methodology that makes multiscale modeling and analysis possible.

Recently, a class of multiscale tree-structured models was introduced in terms of scale-recursive dynamics defined on trees (Chou, 1991, Basseville et al., 1992a,b,c) This class of models is very rich and can be used to describe stochastic processes with multiscale features. Let $\boldsymbol{x}_t$ denote a vector representing the (multivariate) status of the battlefield at some location t within a given scale. Then the multiscale tree-structured model is:

$$\boldsymbol{x}_t = \boldsymbol{A}_t \boldsymbol{x}_{pa(t)} + \boldsymbol{w}_t; \quad t \in T \setminus \{t_0\},$$

where T is a tree with root t_0, and $pa(t)$ denotes a location connected to t but within a coarser scale. Think of $pa(t)$ as a parent of t. The potential observations $\{\boldsymbol{y}_t\}$ are given by

$$\boldsymbol{y}_t = \boldsymbol{C}_t \boldsymbol{x}_t + \boldsymbol{\epsilon}_t; \quad t \in T.$$

Our goal is to obtain the conditional distribution of $\{\boldsymbol{x}_t : t \in T\}$ based on the data $\{\boldsymbol{y}_t\}$. The main advantage of these models over other statistical methods is their association with a fast, recursive, Kalman-filter prediction algorithm (Chou et al., 1994). The algorithm can be used to handle massive amounts of data efficiently, even in the presence of missing data. However, as models for time series or spatial processes at the finest level of the tree, they are somewhat restricted in that they may enforce certain discontinuities in covariance structures. That is, two points that are adjacent in time or space may or may not have the same parent node (or even grandparent node) on the tree. These discontinuities can sometimes lead to blocky artifacts in the resulting predicted values.

There are at least two possible ways that the tree-structured model above can be generalized to describe a more realistic situation in C2. First, we may consider an index set T that is more general than a tree. Second, we may keep the tree structure, but allow for more general nonlinear and non-Gaussian models as follows:

$$\begin{aligned} \boldsymbol{x}_t &= f_t(\boldsymbol{x}_{pa(t)}, \boldsymbol{w}_t); \quad t \in T \setminus \{t_0\}, \\ \boldsymbol{y}_t &= g_t(\boldsymbol{x}_t, \boldsymbol{\epsilon}_t); \quad t \in T. \end{aligned}$$

This paper addresses the first approach.

We consider graphical Markov models for C2, where the nodes of the graph represent random variables and the edges of the graph characterize certain Markov properties (conditional-independence relations). In particular, we propose multiscale graphical models in terms of scale-recursive dynamics defined on acyclic directed graphs. This class of models not only allows us to describe a wide range of spatial-dependence structure, but also can be used to reduce the discontinuities sometimes obtained from tree-structured models.

In Section 2, we describe tree-structured Markov models and its associated Kalman-filter algorithm. In Section 3, we introduce graphical Markov models and develop the generalized-Kalman-filter algorithm for these models. The proposed multiscale graphical models, which are more general than multiscale tree-structured models, are presented in Section 4. Finally, a brief discussion is given in Section 5.

2 Tree-Structured Markov Models

In application areas such as command and control, fast filtering and prediction of data at many different spatial resolutions is an important problem. Gaussian tree-structured Markov models are well adapted to solving the problem in simple scenarios.

Consider a (multivariate) Gaussian random process $\{\boldsymbol{x}_t : t \in T\}$ indexed by the nodes of a tree (T, E), where T is the set of nodes, and E is the set of directed edges. Let t_0 denote the root of the tree, and let $\text{pa}(t)$ denote the parent of a node $t \in T$. Then

$$E = \{(\text{pa}(t), t) : t \in T \setminus \{t_0\}\}.$$

The Gaussian tree-structured Markov model on the tree (T, E) is defined as follows. Assume that the Gaussian process evolves from parents to children in a Markovian manner according to the following state-space model:

$$\boldsymbol{x}_t = \boldsymbol{A}_t \boldsymbol{x}_{pa(t)} + \boldsymbol{w}_t; \quad t \in T \setminus \{t_0\}, \tag{1}$$

$$\boldsymbol{y}_t = \boldsymbol{C}_t \boldsymbol{x}_t + \boldsymbol{\epsilon}_t; \quad t \in T, \tag{2}$$

where $\{\boldsymbol{y}_t : t \in T\}$ are (potential) observations, $\{\boldsymbol{x}_t : t \in T\}$ are hidden (i.e., unobserved), zero-mean, Gaussian state vectors that we would like to predict, $\{\boldsymbol{A}_t : t \in T \setminus \{t_0\}\}$ and $\{\boldsymbol{C}_t : t \in T\}$ are deterministic matrices, $\{\boldsymbol{w}_t : t \in T \setminus \{t_0\}\}$ and $\{\boldsymbol{\epsilon}_t : t \in T\}$ are independent zero-mean Gaussian vectors with covariance matrices given by

$$\boldsymbol{W}_t \equiv \text{var}(\boldsymbol{w}_t); \quad t \in T \setminus \{t_0\},$$
$$\boldsymbol{V}_t \equiv \text{var}(\boldsymbol{\epsilon}_t); \quad t \in T,$$

$\{\boldsymbol{x}_t : t \in T\}$ and $\{\boldsymbol{\epsilon}_t : t \in T\}$ are independent, and $\boldsymbol{x}_{pa(t)}$ and $\boldsymbol{w}_t$ are independent for $t \in T \setminus \{t_0\}$.

For example, we consider a tree-structured model with the hidden process given by,

$$x_{j,k} = x_{j-1,[k/2]} + w_{j,k}; \quad j = J_0 + 1, \dots, J, k \in \mathbb{Z}, \tag{3}$$

where $[k/2]$ denotes the largest integer that is less than or equal to $k/2$, $\text{var}(x_{J_0,k}) = \sigma_0^2$; $k \in \mathbb{Z}$, and $\text{var}(w_{j,k}) = \sigma_j^2$, $j = J_0 + 1, \dots, J$, $k \in \mathbb{Z}$. The associated tree of this model is shown in Figure 1. For $J_0 = 0$, $J = 6$, $\sigma_0^2 = 10$, $\sigma_1^2 = 8$, $\sigma_2^2 = 6$, $\sigma_3^2 = 4$, $\sigma_4^2 = 3$, $\sigma_5^2 = 2$, and $\sigma_6^2 = 1$, the correlation function of the latent process,

$$f(t,s) \equiv \text{corr}(x_{6,t}, x_{6,s}); \quad t, s = 0, 1, \dots, 127,$$

constructed from two nodes $x_{0,0}$ and $x_{0,1}$ at the coarsest scale, is shown in Figure 2. Notice the blocky nature of the correlation function and, while it is not stationary, *en gros* it is nearly so.

The standard Kalman-filter algorithm developed for time series based on a state-space model (e.g., Harvey, 1989) can be generalized to tree-structured models. The derivation of the algorithm has been developed by Chou et al. (1994). In this section, a new and somewhat simpler approach based on Bayes' theorem is used to derive the algorithm.

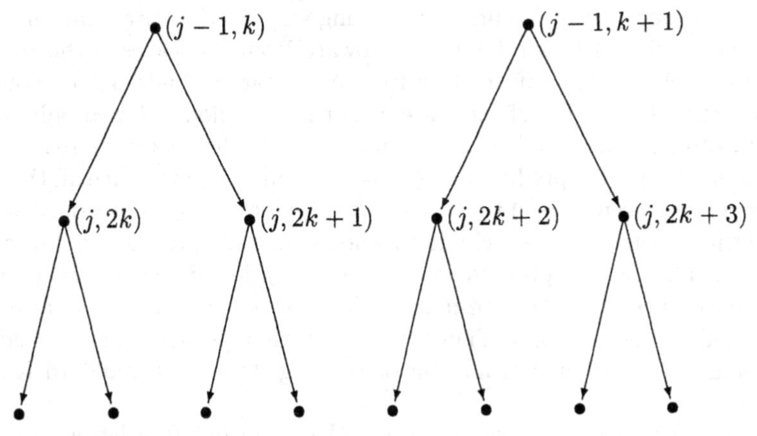

Figure 1. Graphical representation of a tree-structured model.

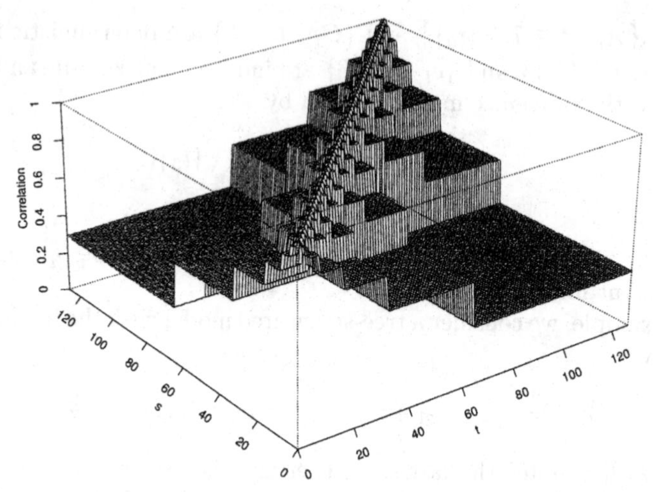

Figure 2. Correlation function $f(t,s) = \mathrm{corr}(x_{6,t}, x_{6,s})$ of $(x_{6,0}, x_{6,1}, \ldots, x_{6,127})'$ for a tree-structured model given by (3).

Kalman filtering (Kalman, 1960, Kalman and Bucy, 1961) in time series is a recursive procedure to obtain the optimal predictors (i.e., conditional expectations) of state vectors based on all the observations that are currently available. Once the end of the series is reached, we can move backward to obtain the optimal predictors of state vectors based on all the data. These two steps, the forward recursion and the backward recursion, are known as filtering and smoothing, respectively.

In a tree-structured model given by (1) and (2), our goal is to obtain the optimal predictors of state vectors based on all the data observed on a tree (T, E). In order to visit all the nodes on T, the generalized Kalman filter for tree-structured models also consists of two steps, the uptree filtering step followed by the downtree smoothing step. In the uptree filtering step, the generalized Kalman filter goes upward from the leaves of the tree and successively collects information from each stage's children (i.e., compute the optimal predictor of the state vector $\boldsymbol{x}_t$ at a node t based only on the data observed at t and its descendents). Once the root t_0 is reached, we obtain the optimal predictor of $\boldsymbol{x}_{t_0}$ based on all the data, since all the other nodes are descendents of the root t_0. We then go back downward by sending information to the root's children, who in turn send to their children, and so on. That is, we proceed in the reverse order of the uptree step and recursively compute the optimal predictor of the state vector $\boldsymbol{x}_t$ at a node t based on all the data. These two steps, though more complicated, are analogous to the filtering and the smoothing step of the standard Kalman filter.

First, we introduce some notation. We use a boldface letter to denote both a vector and a set. Denote $t \prec t'$ if $t' = t$ or t' is a descendent of t.

For $t \in T$,

$$
\begin{aligned}
\gamma_t &\equiv \begin{cases} 1, & \text{if } \boldsymbol{y}_t \text{ is observed,} \\ 0, & \text{otherwise,} \end{cases} \\
\boldsymbol{Y} &\equiv \{\boldsymbol{y}_t : \gamma_t = 1\}, \\
\boldsymbol{Y}_t &\equiv \{\boldsymbol{y}_{t'} : \gamma_{t'} = 1,\ t \prec t'\}, \\
\boldsymbol{Y}_t^* &\equiv \{\boldsymbol{y}_{t'} : \gamma_{t'} = 1, t \prec t', t' \neq t\} = \boldsymbol{Y}_t \setminus \{\boldsymbol{y}_t\}, \\
\boldsymbol{Y}_t^c &\equiv \boldsymbol{Y} \setminus \boldsymbol{Y}_t, \\
\hat{\boldsymbol{x}}_{t_1|t_2} &\equiv E(\boldsymbol{x}_{t_1} \mid \boldsymbol{Y}_{t_2}), \\
\hat{\boldsymbol{x}}^*_{t_1|t_2} &\equiv E(\boldsymbol{x}_{t_1} \mid \boldsymbol{Y}^*_{t_2}), \\
\boldsymbol{\Sigma}_t &\equiv \text{var}(\boldsymbol{x}_t), \\
\boldsymbol{\Sigma}_{t_1,t_2} &\equiv \text{cov}(\boldsymbol{x}_{t_1}, \boldsymbol{x}_{t_2}), \\
\boldsymbol{\Gamma}_{t_1|t_2} &\equiv \text{var}(\boldsymbol{x}_{t_1} \mid \boldsymbol{Y}_{t_2}) = \text{var}(\boldsymbol{x}_{t_1} - \hat{\boldsymbol{x}}_{t_1|t_2}), \\
\boldsymbol{\Gamma}^*_{t_1|t_2} &\equiv \text{var}(\boldsymbol{x}_{t_1} \mid \boldsymbol{Y}^*_{t_2}) = \text{var}(\boldsymbol{x}_{t_1} - \hat{\boldsymbol{x}}^*_{t_1|t_2}), \\
\boldsymbol{\Gamma}_t &\equiv \text{var}(\boldsymbol{x}_t \mid \boldsymbol{Y}) = \text{var}(\boldsymbol{x}_t - \hat{\boldsymbol{x}}_t), \\
\boldsymbol{\Gamma}_{t_1,t_2} &\equiv \text{cov}(\boldsymbol{x}_{t_1}, \boldsymbol{x}_{t_2} \mid \boldsymbol{Y}) = \text{cov}(\boldsymbol{x}_{t_1} - \hat{\boldsymbol{x}}_{t_1}, \boldsymbol{x}_{t_2} - \hat{\boldsymbol{x}}_{t_2}).
\end{aligned}
$$

The uptree filtering algorithm starts with the leaves and proceeds along all upward paths to the root of the tree. At each node t, $\hat{\boldsymbol{x}}_{t|t}$ and $\boldsymbol{\Gamma}_{t|t}$ are obtained recursively. Since the algorithm proceeds in a backward direction, as opposed to the forward direction given by (1), it is necessary to have a backward representation for each $t \in T \setminus \{t_0\}$; that is,

$$
\begin{aligned}
\boldsymbol{x}_{pa(t)} &= E(\boldsymbol{x}_{pa(t)} \mid \boldsymbol{x}_t) + \big(\boldsymbol{x}_{pa(t)} - E(\boldsymbol{x}_{pa(t)} \mid \boldsymbol{x}_t)\big) \\
&\equiv \boldsymbol{B}_t \boldsymbol{x}_t + \boldsymbol{\xi}_t; \qquad\qquad t \in T \setminus \{t_0\},
\end{aligned}
$$

where for $t \in T \setminus \{t_0\}$,

$$
\begin{aligned}
\boldsymbol{B}_t &\equiv \boldsymbol{\Sigma}_{pa(t)} \boldsymbol{A}_t' \boldsymbol{\Sigma}_t^{-1}, \\
\boldsymbol{\xi}_t &\equiv \boldsymbol{x}_{pa(t)} - \boldsymbol{\Sigma}_{pa(t)} \boldsymbol{A}_t' \boldsymbol{\Sigma}_t^{-1} \boldsymbol{x}_t,
\end{aligned}
$$

and

$$
\boldsymbol{R}_t \equiv \text{var}(\boldsymbol{\xi}_t) = \boldsymbol{\Sigma}_{pa(t)} - \boldsymbol{\Sigma}_{pa(t)} \boldsymbol{A}_t' \boldsymbol{\Sigma}_t^{-1} \boldsymbol{A}_t \boldsymbol{\Sigma}_{pa(t)}.
$$

Note that $\boldsymbol{x}_t$ and $\boldsymbol{\xi}_t$ are uncorrelated, since $\boldsymbol{B}_t \boldsymbol{x}_t = E(\boldsymbol{x}_{\text{pa}(t)} \mid \boldsymbol{x}_t)$ and $\boldsymbol{\xi}_t = \boldsymbol{x}_{\text{pa}(t)} - E(\boldsymbol{x}_{\text{pa}(t)} \mid \boldsymbol{x}_t)$ are uncorrelated, for all $t \in T \setminus \{t_0\}$. Also note that $\{\boldsymbol{\Sigma}_t : t \in T \setminus \{t_0\}\}$ can be computed recursively from the forward representation (1) by

$$
\boldsymbol{\Sigma}_t = \boldsymbol{A}_t \boldsymbol{\Sigma}_{pa(t)} \boldsymbol{A}_t' + \boldsymbol{W}_t,
$$

where $\boldsymbol{W}_t \equiv \text{var}(\boldsymbol{w}_t)$; $t \in T \setminus \{t_0\}$.

For a leaf $t \in T$, if $\gamma_t = 0$ (i.e., $\boldsymbol{y}_t$ is not observed), then $\hat{\boldsymbol{x}}_{t|t} = \mathbf{0}$ and $\boldsymbol{\Gamma}_{t|t} = \boldsymbol{\Sigma}_t$, whereas if $\gamma_t = 1$ (i.e., $\boldsymbol{y}_t$ is observed), then by (2),

$$\begin{aligned}\hat{\boldsymbol{x}}_{t|t} &= \operatorname{cov}(\boldsymbol{x}_t, \boldsymbol{y}_t)\big(\operatorname{var}(\boldsymbol{y}_t)\big)^{-1}\boldsymbol{y}_t \\ &= \boldsymbol{\Sigma}_t\boldsymbol{C}_t'(\boldsymbol{C}_t\boldsymbol{\Sigma}_t\boldsymbol{C}_t' + \boldsymbol{V}_t)^{-1}\boldsymbol{y}_t, \\ \boldsymbol{\Gamma}_{t|t} &= \operatorname{var}(\boldsymbol{x}_t) - \operatorname{cov}(\boldsymbol{x}_t, \boldsymbol{y}_t)\big(\operatorname{var}(\boldsymbol{y}_t)\big)^{-1}\operatorname{cov}(\boldsymbol{y}_t, \boldsymbol{x}_t) \\ &= \boldsymbol{\Sigma}_t - \boldsymbol{\Sigma}_t\boldsymbol{C}_t'(\boldsymbol{C}_t\boldsymbol{\Sigma}_t\boldsymbol{C}_t' + \boldsymbol{V}_t)^{-1}\boldsymbol{C}_t\boldsymbol{\Sigma}_t.\end{aligned}$$

Therefore, for a leaf $t \in T$, we have

$$\hat{\boldsymbol{x}}_{t|t} = \gamma_t\boldsymbol{\Sigma}_t\boldsymbol{C}_t'(\boldsymbol{C}_t\boldsymbol{\Sigma}_t\boldsymbol{C}_t' + \boldsymbol{V}_t)^{-1}\boldsymbol{y}_t, \tag{4}$$

$$\boldsymbol{\Gamma}_{t|t} = \boldsymbol{\Sigma}_t - \gamma_t\boldsymbol{\Sigma}_t\boldsymbol{C}_t'(\boldsymbol{C}_t\boldsymbol{\Sigma}_t\boldsymbol{C}_t' + \boldsymbol{V}_t)^{-1}\boldsymbol{C}_t\boldsymbol{\Sigma}_t. \tag{5}$$

For a node $t \in T$ that is not a leaf, let $\{t\alpha_1, t\alpha_2, \ldots, t\alpha_{k_t}\}$ denote all the children of t, where k_t is the number of children of the node t. Suppose that we have computed $\hat{\boldsymbol{x}}_{t\alpha_i|t\alpha_i}$ and $\boldsymbol{\Gamma}_{t\alpha_i|t\alpha_i}$; $i = 1, \ldots, k_t$. Since

$$\boldsymbol{x}_t = \boldsymbol{B}_{t\alpha_i}\boldsymbol{x}_{t\alpha_i} + \boldsymbol{\xi}_{t\alpha_i}; \quad i = 1, \ldots, k_t,$$

it follows that for $i = 1, \ldots, k_t$,

$$\boldsymbol{x}_t \mid \boldsymbol{Y}_{t\alpha_i} \sim \mathrm{N}(\boldsymbol{B}_{t\alpha_i}\hat{\boldsymbol{x}}_{t\alpha_i|t\alpha_i}, \boldsymbol{B}_{t\alpha_i}\boldsymbol{\Gamma}_{t\alpha_i|t\alpha_i}\boldsymbol{B}_{t\alpha_i} + \boldsymbol{R}_{t\alpha_i}).$$

Therefore, for $i = 1, \ldots, k_t$, the following recursions are obtained:

$$\hat{\boldsymbol{x}}_{t|t\alpha_i} = \boldsymbol{B}_{t\alpha_i}\hat{\boldsymbol{x}}_{t\alpha_i|t\alpha_i}, \tag{6}$$

$$\boldsymbol{\Gamma}_{t|t\alpha_i} = \boldsymbol{B}_{t\alpha_i}\boldsymbol{\Gamma}_{t\alpha_i|t\alpha_i}\boldsymbol{B}_{t\alpha_i} + \boldsymbol{R}_{t\alpha_i}. \tag{7}$$

Next we compute $\hat{\boldsymbol{x}}^*_{t|t}$, $\boldsymbol{\Gamma}^*_{t|t}$. Suppose that we have computed $\hat{\boldsymbol{x}}_{t|t\alpha_i}$ and $\boldsymbol{\Gamma}_{t|t\alpha_i}$; $i = 1, \ldots, k_t$. Using Bayes' theorem, we have

$$\begin{aligned}p(\boldsymbol{x}_t \mid \boldsymbol{Y}_t^*) &= p(\boldsymbol{x}_t \mid \boldsymbol{Y}_{t\alpha_1}, \ldots, \boldsymbol{Y}_{t\alpha_{k_t}}) \\ &\propto p(\boldsymbol{Y}_{t\alpha_1}, \ldots, \boldsymbol{Y}_{t\alpha_{k_t}} \mid \boldsymbol{x}_t)p(\boldsymbol{x}_t) \\ &= p(\boldsymbol{x}_t)\prod_{i=1}^{k_t} p(\boldsymbol{Y}_{t\alpha_i} \mid \boldsymbol{x}_t).\end{aligned}$$

Note that the last equality holds, since $\boldsymbol{Y}_{t\alpha_1}, \ldots, \boldsymbol{Y}_{t\alpha_{k_t}}$ are independent, conditional on $\boldsymbol{x}_t$. Hence by using Bayes' theorem again, we have

$$\begin{aligned}p(\boldsymbol{x}_t \mid \boldsymbol{Y}_t^*) &\propto p(\boldsymbol{x}_t)\prod_{i=1}^{k_t}\frac{p(\boldsymbol{x}_t \mid \boldsymbol{Y}_{t\alpha_i})}{p(\boldsymbol{x}_t)} \\ &\propto \exp\left\{-\frac{1}{2}\left[\boldsymbol{x}_t'\boldsymbol{\Sigma}_t^{-1}\boldsymbol{x}_t + \sum_{i=1}^{k_t}\big((\boldsymbol{x}_t - \hat{\boldsymbol{x}}_{t|t\alpha_i})'\boldsymbol{\Gamma}_{t|t\alpha_i}^{-1}(\boldsymbol{x}_t - \hat{\boldsymbol{x}}_{t|t\alpha_i}) - \boldsymbol{x}_t'\boldsymbol{\Sigma}_t^{-1}\boldsymbol{x}_t\big)\right]\right\} \\ &\propto \exp\left\{-\frac{1}{2}\left[\boldsymbol{x}_t'\Big(\boldsymbol{\Sigma}_t^{-1} + \sum_{i=1}^{k_t}(\boldsymbol{\Gamma}_{t|t\alpha_i}^{-1} - \boldsymbol{\Sigma}_t^{-1})\Big)\boldsymbol{x}_t - 2\boldsymbol{x}_t'\sum_{i=1}^{k_t}\boldsymbol{\Gamma}_{t|t\alpha_i}^{-1}\hat{\boldsymbol{x}}_{t|t\alpha_i}\right]\right\}.\end{aligned}$$

Therefore,

$$\hat{\boldsymbol{x}}^*_{t|t} = \boldsymbol{\Gamma}^*_{t|t}\left(\sum_{i=1}^{k_t} \boldsymbol{\Gamma}^{-1}_{t|t\alpha_i} \hat{\boldsymbol{x}}_{t|t\alpha_i}\right), \tag{8}$$

$$\boldsymbol{\Gamma}^*_{t|t} = \left\{\boldsymbol{\Sigma}_t^{-1} + \sum_{i=1}^{k_t}(\boldsymbol{\Gamma}^{-1}_{t|t\alpha_i} - \boldsymbol{\Sigma}_t^{-1})\right\}^{-1}. \tag{9}$$

The final uptree step for each update is to compute $\hat{\boldsymbol{x}}_{t|t}$ and $\boldsymbol{\Gamma}_{t|t}$. Suppose that we have computed $\hat{\boldsymbol{x}}^*_{t|t}$ and $\boldsymbol{\Gamma}^*_{t|t}$. If $\gamma_t = 0$ (i.e., $\boldsymbol{y}_t$ is not observed), then $\hat{\boldsymbol{x}}_{t|t} = \hat{\boldsymbol{x}}^*_{t|t}$ and $\boldsymbol{\Gamma}_{t|t} = \boldsymbol{\Gamma}^*_{t|t}$. If $\gamma_t = 1$ (i.e., $\boldsymbol{y}_t$ is observed), the standard Kalman-filter algorithm can be applied to update $\hat{\boldsymbol{x}}_{t|t}$ and $\boldsymbol{\gamma}_{t|t}$. Using Bayes' theorem again, we have

$$\begin{aligned}
&p(\boldsymbol{x}_t \mid \boldsymbol{Y}_t) = p(\boldsymbol{x}_t \mid \boldsymbol{y}_t, \boldsymbol{Y}^*_t) \\
&\quad \propto p(\boldsymbol{y}_t, \boldsymbol{Y}^*_t \mid \boldsymbol{x}_t) p(\boldsymbol{x}_t) \\
&\quad = p(\boldsymbol{y}_t \mid \boldsymbol{x}_t) p(\boldsymbol{Y}^*_t \mid \boldsymbol{x}_t) p(\boldsymbol{x}_t) \\
&\quad \propto p(\boldsymbol{y}_t \mid \boldsymbol{x}_t) p(\boldsymbol{x}_t \mid \boldsymbol{Y}^*_t) \\
&\quad \propto \exp\{-\tfrac{1}{2}[(\boldsymbol{y}_t - \boldsymbol{C}_t\boldsymbol{x}_t)'\boldsymbol{V}_t^{-1}(\boldsymbol{y}_t - \boldsymbol{C}_t\boldsymbol{x}_t) + (\boldsymbol{x}_t - \hat{\boldsymbol{x}}^*_{t|t})'(\boldsymbol{\Gamma}^*_{t|t})^{-1}(\boldsymbol{x}_t - \hat{\boldsymbol{x}}^*_{t|t})]\} \\
&\quad \propto \exp\{-\tfrac{1}{2}[\boldsymbol{x}_t'(\boldsymbol{C}_t'\boldsymbol{V}_t^{-1}\boldsymbol{C}_t + (\boldsymbol{\Gamma}^*_{t|t})^{-1})\boldsymbol{x}_t - 2\boldsymbol{x}_t'(\boldsymbol{C}_t'\boldsymbol{V}_t^{-1}\boldsymbol{y}_t + (\boldsymbol{\Gamma}^*_{t|t})^{-1}\hat{\boldsymbol{x}}^*_{t|t})]\}.
\end{aligned}$$

Since

$$\left(\boldsymbol{C}_t'\boldsymbol{V}_t^{-1}\boldsymbol{C}_t + (\boldsymbol{\Gamma}^*_{t|t})^{-1}\right)^{-1} = \boldsymbol{\Gamma}^*_{t|t} - \boldsymbol{\Gamma}^*_{t|t}\boldsymbol{C}_t'(\boldsymbol{C}_t\boldsymbol{\Gamma}^*_{t|t}\boldsymbol{C}_t' + \boldsymbol{V}_t)^{-1}\boldsymbol{C}_t\boldsymbol{\Gamma}^*_{t|t},$$

we see that,

$$\hat{\boldsymbol{x}}_{t|t} = \boldsymbol{\Gamma}_{t|t}\left(\gamma_t\boldsymbol{C}_t'\boldsymbol{V}_t^{-1}\boldsymbol{y}_t + (\boldsymbol{\Gamma}^*_{t|t})^{-1}\hat{\boldsymbol{x}}^*_{t|t}\right), \tag{10}$$

$$\boldsymbol{\Gamma}_{t|t} = \boldsymbol{\Gamma}^*_{t|t} - \gamma_t\boldsymbol{\Gamma}^*_{t|t}\boldsymbol{C}_t'(\boldsymbol{C}_t\boldsymbol{\Gamma}^*_{t|t}\boldsymbol{C}_t' + \boldsymbol{V}_t)^{-1}\boldsymbol{C}_t\boldsymbol{\Gamma}^*_{t|t}. \tag{11}$$

Thus, the uptree filtering algorithm can be obtained recursively by (6) through (11) with the initial predictors and the prediction variances of the state vectors at the leaves given by (4) and (5). The algorithm stops at the root t_0, where we obtain

$$\hat{\boldsymbol{x}}_{t_0} = \hat{\boldsymbol{x}}_{t_0|t_0},$$
$$\boldsymbol{\Gamma}_{t_0} = \boldsymbol{\Gamma}_{t_0|t_0}.$$

The downtree smoothing algorithm moves downward from the root to the leaves of the tree, allowing $\hat{\boldsymbol{x}}_t$ and $\boldsymbol{\Gamma}_t$ (i.e., the optimal predictor of $\boldsymbol{x}_t$ and its prediction variance based on all the data) to be computed recursively for $t \in T$. To do this, we first compute the following conditional density for each $t \in T \setminus \{t_0\}$:

$$\begin{aligned} p(\boldsymbol{x}_t, \boldsymbol{x}_{\mathrm{pa}(t)} \mid \boldsymbol{Y}) &= p(\boldsymbol{x}_{\mathrm{pa}(t)} \mid \boldsymbol{Y}) p(\boldsymbol{x}_t \mid \boldsymbol{x}_{\mathrm{pa}(t)}, \boldsymbol{Y}) \\ &= p(\boldsymbol{x}_{\mathrm{pa}(t)} \mid \boldsymbol{Y}) p(\boldsymbol{x}_t \mid \boldsymbol{x}_{\mathrm{pa}(t)}, \boldsymbol{Y}_t, \boldsymbol{Y}_t^c) \\ &= p(\boldsymbol{x}_{\mathrm{pa}(t)} \mid \boldsymbol{Y}) \frac{p(\boldsymbol{x}_t, \boldsymbol{Y}_t \mid \boldsymbol{x}_{\mathrm{pa}(t)}, \boldsymbol{Y}_t^c)}{p(\boldsymbol{Y}_t \mid \boldsymbol{x}_{\mathrm{pa}(t)}, \boldsymbol{Y}_t^c)}. \end{aligned} \tag{12}$$

Since $(\boldsymbol{x}_t', \boldsymbol{Y}_t')'$ and $\boldsymbol{Y}_t^c$ are independent, conditional on $\boldsymbol{x}_{\mathrm{pa}(t)}$, we have from (12),

$$\begin{aligned} p(\boldsymbol{x}_t, \boldsymbol{x}_{\mathrm{pa}(t)} \mid \boldsymbol{Y}) &= p(\boldsymbol{x}_{\mathrm{pa}(t)} \mid \boldsymbol{Y}) \frac{p(\boldsymbol{x}_t, \boldsymbol{Y}_t \mid \boldsymbol{x}_{\mathrm{pa}(t)})}{p(\boldsymbol{Y}_t \mid \boldsymbol{x}_{\mathrm{pa}(t)})} \\ &= p(\boldsymbol{x}_{\mathrm{pa}(t)} \mid \boldsymbol{Y}) p(\boldsymbol{x}_t \mid \boldsymbol{x}_{\mathrm{pa}(t)}, \boldsymbol{Y}_t) \\ &= p(\boldsymbol{x}_{\mathrm{pa}(t)} \mid \boldsymbol{Y}) \frac{p(\boldsymbol{x}_t, \boldsymbol{x}_{\mathrm{pa}(t)} \mid \boldsymbol{Y}_t)}{p(\boldsymbol{x}_{\mathrm{pa}(t)} \mid \boldsymbol{Y}_t)} \\ &= p(\boldsymbol{x}_{\mathrm{pa}(t)} \mid \boldsymbol{Y}) \frac{p(\boldsymbol{x}_t \mid \boldsymbol{Y}_t) p(\boldsymbol{x}_{\mathrm{pa}(t)} \mid \boldsymbol{x}_t, \boldsymbol{Y}_t)}{p(\boldsymbol{x}_{\mathrm{pa}(t)} \mid \boldsymbol{Y}_t)} \\ &= p(\boldsymbol{x}_{\mathrm{pa}(t)} \mid \boldsymbol{Y}) \frac{p(\boldsymbol{x}_t \mid \boldsymbol{Y}_t) p(\boldsymbol{x}_{\mathrm{pa}(t)} \mid \boldsymbol{x}_t)}{p(\boldsymbol{x}_{\mathrm{pa}(t)} \mid \boldsymbol{Y}_t)} \\ &\propto \frac{p(\boldsymbol{x}_{\mathrm{pa}(t)} \mid \boldsymbol{Y})}{p(\boldsymbol{x}_{\mathrm{pa}(t)} \mid \boldsymbol{Y}_t)} \exp\{-\tfrac{1}{2}[(\boldsymbol{x}_t - \hat{\boldsymbol{x}}_{t|t})' \boldsymbol{\Gamma}_{t|t}^{-1} (\boldsymbol{x}_t - \hat{\boldsymbol{x}}_{t|t}) \\ &\qquad + (\boldsymbol{x}_{\mathrm{pa}(t)} - \boldsymbol{B}_t \boldsymbol{x}_t)' \boldsymbol{R}_t^{-1} (\boldsymbol{x}_{\mathrm{pa}(t)} - \boldsymbol{B}_t \boldsymbol{x}_t)]\}. \end{aligned}$$

Evaluating the conditional density above at $\boldsymbol{x}_{\mathrm{pa}(t)} = \hat{\boldsymbol{x}}_{\mathrm{pa}(t)}$ yields

$$\begin{aligned} &p(\boldsymbol{x}_t, \boldsymbol{x}_{\mathrm{pa}(t)} = \hat{\boldsymbol{x}}_{\mathrm{pa}(t)} \mid \boldsymbol{Y}) \\ &\quad \propto \exp\{-\tfrac{1}{2}[(\boldsymbol{x}_t - \hat{\boldsymbol{x}}_{t|t})' \boldsymbol{\Gamma}_{t|t}^{-1} (\boldsymbol{x}_t - \hat{\boldsymbol{x}}_{t|t}) + (\hat{\boldsymbol{x}}_{\mathrm{pa}(t)} - \boldsymbol{B}_t \boldsymbol{x}_t)' \boldsymbol{R}_t^{-1} (\hat{\boldsymbol{x}}_{\mathrm{pa}(t)} - \boldsymbol{B}_t \boldsymbol{x}_t)]\} \\ &\quad \propto \exp\{-\tfrac{1}{2}[\boldsymbol{x}_t' (\boldsymbol{\Gamma}_{t|t}^{-1} + \boldsymbol{B}_t' \boldsymbol{R}_t^{-1} \boldsymbol{B}_t) \boldsymbol{x}_t - 2\boldsymbol{x}_t' (\boldsymbol{\Gamma}_{t|t}^{-1} \hat{\boldsymbol{x}}_{t|t} + \boldsymbol{B}_t' \boldsymbol{R}_t^{-1} \hat{\boldsymbol{x}}_{\mathrm{pa}(t)})]\}. \end{aligned}$$

We obtain the recursion,

$$\hat{\boldsymbol{x}}_t = (\boldsymbol{\Gamma}_{t|t}^{-1} + \boldsymbol{B}_t' \boldsymbol{R}_t^{-1} \boldsymbol{B}_t)^{-1} (\boldsymbol{\Gamma}_{t|t}^{-1} \hat{\boldsymbol{x}}_{t|t} + \boldsymbol{B}_t' \boldsymbol{R}_t^{-1} \hat{\boldsymbol{x}}_{\mathrm{pa}(t)}). \tag{13}$$

Also,

$$\begin{aligned} &(\boldsymbol{\Gamma}_{t|t}^{-1} + \boldsymbol{B}_t' \boldsymbol{R}_t^{-1} \boldsymbol{B}_t)^{-1} \boldsymbol{B}_t' \boldsymbol{R}_t^{-1} \\ &\quad = \{\boldsymbol{\Gamma}_{t|t} - \boldsymbol{\Gamma}_{t|t} \boldsymbol{B}_t' (\boldsymbol{R}_t + \boldsymbol{B}_t \boldsymbol{\Gamma}_{t|t} \boldsymbol{B}_t')^{-1} \boldsymbol{B}_t \boldsymbol{\Gamma}_{t|t}\} \boldsymbol{B}_t' \boldsymbol{R}_t^{-1} \\ &\quad = \boldsymbol{\Gamma}_{t|t} \boldsymbol{B}_t' \boldsymbol{R}_t^{-1} - \boldsymbol{\Gamma}_{t|t} \boldsymbol{B}_t' (\boldsymbol{R}_t + \boldsymbol{B}_t \boldsymbol{\Gamma}_{t|t} \boldsymbol{B}_t')^{-1} \boldsymbol{B}_t \boldsymbol{\Gamma}_{t|t} \boldsymbol{B}_t' \boldsymbol{R}_t^{-1} \\ &\quad = \boldsymbol{\Gamma}_{t|t} \boldsymbol{B}_t' \boldsymbol{R}_t^{-1} \\ &\qquad - \boldsymbol{\Gamma}_{t|t} \boldsymbol{B}_t' (\boldsymbol{R}_t + \boldsymbol{B}_t \boldsymbol{\Gamma}_{t|t} \boldsymbol{B}_t')^{o} te{-}1 (\boldsymbol{B}_t \boldsymbol{\Gamma}_{t|t} \boldsymbol{B}_t' + \boldsymbol{R}_t - \boldsymbol{R}_t) \boldsymbol{R}_t^{-1} \\ &\quad = \boldsymbol{\Gamma}_{t|t} \boldsymbol{B}_t' \boldsymbol{R}_t^{-1} - \{\boldsymbol{\Gamma}_{t|t} \boldsymbol{B}_t' \boldsymbol{R}_t^{-1} - \boldsymbol{\Gamma}_{t|t} \boldsymbol{B}_t' (\boldsymbol{R}_t + \boldsymbol{B}_t \boldsymbol{\Gamma}_{t|t} \boldsymbol{B}_t')^{-1}\} \\ &\quad = \boldsymbol{\Gamma}_{t|t} \boldsymbol{B}_t' \boldsymbol{\Gamma}_{\mathrm{pa}(t)|t}^{-1}, \end{aligned}$$

where the last equality follows from (7). Therefore, (13) can also be written as

$$\begin{aligned}
\hat{\boldsymbol{x}}_t &= (\boldsymbol{\Gamma}_{t|t}^{-1} + \boldsymbol{B}_t'\boldsymbol{R}_t^{-1}\boldsymbol{B}_t)^{-1}(\boldsymbol{\Gamma}_{t|t}^{-1}\hat{\boldsymbol{x}}_{t|t} + \boldsymbol{B}_t'\boldsymbol{R}_t^{-1}\hat{\boldsymbol{x}}_{\mathrm{pa}(t)}) \\
&= (\boldsymbol{\Gamma}_{t|t}^{-1} + \boldsymbol{B}_t'\boldsymbol{R}_t^{-1}\boldsymbol{B}_t)^{-1} \\
&\quad \times (\boldsymbol{\Gamma}_{t|t}^{-1}\hat{\boldsymbol{x}}_{t|t} + \boldsymbol{B}_t'\boldsymbol{R}_t^{-1}\boldsymbol{B}_t\hat{\boldsymbol{x}}_{t|t} + \boldsymbol{B}_t'\boldsymbol{R}_t^{-1}\hat{\boldsymbol{x}}_{\mathrm{pa}(t)} - \boldsymbol{B}_t'\boldsymbol{R}_t^{-1}\boldsymbol{B}_t\hat{\boldsymbol{x}}_{t|t}) \\
&= \hat{\boldsymbol{x}}_{t|t} + (\boldsymbol{\Gamma}_{t|t}^{-1} + \boldsymbol{B}_t'\boldsymbol{R}_t^{-1}\boldsymbol{B}_t)^{-1}(\boldsymbol{B}_t'\boldsymbol{R}_t^{-1}\hat{\boldsymbol{x}}_{\mathrm{pa}(t)} - \boldsymbol{B}_t'\boldsymbol{R}_t^{-1}\boldsymbol{B}_t\hat{\boldsymbol{x}}_{t|t}) \\
&= \hat{\boldsymbol{x}}_{t|t} + (\boldsymbol{\Gamma}_{t|t}^{-1} + \boldsymbol{B}_t'\boldsymbol{R}_t^{-1}\boldsymbol{B}_t)^{-1}(\boldsymbol{B}_t'\boldsymbol{R}_t^{-1}\hat{\boldsymbol{x}}_{\mathrm{pa}(t)} - \boldsymbol{B}_t'\boldsymbol{R}_t^{-1}\hat{\boldsymbol{x}}_{\mathrm{pa}(t)|t}) \\
&= \hat{\boldsymbol{x}}_{t|t} + (\boldsymbol{\Gamma}_{t|t}^{-1} + \boldsymbol{B}_t'\boldsymbol{R}_t^{-1}\boldsymbol{B}_t)^{-1}\boldsymbol{B}_t'\boldsymbol{R}_t^{-1}(\hat{\boldsymbol{x}}_{\mathrm{pa}(t)} - \hat{\boldsymbol{x}}_{\mathrm{pa}(t)|t}) \\
&= \hat{\boldsymbol{x}}_{t|t} + \boldsymbol{\Gamma}_{t|t}\boldsymbol{B}_t'\boldsymbol{\Gamma}_{\mathrm{pa}(t)|t}^{-1}(\hat{\boldsymbol{x}}_{\mathrm{pa}(t)} - \hat{\boldsymbol{x}}_{\mathrm{pa}(t)|t}),
\end{aligned} \tag{14}$$

where $\hat{\boldsymbol{x}}_{\mathrm{pa}(t)|t}$ and $\Gamma_{\mathrm{pa}(t)|t}$ have been computed from (6) and (7), respectively, and $\hat{\boldsymbol{x}}_{t|t}$ and $\Gamma_{t|t}$ have been computed from (10) and (11), respectively, in the uptree filtering step. From (14), the mean squared prediction error (prediction variance) of $\hat{\boldsymbol{x}}_t$ can be obtained recursively by

$$\begin{aligned}
\boldsymbol{\Gamma}_t &= \mathrm{var}(\boldsymbol{x}_t - \hat{\boldsymbol{x}}_t) \\
&= \mathrm{var}(\boldsymbol{x}_t) - \mathrm{var}(\hat{\boldsymbol{x}}_t) \\
&= \mathrm{var}(\boldsymbol{x}_t) - \mathrm{var}\big(\hat{\boldsymbol{x}}_{t||t} + \boldsymbol{\Gamma}_{t|t}\boldsymbol{B}_t'\boldsymbol{\Gamma}_{\mathrm{pa}(t)|t}^{-1}(\hat{\boldsymbol{x}}_{\mathrm{pa}(t)} - \hat{\boldsymbol{x}}_{\mathrm{pa}(t)|t})\big) \\
&= \mathrm{var}(\boldsymbol{x}_t) - \big\{\mathrm{var}(\hat{\boldsymbol{x}}_{t|t}) + \mathrm{var}\big(\boldsymbol{\Gamma}_{t|t}\boldsymbol{B}_t'\boldsymbol{\Gamma}_{\mathrm{pa}(t)|t}^{-1}(\hat{\boldsymbol{x}}_{\mathrm{pa}(t)} - \hat{\boldsymbol{x}}_{pa(t)|t})\big)\big\} \\
&= \mathrm{var}(\boldsymbol{x}_t) - \mathrm{var}(\hat{\boldsymbol{x}}_{t|t}) \\
&\qquad - \boldsymbol{\Gamma}_{t|t}\boldsymbol{B}_t'\boldsymbol{\Gamma}_{\mathrm{pa}(t)|t}^{-1}\,\mathrm{var}(\hat{\boldsymbol{x}}_{\mathrm{pa}(t)} - \hat{\boldsymbol{x}}_{\mathrm{pa}(t)|t})\boldsymbol{\Gamma}_{\mathrm{pa}(t)|t}^{-1}\boldsymbol{B}_t\boldsymbol{\Gamma}_{t|t} \\
&= \boldsymbol{\Gamma}_{t|t} + \boldsymbol{\Gamma}_{t|t}\boldsymbol{B}_t'\boldsymbol{\Gamma}_{\mathrm{pa}(t)|t}^{-1}\{\mathrm{var}(\hat{\boldsymbol{x}}_{\mathrm{pa}(t)|t}) - \mathrm{var}(\hat{\boldsymbol{x}}_{\mathrm{pa}(t)})\}\boldsymbol{\Gamma}_{\mathrm{pa}(t)|t}^{-1}\boldsymbol{B}_t\boldsymbol{\Gamma}_{t|t} \\
&= \boldsymbol{\Gamma}_{t|t} + \boldsymbol{\Gamma}_{t|t}\boldsymbol{B}_t'\boldsymbol{\Gamma}_{\mathrm{pa}(t)|t}^{-1}(\boldsymbol{\Gamma}_{\mathrm{pa}(t)} - \boldsymbol{\Gamma}_{\mathrm{pa}(t)|t})\boldsymbol{\Gamma}_{\mathrm{pa}(t)|t}^{-1}\boldsymbol{B}_t\boldsymbol{\Gamma}_{t|t},
\end{aligned} \tag{15}$$

where the second equality holds since $\boldsymbol{x} - \hat{\boldsymbol{x}}_t$ and $\hat{\boldsymbol{x}}_t$ are uncorrelated, and the last equality holds since

$$\begin{aligned}
&\mathrm{var}(\hat{\boldsymbol{x}}_{\mathrm{pa}(t)|t}) - \mathrm{var}(\hat{\boldsymbol{x}}_{\mathrm{pa}(t)}) \\
&\qquad = \big(\mathrm{var}(\boldsymbol{x}_{\mathrm{pa}(t)}) - \mathrm{var}(\hat{\boldsymbol{x}}_{\mathrm{pa}(t)})\big) - \big(\mathrm{var}(\boldsymbol{x}_{\mathrm{pa}(t)}) - \mathrm{var}(\hat{\boldsymbol{x}}_{\mathrm{pa}(t)|t})\big) \\
&\qquad = \boldsymbol{\Gamma}_{\mathrm{pa}(t)} - \boldsymbol{\Gamma}_{\mathrm{pa}(t)|t}.
\end{aligned}$$

Note that $\boldsymbol{\Gamma}_{\mathrm{pa}(t)|t}$ and $\boldsymbol{\Gamma}_{t|t}$ have been computed from (7) and (11), respectively, in the uptree filtering step.

Thus, the generalized Kalman filter for tree-structured models can be carried out recursively by uptree filtering: use (6) through (11) with the initial predictors and the prediction variances of the state vectors at the leaves given by (4) and (5); follow this by downtree smoothing: use (14) and (15) to obtain the optimal predictor $\hat{\boldsymbol{x}}_t$ and the prediction variance $\boldsymbol{\Gamma}_t$, for all $t \in T$.

Luettgen and Willsky (1995) show that the prediction errors $\hat{\boldsymbol{x}}_t - \boldsymbol{x}_t$, $t \in T$, also follow a multiscale tree-structured model. That is,

$$\hat{\boldsymbol{x}}_t - \boldsymbol{x}_t = \boldsymbol{G}_t(\hat{\boldsymbol{x}}_{\mathrm{pa}(t)} - \boldsymbol{x}_{\mathrm{pa}(t)}) + \boldsymbol{\xi}_t; \quad t \in T, \tag{16}$$

where

$$\boldsymbol{G}_t \equiv \boldsymbol{\Gamma}_{t|t}\boldsymbol{B}_t'\boldsymbol{\Gamma}^{-1}_{\mathrm{pa}(t|t)}; \quad t \in T \setminus \{t\},$$

$\hat{\boldsymbol{x}}_{\mathrm{pa}(t)} - \boldsymbol{x}_{\mathrm{pa}(t)}$ and $\{\boldsymbol{\xi}_t : t \in T_t^c\}$ are independent for $t \in T \setminus \{t_0\}$, and

$$\mathrm{var}(\boldsymbol{\xi}_t) = \boldsymbol{R}_t = \boldsymbol{\Sigma}_{\mathrm{pa}(t)} - \boldsymbol{\Sigma}_{\mathrm{pa}(t)}\boldsymbol{A}_t'\boldsymbol{\Sigma}_t^{-1}\boldsymbol{A}_t\boldsymbol{\Sigma}_{\mathrm{pa}(t)}; \quad t \in T.$$

Notice that $\boldsymbol{G}_t$, $t \in T$, can be computed in the uptree filtering step. From (16), for any $t, t' \in T$, we can obtain the prediction covariance between any two variables as:

$$\begin{aligned}\boldsymbol{\Gamma}_{t,t'} &= \mathrm{cov}(\hat{\boldsymbol{x}}_t - \boldsymbol{x}_t, \hat{\boldsymbol{x}}_{t'} - \boldsymbol{x}_{t'}) \\ &= \boldsymbol{G}_{t_1}\boldsymbol{G}_{t_2}\cdots\boldsymbol{G}_t\,\mathrm{var}(\hat{x}_{\mathrm{an}(t,t')} - x_{\mathrm{an}(t,t')})(\boldsymbol{G}_{t_1'}\boldsymbol{G}_{t_2'}\cdots\boldsymbol{G}_{t'})',\end{aligned}$$

where $(\mathrm{an}(t,t'), t_1, t_2, \ldots, t)$ and $(\mathrm{an}(t,t'), t_1', t_2', \ldots, t')$ are the paths from $\mathrm{an}(t,t')$ (the first common ancestor of t and t') to t and t', respectively. In particular, we have

$$\boldsymbol{\Gamma}_{t,\mathrm{pa}(t)} = \boldsymbol{G}_t\boldsymbol{\Gamma}_{\mathrm{pa}(t)}.$$

3 Graphical Markov Models

The results in Section 2 demonstrate how recursive filtering yields optimal prediction of the x-process at all resolutions of the tree. Some data structures will dictate the use of acyclic directed graphical models, which represent a generalization of the tree-structured models. This extra flexibility hints at graphical models' usefulness in command and control problems.

Graphical Markov models, first introduced by Wright (1921, 1934), use graphs to represent possible dependencies among random variables. Graphs can be either undirected (edges represented by lines), directed (edges represented by arrows), or a mixture of the two. In graphical Markov models, the nodes of the graph represent random variables, and the edges of the graph characterize certain Markov properties (conditional independence relations). These models have been applied in a wide variety of areas, including undirected graphical models (also known as Markov random fields) in spatial statistics (Besag, 1974, Isham, 1981, Cressie, 1993), and acyclic directed graphical models in image analysis (Davidson and Cressie, 1993, Fieguth et al., 1995, Fosgate et al., 1997, Cressie and Davidson, 1998), categorical data analysis (Darroch et al., 1980, Edwards and Kreiner, 1983, Wermuth and Lauritzen, 1983), influence diagrams (Howard and Matheson, 1981, Shachter, 1986, Smith, 1989), and probabilistic expert systems

(Lauritzen and Spiegelhalter, 1988, Pearl, 1988, Spiegelhalter et al., 1993). The books by Pearl (1988), Whittaker (1990), Edwards (1995), and Lauritzen (1996) are good references for graphical Markov models and their applications.

In this section, we shall introduce graphs and graphical Markov models. The emphasis throughout is on Gaussian graphical models, which are also called covariance-selection models (Dempster, 1972). We shall develop a fast optimal-prediction algorithm for Gaussian graphical models.

3.1 Graphs

A graph G is made up of the pair (V, E), where V denotes a finite set of nodes (or vertices), and $E \subset V \times V$ denotes the set (possibly empty) of edges between two distinct nodes. An edge $(v, v') \in E$ such that $(v', v) \in E$ is called an undirected edge and is represented as a line between v and v' in our figures. An edge $(v, v') \in E$ such that $(v', v) \notin E$ is called a directed edge and is represented by an arrow from v to v' in our figures. The subset $A \subset V$ induces a subgraph $G_A = (A, E_A)$, where $E_A = E \cap (A \times A)$.

For both directed and undirected graphs, we define a path of length k from v_0 to v_k to be a sequence of nodes $v_0, v_1, \ldots, v_k$, $k \geq 1$, such that (v_i, v_{i+1}) is an edge for each $i = 0, \ldots, k-1$. A directed path $v_0, v_1, \ldots, v_k$ is a path such that (v_i, v_{i+1}) is a directed edge for each $i = 0, \ldots, k-1$. A cycle of length k in both directed and undirected graphs is a path $(v_0, v_1, \ldots, v_k)$ such that $v_0 = v_k$. A graph G is said to be connected if there is a path between any pair of nodes.

If E contains only undirected edges, then the graph G is called an undirected graph. If E contains only directed edges, then the graph G is called a directed graph. An acyclic directed graph is a directed graph that has no cycles in it. A graph with both directed and undirected edges but with no directed cycle is called a chain graph. We shall consider only undirected graphs and acyclic directed graphs.

Two nodes $v, v' \in V$ are said to be adjacent (or neighbors) if either $(v, v') \in E$ or $(v', v) \in E$. For $A \subset V$, we define the boundary of A to be

$$\mathrm{bd}(A) \equiv \{v' \in V \setminus A : (v, v') \text{ or } (v', v) \in E, \text{ where } v \in A\}.$$

The closure of $A \subset V$ is $\mathrm{cl}(A) \equiv A \cup \mathrm{bd}(A)$.

For a directed edge (v, v'), v is said to be a parent of v', and v' is said to be a child of v. A node v of a directed graph is said to be terminal if it has no children. We say that v' is a descendant of v if there exists a directed path from v to v'.

An undirected tree is a connected undirected graph with no cycles. A directed tree is a connected acyclic directed graph such that there is a node $v_0 \in V$ that has no parent, called the root, and each $v \in V \setminus \{v_0\}$ has exactly one parent node. Note that, given an undirected tree, we can construct a directed tree by first choosing any node of the tree as a root, and directing all edges away from this root.

A graphical Markov model is defined by a collection of conditional independence relations among a set of random variables (vectors) in the form of a graph, where the set of random variables (vectors) are indexed by the nodes of the graph. For example, tree-structured models described in Section 2 are a special class of graphical Markov models. The generalized-Kalman-filter algorithm we shall derive does not work directly on the graph, but on an associated tree, called a junction tree. Each node on a junction tree contains clusters of variables obtained from the graphical structure.

Let $G = (V, E)$ be a graph. We consider a (multivariate) stochastic process $\{\boldsymbol{x}_v : v \in V\}$ with V as an index set. For $A \subset V$, denote $\boldsymbol{x}_A \equiv \{\boldsymbol{x}_v : v \in A\}$. For $A, B \subset V$, denote $p(\boldsymbol{x}_A)$ as the probability density of $\boldsymbol{x}_A$, and denote $p(\boldsymbol{x}_A \mid \boldsymbol{x}_B)$ as the conditional probability density of $\boldsymbol{x}_A$ given $\boldsymbol{x}_B$. For disjoint sets $A, B, C \subset V$, denote the conditional independence relation, $\boldsymbol{x}_A$ conditionally independent of $\boldsymbol{x}_B$ given $\boldsymbol{x}_C$, as

$$\boldsymbol{x}_A \perp \boldsymbol{x}_B \mid \boldsymbol{x}_C;$$

that is,

$$p(\boldsymbol{x}_A, \boldsymbol{x}_B \mid \boldsymbol{x}_C) = p(\boldsymbol{x}_A \mid \boldsymbol{x}_C)p(\boldsymbol{x}_B \mid \boldsymbol{x}_C).$$

Without loss of generality, we assume throughout the section that G is connected, since all the results can likewise be applied to a disconnected graph by applying them successively to each connected component.

A graph is complete if all pairs of distinct nodes are joined by an (directed or undirected) edge. A subset is complete if it induces a complete subgraph. A complete subset that is not a proper (genuine) subset of any other complete subset is called a clique.

For $A, B, S \subset V$, we say S separates A and B if all paths from an element $a \in A$ to an element $b \in B$ intersect S. A pair (A, B) of nonempty subsets of $V = A \cup B$ is said to form a decomposition of G if $A \cap B$ is complete, and $A \cap B$ separates $A \setminus B$ and $B \setminus A$. If the sets A and B are both proper subsets of V, the decomposition is proper. An undirected graph G is said to be decomposable if it is complete, or if there exists a proper decomposition (A, B) into decomposable subgraphs G_A and G_B.

A chord in a cycle is an edge connecting two nonconsecutive nodes in the cycle. An undirected graph is chordal (or triangulated) if every cycle of length greater than three has a chord. A well known result states that an undirected graph is decomposable if and only if it is chordal (Golumbic, 1980, Lauritzen et al., 1984, Whittaker, 1990).

For an acyclic directed graph G, we define its moral graph G^m as the undirected graph obtained from G by "marrying" parents, that is, by placing undirected edges between all parents of each node and then dropping the directions from all directed edges.

3.2 Graphical Markov Models Over Undirected Graphs

Let $G = (V, E)$ be an undirected graph, and $\boldsymbol{x}_V \equiv \{\boldsymbol{x}_v : v \in V\}$ be a stochastic process with index set V. Associated with this graphical structure, there are at least three Markov properties for $\boldsymbol{x}_V$.

Definition 1. A stochastic process $\boldsymbol{x}_V \equiv \{\boldsymbol{x}_v : v \in V\}$ on an undirected graph $G = (V, E)$ is said to have the local Markov property if, for any $v \in V$,

$$\boldsymbol{x}_v \perp \boldsymbol{x}_{V\setminus \mathrm{cl}(v)} \mid \boldsymbol{x}_{\mathrm{bd}(v)};$$

it is said to have the pairwise Markov property if, for any nonadjacent nodes (v_1, v_2),

$$\boldsymbol{x}_{v_1} \perp \boldsymbol{x}_{v_2} \mid \boldsymbol{x}_{V\setminus\{v_1,v_2\}};$$

it is said to have the global Markov property if, for any disjoint subsets $A, B, S \subset V$ such that S separates A from B in G,

$$\boldsymbol{x}_A \perp \boldsymbol{x}_B \mid \boldsymbol{x}_S.$$

We call $\boldsymbol{x}_V$ a graphical Markov model over G if any one of the Markov properties above is satisfied.

Note that by the definition above, it is clear that the global Markov property implies the local Markov property, and the local Markov property implies the pairwise Markov property. Though the three Markov properties are in general different (some discussion can be found in Speed, 1979), and thus may lead to different graphical Markov models, they are equivalent under quite general conditions. For example, the following proposition is due to Lauritzen et al. (1990).

Proposition 1. *Suppose that $\boldsymbol{x}_V$ is a stochastic process with index set V. Let $G = (V, E)$ be an undirected graph, and let $p(\boldsymbol{x}_V)$ be the joint probability density of $\boldsymbol{x}_V$ with respect to a product measure. Then $\boldsymbol{x}_V$ satisfies the global Markov property over G if $p(\boldsymbol{x}_V)$ factorizes, namely*

$$p(\boldsymbol{x}_V) = \prod_{A \in \mathcal{A}} \Psi_A(\boldsymbol{x}_A),$$

for some nonnegative functions $\{\Psi_A : A \in \mathcal{A}\}$, where $\mathcal{A}$ is a collection of complete subsets of V.

Note that a result known as the Hammersley–Clifford theorem (Hammersley and Clifford, 1971) shows that when $\boldsymbol{x}_V$ satisfies the positivity condition (i.e., the support of $\boldsymbol{x}_V$ is the Cartesian product of the support of each individual component of $\boldsymbol{x}_V$), any one of the three Markov properties implies that $p(\boldsymbol{x}_V)$ factorizes. Therefore, the three Markov properties are equivalent under the positivity condition. Since the Gaussian processes we are focusing on obviously satisfy the positivity condition, we shall consider only the global Markov property throughout Section 3.

A graphical model over an undirected graph G is called a decomposable graphical model (Frydenberg and Lauritzen, 1989) if G is decomposable. This class of models is crucial for the rest of this section, and is the key to deriving a fast optimal-prediction algorithm for Gaussian graphical models. An important structure associated with computational aspects of decomposable graphs is a junction tree defined as follow:

Definition 2. Let $\mathcal{H}_J$ denote the set of cliques of an undirected graph $G = (V, E)$. Then $\mathcal{T}_J = (\mathcal{H}_J, \mathcal{E}_J)$ is called a junction tree for G, where $\mathcal{E}_J$ is the set of edges of $\mathcal{T}_J$, if $\mathcal{T}_J$ is an undirected tree and $\mathcal{T}_J$ satisfies the clique-intersection property (i.e., for every two cliques $K, K' \in \mathcal{H}_J$, all cliques of the undirected tree on the path between the clique K and the clique K' contain all the elements of $K \cap K'$).

Buneman (1974) and Gavril (1974) independently proved that a connected undirected graph G is decomposable if and only if there exists a junction tree for G (i.e., there is an associated tree $\mathcal{T}_J = (\mathcal{H}_J, \mathcal{E}_J)$ for which the clique-intersection property holds). Pearl (1988) showed that, given a junction-tree representation for a decomposable graphical model $\{\boldsymbol{x}_v : v \in V\}$, the joint probability density can be written as

$$p(\boldsymbol{x}_V) = \frac{\prod_{K_i \in \mathcal{H}_J} p(\boldsymbol{x}_{K_i})}{\prod_{(K_i, K_j) \in \mathcal{E}_J} p(\boldsymbol{x}_{K_i \cap K_j})}. \tag{17}$$

For any two cliques $K_i, K_j \in \mathcal{H}_J$ such that $(K_i, K_j) \in \mathcal{E}_J$, their intersection $K_i \cap K_j$ is called their separator.

Though many undirected graphs of interest may not be decomposable, one may always identify a decomposable cover G' of an undirected graph G, by an addition of appropriate edges, called a filling-in. For example, the complete graph is a decomposable cover for any undirected graph. However, by filling in new edges, we may obscure certain conditional independencies. Now because filling in does not introduce any new conditional independence relations, the underlying probability distribution is unaffected. A filling-in is said to be minimal if it contains a minimal number of edges. Tarjan and Yannakakis (1984) give various methods for testing whether an undirected graph is decomposable. Algorithms for finding minimal decompositions can be found in Tarjan (1985) and Leimer (1993).

3.3 Graphical Markov Models Over Acyclic Directed Graphs

Consider the same setup as in Section 3.2, except that now $G = (V, E)$ is assumed to be an acyclic directed graph, where V is the set of nodes and E is the set of directed edges.

Definition 3. We say that $\boldsymbol{x}_V$ is a graphical Markov model over an acyclic directed graph $G = (V, E)$ if each $\boldsymbol{x}_v$, $v \in V$, is conditionally independent of its nondescendents given its parents, where the nondescendents of v are those $v' \in V$ such that there is no directed path from v to v'.

Note that acyclic directed graphical models are also referred to in the literature as recursive graphical models, Bayesian networks, belief networks, causal networks, influence diagrams, causal Markov models, and directed Markov fields. In the spatial context, these models are called partially ordered Markov models (Cressie and Davidson, 1998, Davidson et al., 1999).

Assume that $p(\boldsymbol{x}_V)$ is the joint probability density of $\boldsymbol{x}_V$ with respect to a product measure. It follows that

$$p(\boldsymbol{x}_V) = p(\boldsymbol{x}_{V_0}) \prod_{v \in V \setminus V_0} p(\boldsymbol{x}_v \mid \boldsymbol{x}_{pa(v)}), \tag{18}$$

where $V_0 \equiv \{v : pa(v) = \emptyset\}$ is the set of minimal elements.

The following lemma enables us to transform an acyclic directed graphical model to an undirected graphical model.

Lemma 1. *Suppose that $\boldsymbol{x}_V$ is a graphical Markov model over an acyclic directed graph G. Then $\boldsymbol{x}_V$ satisfies the global Markov property over the moral graph G^m.*

Proof. The result follows directly from Proposition 1 and (18), since for each $v \in V$, the set $v \cup \text{pa}(v)$ is complete in G^m. □

Note that by filling in appropriate undirected edges, the moral graph G^m can then be converted into a decomposable graph G', as described at the end of Section 3.2. Therefore, we can make any acyclic directed graph decomposable by marrying parents, making all directed edges undirected, and adding new undirected edges.

3.4 Generalized Kalman Filter for Gaussian Graphical Models

We shall now derive the generalized-Kalman-filter algorithm for Gaussian graphical models over an undirected graph or an acyclic directed graph. Recall from Section 3.2 and Section 3.3 that any graphical model over an undirected graph or an acyclic directed graph G can be represented as a graphical model over a decomposable graph G', which is defined at the end of Section 3.2. Also, we have the following proposition:

Proposition 2. *Let $G = (V, E)$ be a connected decomposable graph and $\mathcal{T}_J = (\mathcal{H}_J, \mathcal{E}_J)$be a junction tree for G. Suppose that $\boldsymbol{x}_V \equiv \{\boldsymbol{x}_v : v \in V\}$ satisfies the global Markov property over G. Then the stochastic process $\{\boldsymbol{x}_H : H \in \mathcal{H}_J\}$ satisfies the global Markov property over $\mathcal{T}_J$, where $\boldsymbol{x}_H \equiv \{\boldsymbol{x}_v : v \in H\}$.*

Proof. Let $\mathcal{A}$, $\mathcal{B}$, and $\mathcal{S}$ be disjoint subsets of $\mathcal{H}_J$ such that $\mathcal{S}$ separates $\mathcal{A}$ from $\mathcal{B}$ in $\mathcal{T}_J$. Define $A \equiv \{v \in H : H \in \mathcal{A}\}$, $B \equiv \{v \in H : H \in \mathcal{B}\}$, and $S \equiv \{v \in H : H \in \mathcal{S}\}$. By Definition 1, it is enough to show that S separates $A \backslash S$ from $B \backslash S$ in V. Suppose that S does not separate $A \backslash S$ from $B \setminus S$ in V. Then there exists a path $(v_0, v_1, \ldots, v_k)$ in G with $v_0 \in A \setminus S$, $v_k \in B \setminus S$, and $v_i \in V \setminus S$, $i = 0, 1, \ldots k-1$. Hence, we can find $H_1 \in \mathcal{A}$, $H_k \in \mathcal{B}$, and $H_i \in \mathcal{H}_J \setminus \mathcal{S}$, $i = 2, \ldots, k-1$, such that $\{v_{i-1}, v_i\} \subset H_i$, $i = 1, \ldots, k$. For each $i = 2, \ldots, k$, since $(H_{i-1} \cap H_i) \setminus S \neq \emptyset$, it follows from the clique-intersection property that the path from H_{i-1} to H_i in $\mathcal{T}_J$ does not intersect $\mathcal{S}$. Therefore, the path from $H_1 \in \mathcal{A}$ to $H_k \in \mathcal{B}$ does not intersect $\mathcal{S}$. This contradicts the assumption that $\mathcal{S}$ separates $\mathcal{A}$ from $\mathcal{B}$ in $\mathcal{T}_J$. □

By choosing a node H_0 of the tree $\mathcal{T}_J$ as a root, and directing all edges away from this root, we obtain a directed tree $\mathcal{T}_D = (\mathcal{H}_J, \mathcal{E}_D)$, where $\mathcal{E}_D$ is the set of the corresponding directed edges of $\mathcal{E}_J$. Since from Proposition 2, $\{\boldsymbol{x}_H : H \in \mathcal{H}_J\}$ satisfies the global Markov property over $\mathcal{T}_J$, it follows from Definition 3 that $\{\boldsymbol{x}_H : H \in \mathcal{H}_J\}$ is a graphical Markov model over the directed tree $\mathcal{T}_D$.

Let $G = (V, E)$ be a connected decomposable graph and let $\boldsymbol{x}_V \equiv \{\boldsymbol{x}_v : v \in V\}$ be a zero-mean Gaussian graphical model over G; that is, $\boldsymbol{x}_V$ is a Gaussian process with index set V, and it satisfies the global Markov property over G (Definition 1). We assume that the data are observed, perhaps incompletely, according to

$$\boldsymbol{y}_v = \boldsymbol{C}_v \boldsymbol{x}_v + \boldsymbol{\epsilon}_v; \quad v \in V, \tag{19}$$

where $\{\boldsymbol{C}_v : v \in V\}$ are deterministic matrices, and $\{\boldsymbol{\epsilon}_v : v \in V\}$ is a zero-mean Gaussian white-noise process independent of $\boldsymbol{x}_V$.

For each $H \in \mathcal{H}_J$, denote $pa(H)$ as the parent of H in $\mathcal{T}_D$. Let $\boldsymbol{y}_H \equiv \{\boldsymbol{y}_v : v \in H\}$ and $\boldsymbol{\epsilon}_H \equiv \{\boldsymbol{\epsilon}_v : v \in H\}$. We have,

$$\boldsymbol{x}_H = \boldsymbol{A}_H \boldsymbol{x}_{pa(H)} + \boldsymbol{\eta}_H; \quad H \in \mathcal{H}_J \setminus \{H_0\}, \tag{20}$$

$$\boldsymbol{y}_H = \boldsymbol{C}_H \boldsymbol{x}_H + \boldsymbol{\epsilon}_H; \quad H \in \mathcal{H}_J, \tag{21}$$

for some matrices $\{\boldsymbol{A}_H\}$ and $\{\boldsymbol{C}_H\}$, where (21) is obtained from (19), and $\{\boldsymbol{\eta}_H : H \in \mathcal{H}_J \setminus \{H_0\}\}$ are zero-mean, independent Gaussian vectors with

$$\text{var}(\boldsymbol{\eta}_H) = \text{var}(\boldsymbol{x}_H \mid \boldsymbol{x}_{pa(H)}); \quad H \in \mathcal{H}_J \setminus \{H_0\}.$$

We therefore obtain a Gaussian tree-structured Markov model given by (20) and (21). Thus, the generalized-Kalman-filter algorithm constructed in Section 2 can now be used to compute the optimal predictors and their prediction variances for the state vectors $\{\boldsymbol{x}_H : H \in \mathcal{H}_J\}$. In particular, we obtain the optimal predictors and their prediction variances for the state vectors $\{\boldsymbol{x}_v : v \in V\}$.

For computational convenience, nodes that have only one child on the directed tree $\mathcal{T}_D$ can be removed to simplify the tree structure without affecting the global Markov property. When a node H is removed, the edges between H and its parent and between H and its children are also removed, and new directed edges from the parent of H to each child of H are added. Moreover, for a terminal node H on the directed tree $\mathcal{T}_D$, we can reduce the size of H by replacing H by $H \setminus \mathrm{pa}(H)$. In both cases, the global Markov property of $\{\boldsymbol{x}_H : H \in \mathcal{H}_J\}$ over $\mathcal{T}_J$ guarantees that we still have a tree-structured model over the new reduced directed tree $\mathcal{T}_R$.

Note that nodes can be removed only if no information is lost. A node H with no data observed can also be removed if we are not interested in obtaining the optimal predictor for $\boldsymbol{x}_H$, or we can obtain the optimal predictor for $\boldsymbol{x}_H$ through other nodes.

3.5 Example

We consider a Gaussian graphical model over an acryclic directed graph G shown as in Figure 3(a). We assumed that the data are observed only at the terminal nodes by

$$y_k = x_k + \varepsilon_k; \quad k = 1, \ldots, 15,$$

where $\{y_1, \ldots, y_{15}\}$ are observed data, and $\{\varepsilon_1, \ldots, \varepsilon_{15}\}$ is a Gaussian white-noise process independent of $\{x_1, \ldots, x_{15}\}$. The goal is to predict $\{x_1, \ldots, x_{15}\}$ based on the data $\{y_1, \ldots, y_{15}\}$.

The corresponding moral graph G^m is shown in Figure 3(b). A decomposable graph G', obtained by filling in certain undirected edges from G^m, is

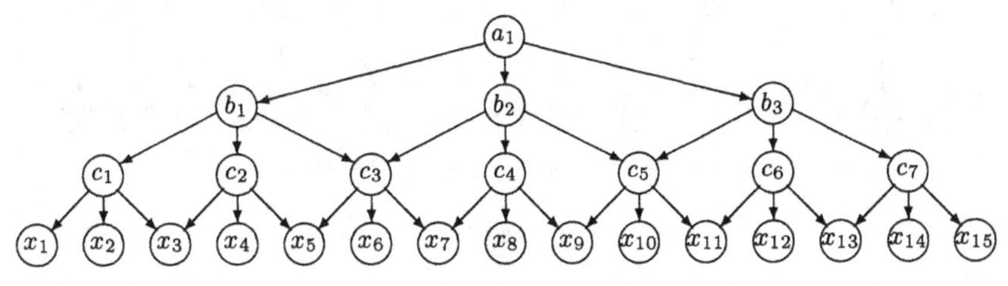

(a) An acyclic directed graph G

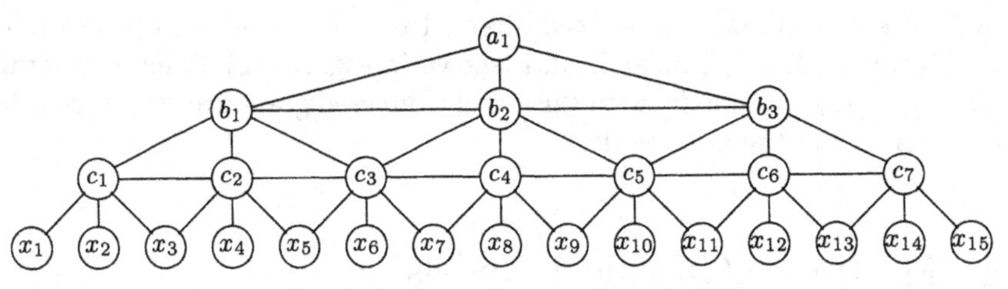

(b) The corresponding moral graph G^m

Figure 3.

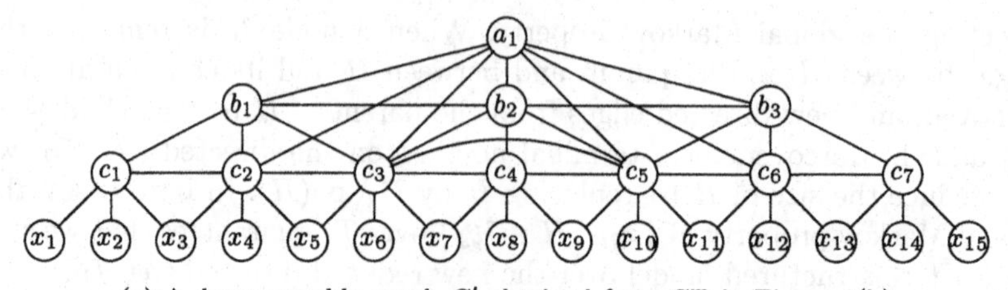

(a) A decomposable graph G' obtained from G^m in Figure 1(b)

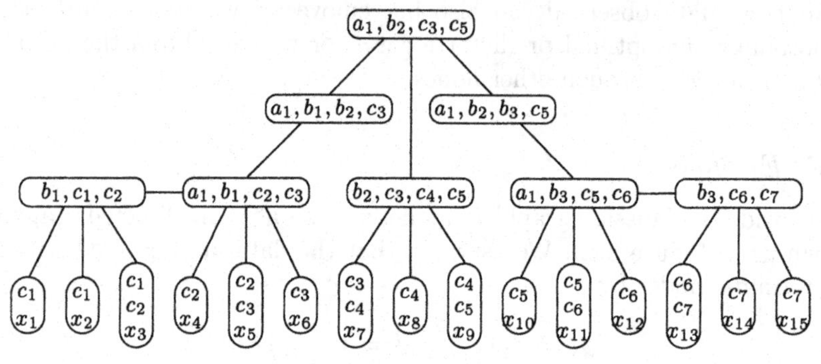

(b) A corresponding junction tree $\mathcal{T}_J$ of G'

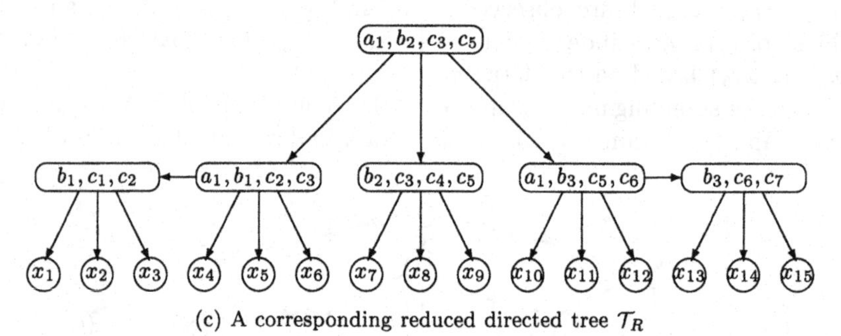

(c) A corresponding reduced directed tree $\mathcal{T}_R$

Figure 4.

shown in Figure 4(a). A junction tree $\mathcal{T}_J$ composed of cliques of G' is shown in Figure 4(b). Finally, a reduced directed tree $\mathcal{T}_R$, to which the generalized Kalman filter will be applied, is shown in Figure 4(c). A more general Gaussian graphical model with this acyclic directed graphical structure will be described in the next section.

4 Multiscale Graphical Models

In Section 2, we introduced tree-structured models with natural multiscale structures. The main advantage of these models is that associated with them there is a fast optimal prediction algorithm that can be used to handle large amounts of data. However, to represent a time series or a spatial process by the variables in the finest level of a tree, these models are somewhat restricted in that they may enforce certain discontinuities in covariance structures (see Figure 2). That is, two points that are adjacent in time or space may or may not have the same parent node on the tree, and even their parents may come from different parents. These discontinuities sometimes lead to blocky artifacts in the resulting predicted values. To fix this problem, we propose more general multiscale models based on an acyclic directed graphical structure.

We consider a series of hidden zero-mean Gaussian stochastic processes $\{x_{j,k} : k \in \mathbb{Z}^d\}$ with scales $j = J_0, \ldots, J$. The stochastic processes evolve in a Markov manner from coarser (smaller j) to finer (larger j) scales based on

$$x_{j,k} \equiv \sum_{l \in M_{j,k}} a_{j,k,l} x_{j-1,l} + w_{j,k}; \quad j = J_0 + 1, \ldots, J, k \in \mathbb{Z}^d, \tag{22}$$

where $\{M_{j,k}\}$ are finite subsets of $\mathbb{Z}$, $\{a_{j,k,l}\}$ are constants,

$$x_{J_0,k} \sim \mathrm{N}(\mathbf{0}, \sigma_{J_0}^2), \qquad \text{independently for } k \in \mathbb{Z}^d,$$
$$w_{j,k} \sim \mathrm{N}(\mathbf{0}, \sigma_j^2), \qquad \text{independently for } j = J_0 + 1, \ldots, J,\ k \in \mathbb{Z},$$

and $\{x_{J_0,k} : k \in \mathbb{Z}^d\}$ and $\{w_{j,k} : j = J_0+1, \ldots, J, k \in \mathbb{Z}^d\}$ are independent.

Let D_J be a finite subset of $\mathbb{Z}^d$, and let

$$D_{j-1} \equiv \{(j-1, l) : ((j-1, l), (j, k)) \in F, k \in D_j\}; \quad j = J_0 + 1, \ldots, J,$$

where

$$F \equiv \{((j-1, l), (j, k)) : j = J_0 + 1, \ldots, J, l \in M_{j,k}, k, l \in \mathbb{Z}^d\}.$$

For each scale $j = J_0, \ldots, J$, we assume that the data are observed (perhaps incompletely) according to

$$y_{j,k} = x_{j,k} + \varepsilon_{j,k}; \quad k \in D_j, \tag{23}$$

where $\{\varepsilon_{j,k}\}$ is a zero-mean Gaussian white-noise process, independent of $\{x_{j,k}\}$, with variance σ_ε^2. Define, for $k \in D_j$, $j = J_0, \ldots, J$,

$$\gamma_{j,k} \equiv \begin{cases} 1, & \text{if } y_{j,k} \text{ is observed,} \\ 0, & \text{otherwise.} \end{cases}$$

The goal is to predict $\{x_{j,k} : k \in D_j, j = J_0, \ldots, J\}$ based on the data $\boldsymbol{Y} \equiv \{y_{j,k} : \gamma_{j,k} = 1\}$. Notice that if each $M_{j,k}$ contains only a single element, we obtain a tree-structured model.

Write

$$\begin{aligned} \boldsymbol{X}_j &\equiv (x_{j,k} : k \in D_j)'; \quad j = J_0, \dots, J, \\ \boldsymbol{X} &\equiv (\boldsymbol{X}'_{J_0}, \dots, \boldsymbol{X}'_J)', \\ \boldsymbol{w}_j &\equiv (w_{j,k} : k \in D_j)'; \quad j = J_0 + 1, \dots, J, \\ \boldsymbol{\epsilon} &\equiv (\varepsilon_{j,k} : \gamma_{j,k} = 1)'. \end{aligned}$$

Then, (22) and (23) can be written as

$$\boldsymbol{X}_j = \boldsymbol{A}_j \boldsymbol{X}_{j-1} + \boldsymbol{w}_j; \quad j = J_0 + 1, \dots, J, \tag{24}$$

$$\boldsymbol{Y} = \boldsymbol{C}\boldsymbol{X} + \boldsymbol{\epsilon}, \tag{25}$$

for some matrices $\boldsymbol{A}_{J_0+1}, \dots, \boldsymbol{A}_J$ and for some matrix $\boldsymbol{C}$.

Note that from (24) and (25), we have the following representation:

$$\boldsymbol{X}_j = (\boldsymbol{A}_j \dots \boldsymbol{A}_{J_0+1})\boldsymbol{X}_{J_0} + (\boldsymbol{A}_j \cdots \boldsymbol{A}_{J_0+2})\boldsymbol{w}_{J_0+1} + \cdots + \boldsymbol{A}_j \boldsymbol{w}_{j-1} + \boldsymbol{w}_j,$$

for $j = J_0 + 1, \dots, J$. Therefore, for $j = J_0 + 1, \dots, J$,

$$\begin{aligned} \mathrm{var}(\boldsymbol{X}_j) = \sigma^2_{J_0} &(\boldsymbol{A}_j \cdots \boldsymbol{A}_{J_0+1})(\boldsymbol{A}_j \cdots \boldsymbol{A}_{J_0+1})' \\ &+ \sigma^2_{J_0+1}(\boldsymbol{A}_j \dots \boldsymbol{A}_{J_0+2})(\boldsymbol{A}_j \cdots \boldsymbol{A}_{J_0+2})' + \cdots \\ &+ \sigma^2_{j-1}\boldsymbol{A}_j \boldsymbol{A}'_j + \sigma^2_j \boldsymbol{I}. \end{aligned} \tag{26}$$

4.1 Multiscale Graphical Models for One-Dimensional Processes

We give here two examples of multiscale graphical models for one-dimensional processes. For the first example, we consider a multiscale graphical model with the hidden processes given by,

$$x_{j,2k} = x_{j-1,k} + w_{j,2k}; \qquad j = J_0 + 1, \dots, J, \ k \in \mathbb{Z}, \tag{27}$$

$$x_{j,2k+1} = \frac{1}{1 + \alpha_j}(x_{j-1,k} + \alpha_j x_{j-1,k+1}) + w_{j,2k+1}; \qquad j = J_0 + 1, \dots, J, \ k \in \mathbb{Z}, \tag{28}$$

where $\alpha_j > 0$, $j = J_0 + 1, \dots, J$, are constants. The associated acyclic directed graph of this model is shown in Figure 5. Note that because each node has more than one parent, this is not a tree. If we assume that the data are observed (perhaps incompletely) according to (23), then we can apply the generalized-Kalman-filter algorithm (Section 3.4) to these models to compute the optimal predictors of $\{x_{j,k}\}$. Recall that in Section 3.5, we have given an example of a Gaussian graphical model whose graph structure is precisely that implied by (27) and (28) for construction of the algorithm.

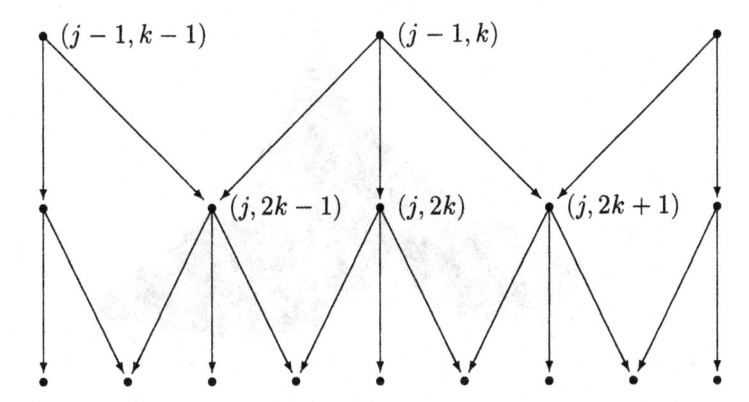

Figure 5. Graphical representation of first multiscale graphical model.

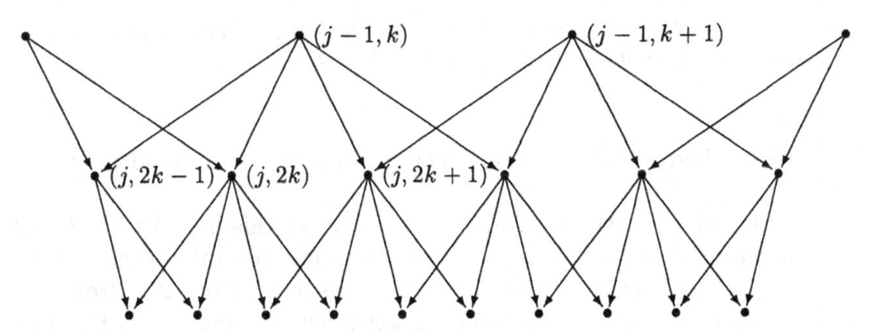

Figure 6. Graphical representation of second multiscale graphical model.

For the second example, we consider a multiscale graphical model with the hidden processes given by,

$$x_{j,2k} = \frac{1}{1+\alpha_j}(x_{j-1,k-1} + \alpha_j x_{j-1,k}) + w_{j,2k}; \quad j = J_0+1,\ldots,J, k \in \mathbb{Z}, \tag{29}$$

$$x_{j,2k+1} = \frac{1}{1+\alpha_j}(\alpha_j x_{j-1,k} + x_{j-1,k+1}) + w_{j,2k+1}; \quad j = J_0+1,\ldots,J, k \in \mathbb{Z}, \tag{30}$$

where $\alpha_j > 0$, $j = J_0, \ldots, J$, are constants. The associated acyclic directed graph of this model is shown in Figure 6. For $J_0 = 0$, $J = 6$, $\sigma_0^2 = 10$, $\sigma_1^2 = 8$, $\sigma_2^2 = 6$, $\sigma_3^2 = 4$, $\sigma_4^2 = 3$, $\sigma_5^2 = 2$, $\sigma_6^2 = 1$, and $\alpha_1 = \alpha_2 = \cdots = \alpha_5 = 2$, the correlation function of the latent process,

$$f(t,s) \equiv \mathrm{corr}(x_{6,t}, x_{6,s}); \quad t,s = 0,1,\ldots,127,$$

constructed from two nodes $x_{0,0}$ and $x_{0,1}$ at the coarsest scale, is shown in Figure 7. The correlation function is not stationary but the figure shows that it is nearly so. Notice there is no blocky structure of the sort seen in

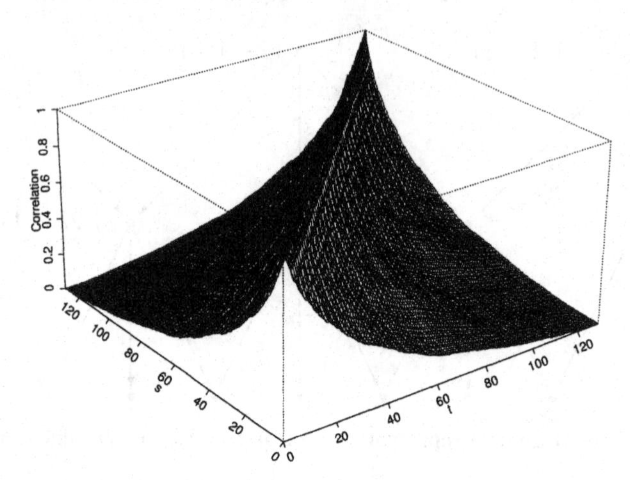

Figure 7. Correlation function $f(t,s) = \mathrm{corr}(x_{6,t}, x_{6,s})$ of $(x_{6,0}, x_{6,1}, \ldots, x_{6,127})'$ for a multiscale graphical model given by (29) and (30).

Figure 2, where the correlation function for a tree-structured model was given.

Figure 8 shows several correlation functions of this model with $J_0 = 0$, $J = 6$, and $\alpha_1 = \alpha_2 = \cdots = \alpha_5 = 2$, where each is generated from only one scale (i.e., only one of $\{\sigma_0^2, \sigma_1^2, \ldots, \sigma_6^2\}$ is non-zero). Note that from (26), a linear combination of these correlation functions, plus the white-noise correlation function, generate the correlation function shown in Figure 7. It is easy to see from these multiscale features how different covariance structure can be constructed from this model.

4.2 *Multiscale Graphical Models for Two-Dimensional Processes*

Consider a two-dimensional multiscale graphical model with the corresponding acyclic directed graph having a graphical structure illustrated in Figure 9. From the scale J_0 to the scale J_1, a non-tree acyclic directed graph is assumed (Figure 9(a)) but from the scale $J_1 + 1$ to the scale J, a quadtree structure (Figure 9(b)) is assumed, where $J_0 \leq J_1 \leq J$. Specifically, for $j = J_0 + 1, \ldots, J_1$, $k_1, k_2 \in \mathbb{Z}$, we assume

$$x_{j,2k_1,2k_2} = x_{j-1,k_1,k_2} + w_{j,2k_1,2k_2},$$

$$x_{j,2k_1+1,2k_2} = \frac{1}{1+\alpha_{j,1}}(x_{j-1,k_1,k_2} + \alpha_{j,1} x_{j-1,k_1+1,k_2}) + w_{j,2k_1+1,2k_2},$$

$$x_{j,2k_1,2k_2+1} = \frac{1}{1+\alpha_{j,2}}(x_{j-1,k_1,k_2} + \alpha_{j,2} x_{j-1,k_1,k_2+1}) + w_{j,2k_1,2k_2+1},$$

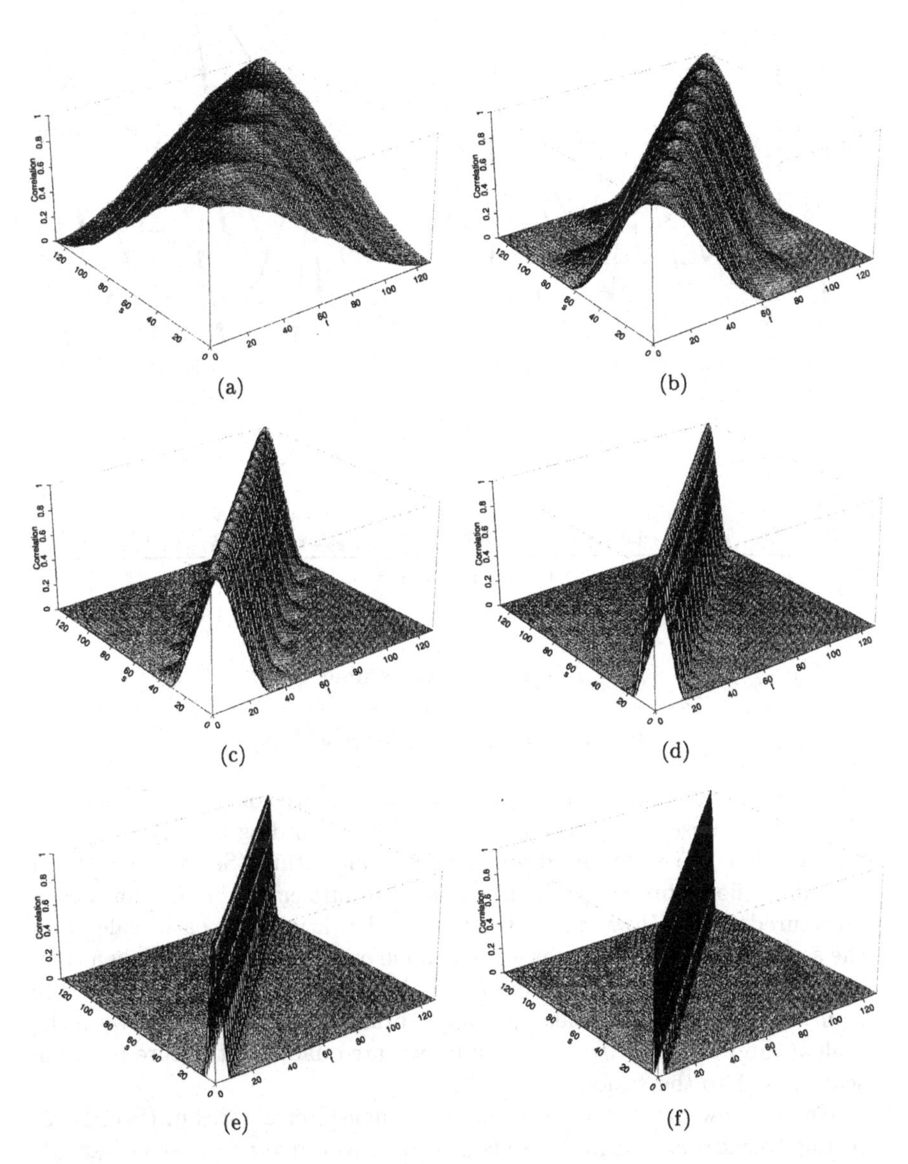

Figure 8. Correlation functions $f(t, s) = \text{corr}(x_{6,t}, x_{6,s})$ of $\{x_{6,t} : t \in \mathbb{Z}\}$ for the model given by (29) and (30)) with $J_0 = 0$, $J = 6$, $\alpha_j = 2$, $j = 1, \ldots, 6$, and the only nonzero $\{\sigma_j^2 : j = 0, \ldots, 6\}$ is (a) $\sigma_0^2 = 1$; (b) $\sigma_1^2 = 1$; (c) $\sigma_2^2 = 1$; (d) $\sigma_3^2 = 1$; (e) $\sigma_4^2 = 1$; (f) $\sigma_5^2 = 1$.

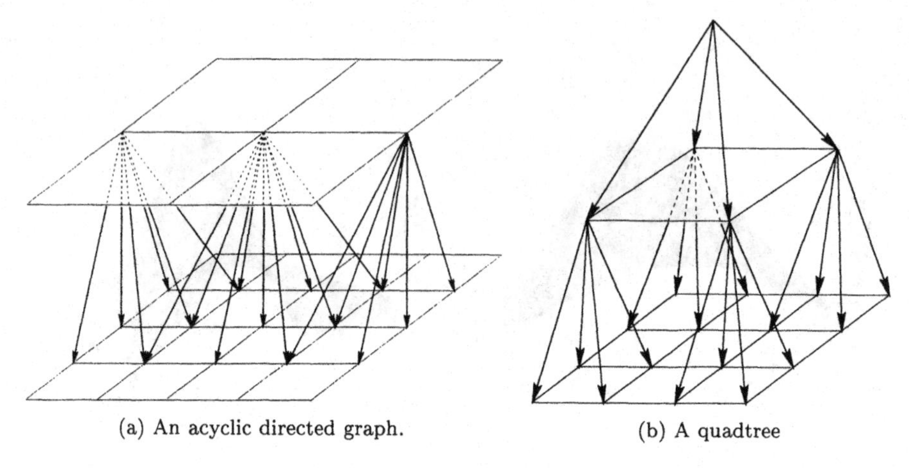

(a) An acyclic directed graph. (b) A quadtree

Figure 9.

$$x_{j,2k_1+1,2k_2+1} = \frac{x_{j-1,k_1,k_2} + \alpha_{j,3}x_{j-1,k_1+1,k_2} + \alpha_{j,4}x_{j-1,k_1,k_2+1} + \alpha_{j,5}x_{j-1,k_1+1,k_2+1}}{1 + \alpha_{j,3} + \alpha_{j,4} + \alpha_{j,5}} + w_{j,2k_1+1,2k_2+1},$$

and for $j = J_1 + 1, \ldots, J, k_1, k_2 \in \mathbb{Z}$, we assume

$$x_{j,k_1,k_2} = x_{j-1,[k_1/2],[k_2/2]} + w_{j,k_1,k_2},$$

where $\alpha_{j,m} > 0$, $j = J_0 + 1, \ldots, J_1, m = 1, \ldots, 5$, are constants. Recall that the data are observed (perhaps incompletely) according to (23).

Note that the generalized-Kalman-filter algorithm (Section 3.4) for a two-dimensional process is usually difficult to implement, except for a tree-structured model. However, by treating all the variables at each scale, from the scale J_0 to the scale J_1, as a single node on the graph, we obtain a tree-structured model that consists of a multivariate Markov-chain structure from the scale J_0 to the scale J_1, a multi-branch tree structure between the scale J_1 and the scale $J_1 + 1$, and a univariate quadtree structure from the scale $J_1 + 1$ to the scale J.

We can now apply the generalized-Kalman-filter algorithm (Section 2) to this tree-structured model to compute the optimal predictors of $\{x_{j,k}\}$. Note that the generalized-Kalman-filter algorithm involves multivariate updates for state vectors from the scale J_0 to the scale J_1. However, these multivariate computations are not all that computationally intensive, because the number of variables in coarser scales are relatively small. In fact, the number of variables in each scale is around one fourth the number of variables at the next finer scale. Hence, the generalized-Kalman-filter algorithm is still computationally efficient for this model.

5 Discussion

While the theory of multiscale graphical modeling is quite general, our examples in this article are restricted to zero-mean Gaussian distributions with linear autoregressive structure. Our future research will be devoted to generalizing the filtering algorithms to allow for non-Gaussian distributions with non-zero means. This will be a challenging problem, since recall that model specification consists of a data step (noise model for $\boldsymbol{y}$) and a process step (graphical model for $\boldsymbol{x}$). It is not yet clear how (moments of) the posterior distribution of $\boldsymbol{x}$ given $\boldsymbol{y}$ can be obtained from a fast Kalman-filter-like algorithm.

Currently, if data exhibit obvious spatial trend and obvious non Gaussianity, our recommendation is to try to transform the data (e.g., using a Box-Cox family of transformations) and then detrend (e.g., median polish). The generalized-Kalman-filter algorithm may then be applied to the residuals.

Incorporation of time into the graphical Markov model is, in principle, straightforward. Any autoregressive temporal structure on the spatio-temporal process $\{\boldsymbol{x}_\tau\}$, where τ indexes time, can be conceptualized as a tree with an $\boldsymbol{x}_\tau$ at each node. Since those nodes are themselves trees or more general graphical structures, the resulting graph with individual random variables at each node is of the general type considered in this article. Ideally, it is a tree, for which the algorithms of Section 2 are immediately applicable.

Another generalization of these models is to deal with the case where data are multivariate. Again, this should be straightforward; instead of individual random variables at each node, there are random vectors. Within the linear Gaussian modeling framework, this results in bigger vectors and matrices, but the algorithms are conceptually the same. The one new feature will be inversions of matrices of order determined by the number of variables in the original multivariate random vector. This does have the potential to slow up the algorithm considerably if the order becomes too large.

Finally, in applications to command and control, one is often interested in nonlinear functions of $\boldsymbol{x}$, for example, $x_J^\dagger \equiv \max\{x_{J,k} : k \in \mathbb{Z}\}$ is the largest feature in the region of interest, at the finest resolution. Loss functions other than squared error (Section 1) are needed to extract nonlinear functions; their development and implementation in spatio-temporal contexts, will be a topic for future research.

Acknowledgments: The authors would like to thank Marc Moore for helpful comments. This research was supported by the Office of Naval Research under grant no. N00014-99-1-0214.

6 References

Basseville, M., A. Benveniste, K.C. Chou, S.A. Golden, R. Nikoukhah, and A.S. Willsky (1992a). Modeling and estimation of multiresolution stochastic processes. *IEEE Transactions on Information Theory 38*, 766–784.

Basseville, M., A. Benveniste, and A.S. Willsky (1992b). Multiscale autoregressive processes I. Schur–Levinson parametrizations. *IEEE Transactions on Signal Processing 40*, 1915–1934.

Basseville, M., A. Benveniste, and A.S. Willsky (1992c). Multiscale autoregressive processes. II. Lattice structures for whitening and modeling. *IEEE Transactions on Signal Processing 40*, 1935–1954.

Besag, J. (1974). Spatial interaction and the statistical analysis of lattice systems (with discussion). *Journal of the Royal Statistical Society. Series B. Methodological 36*, 192–236.

Buneman, P (1974). A characterization of rigid circuit graphs. *Discrete Mathematics 9*, 205–212.

Chou, K.C. (1991). *Multiscale systems, Kalman filters, and Riccati equations.* Ph.D. thesis, Dept. of Electrical Engineering and Computer Science, M.I.T., Cambridge, MA.

Chou, K.C., A.S. Willsky, and R. Nikoukhah (1994). Multiscale systems, Kalman filters, and Riccati equations. *IEEE Transactions on Automatic Control 39*, 479–492.

Cressie, N.A.C. (1993). *Statistics for Spatial Data* (revised ed.). Wiley Series in Probability and Mathematical Statistics: Applied Probability and Statistics. New York: Wiley.

Cressie, N.A.C. and J.L. Davidson (1998). Image analysis with partially ordered Markov models. *Computational Statistics & Data Analysis 29*, 1–26.

Darroch, J.N., S.L. Lauritzen, and T.P. Speed (1980). Markov fields and log-linear interaction models for contingency tables. *The Annals of Statistics 8*, 522–539.

Davidson, J.L. and N.A.C. Cressie (1993). Markov pyramid models in images analysis. In E. Dougherty, P. Gader, and J. Serra (Eds.), *Image Algebra and Morphological Image Processing. IV*, Volume 2030 of *Proceedings of SPIE*, pp. 179–190. San Diego, CA, 1993: SPIE—The International Society for Optical Engineering, Bellingham, WA.

Davidson, J.L., N. Cressie, and X. Hua (1999). Texture synthesis and pattern recognition for partially ordered Markov models. *Pattern Recognition 32*, 1475–1505.

Dempster, A.P. (1972). Covariance selection. *Biometrics 28*, 157–175.

Edwards, D. (1995). *Introduction to Graphical Modelling.* Springer Texts in Statistics. New York: Springer.

Edwards, D. and S. Kreiner (1983). The analysis of contingency tables by graphical models. *Biometrika 70*, 553–565.

Fieguth, P.W., W.C. Karl, A.S. Willsky, and C. Wunsch (1995). Multiresolution optimal interpolation and statistical analysis of TOPEX/POSEIDON satellite altimetry. *IEEE Transactions on Geoscience & Remote Sensing 33*, 280–292.

Fosgate, C.H., H. Krim, W.W. Irving, W.C. Karl, and A.S. Willsky (1997). Multiscale segmentation and anomaly enhancement of SAR imagery. *IEEE Transactions on Image Processing 6*, 7–20.

Frydenberg, M. and S.L. Lauritzen (1989). Decomposition of maximum likelihood in mixed interaction models. *Biometrika 76*, 539–555.

Gavril, F. (1974). The intersection graphs of subtrees in trees are exactly the chordal graphs. *Journal of Combinatorial Theory. Series B 16*, 47–56.

Golumbic, M.C. (1980). *Algorithmic Graph Theory and Perfect Graphs.* Computer Science and Applied Mathematics. New York: Academic Press.

Hammersley, J.M. and P. Clifford (1971). Markov fields on finite graphs and lattices. Oxford University.

Harvey, A.C. (1989). *Forecasting, Structural Time Series Models and the Kalman Filter.* Cambridge: Cambridge University Press.

Howard, R.A. and J.E. Matheson (1981). Influence diagrams. In R. Howard and J. Matheson (Eds.), *Readings on the Principles and Applications of Decision Analysis*, Volume II, pp. 719–762. Menlo Park, CA: Strategic Decisions Group.

Isham, V. (1981). An introduction to spatial point processes and Markov random fields. *International Statistical Review 49*, 21–43.

Kalman, R.E. (1960). A new approach to linear filtering and prediction problems. *Journal of Basic Engineering 82*, 34–45.

Kalman, R.E. and R.S. Bucy (1961). New results in linear filtering and prediction theory. *Journal of Basic Engineering 83*, 95–108.

Lauritzen, S.L. (1996). *Graphical Models.* Oxford Statistical Science Series. New York: Oxford University Press.

Lauritzen, S.L., A.P. Dawid, B.N. Larson, and H.G. Leimer (1990). Independence properties of directed Markov fields. *Networks 20*, 491–505.

Lauritzen, S.L., T. P. Speed, and K. Vijayan (1984). Decomposable graphs and hypergraphs. *Australian Mathematical Society. Journal. Series A. Pure Mathematics and Statistics 36*, 12–29.

Lauritzen, S.L. and D.J. Spiegelhalter (1988). Local computations with probabilities on graphical structures and their application to expert systems (with discussion). *Journal of the Royal Statistical Society. Series B. Methodological 50*, 157–224.

Leimer, H.-G. (1993). Optimal decomposition by clique separators. *Discrete Mathematics 113*, 99–123.

Luettgen, M.R. and A.S. Willsky (1995). Multiscale smoothing error models. *IEEE Transactions on Automatic Control 40*, 173–175.

Paté-Cornell, M.E. and P.S. Fischbeck (1995). Probabilistic interpretation of command and control signals: Bayesian updating of the probability of nuclear attack. *Reliability Engineering and System Safety 47*, 27–36.

Pearl, J. (1988). *Probabilistic Reasoning in Intelligent Systems: Networks of Plausible Inference.* The Morgan Kaufmann Series in Representation and Reasoning. San Mateo, CA: Morgan Kaufmann.

Shachter, R.D. (1986). Evaluating influence diagrams. *Operations Research 34*, 871–882.

Smith, J.Q. (1989). Influence diagrams for statistical modelling. *The Annals of Statistics 17*, 654–672.

Speed, T.P. (1979). A note on nearest-neighbour Gibbs and Markov probabilities. *Sankhyā. The Indian Journal of Statistics. Series A 41*, 184–197.

Spiegelhalter, D.J., A.P. Dawid, S.L. Lauritzen, and R.G. Cowell (1993). Bayesian analysis in expert systems (with discussion). *Statistical Science 8*, 219–283.

Tarjan, R.E. (1985). Decomposition by clique separators. *Discrete Mathematics 55*, 221–232.

Tarjan, R.E. and M. Yannakakis (1984). Simple linear-time algorithms to test chordality of graphs, test acyclicity of hypergraphs, and selectively reduce acyclic hypergraphs. *SIAM Journal on Computing 13*, 566–579.

Wermuth, N. and S.L. Lauritzen (1983). Graphical and recursive models for contingency tables. *Biometrika 70*, 537–552.

Whittaker, J. (1990). *Graphical Models in Applied Multivariate Statistics.* Chichester, UK: Wiley.

Wright, S. (1921). Correlation and causation. *Journal of Agricultural Research 20*, 557–585.

Wright, S. (1934). The method of path coefficients. *The Annals of Mathematical Statistics 5*, 161–215.

6

Unsupervised Image Segmentation Using a Telegraph Parameterization of Pickard Random Field

Jérôme Idier, Yves Goussard, and Andrea Ridolfi

ABSTRACT This communication presents a nonsupervised three-dimensional segmentation method based upon a discrete-level unilateral Markov field model for the labels and conditionaly Gaussian densities for the observed voxels. Such models have been shown to yield numerically efficient algorithms, for segmentation and for estimation of the model parameters as well. Our contribution is twofold. First, we deal with the degeneracy of the likelihood function with respect to the parameters of the Gaussian densities, which is a well-known problem for such mixture models. We introduce a bounded *penalized* likelihood function that has been recently shown to provide a consistent estimator in the simpler cases of independent Gaussian mixtures. On the other hand, implementation with EM reestimation formulas remains possible with only limited changes with respect to the standard case. Second, we propose a *telegraphic* parameterization of the unilateral Markov field. On a theoretical level, this parameterization ensures that some important properties of the field (e.g., stationarity) do hold. On a practical level, it reduces the computational complexity of the algorithm used in the segmentation and parameter estimation stages of the procedure. In addition, it decreases the number of model parameters that must be estimated, thereby improving convergence speed and accuracy of the corresponding estimation method.

1 Introduction

In this paper, we present a method for segmenting images modeled as N-ary Markov random fields (MRFs). Such image representations have proved useful for segmentation because they can explicitly model important features of actual images, such as the presence of homogeneous regions separated by sharp discontinuities. However, Markov-based segmentation methods are often computationally intensive and therefore difficult to apply in a three-dimensional (3D) context. In addition, specification of the MRF parameter values is often difficult to perform. This can be done in a heuristic manner, but such an approach is strongly application-dependent and becomes very burdensome for complex models (i.e., large neighborhoods and large number of levels). Deriving *unsupervised* methods in which the MRF parameters are estimated from the observed data is more satisfactory, but such a task generally requires approximations in order to be mathematically feasible (Besag, 1986), and the corresponding amount of computation is generally much higher than for a segmentation operation.

In order to overcome such difficulties, Devijver and Dekesel (1988) proposed an unsupervised segmentation approach based on a hidden Markov model (HMM) that belongs to a special class of *unilateral* MRFs: Pickard random fields (PRFs). The PRF is observed through an independent Gaussian process and the labels as well as the model parameters are estimated using maximum likelihood techniques. Because of the specific properties of PRF models, a significant reduction of the computational burden is achieved, and application of such methods to 3D problems can be envisioned. However, three kinds of difficulties remain: firstly, from a theoretical standpoint, the estimated MRF parameters are not necessarily consistent with the assumed stationarity of the model; secondly, the likelihood function of the observation parameters presents attractive singular points, as it is well-known in Gaussian mixture identification problems (Nádas, 1983), and this hinders the convergence of the estimation procedure; thirdly, the convergence of the estimation procedure is made even more difficult by the fairly large number of parameters that need to be estimated.

Here, we present a segmentation method that extends the technique introduced by Devijver and Dekesel and corrects some of its deficiencies. First, the method is based upon a parsimonious parameterization of PRFs, referred to as a *telegraph model*, which simplifies the parameter estimation procedure, speeds up its convergence and ensures that some necessary conditions (such as marginal stationarity of the rows and columns) are fulfilled. Second, the singularity of the likelihood function of the parameters is dealt with by using a well-behaved *penalized* likelihood function that lends itself to the derivation of an efficient maximization procedure. Therefore, the resulting unsupervised segmentation method presents a safe overall convergence and exhibits a moderate amount of computation, which makes it suitable to process large 3D images as illustrated in the sequel.

2 Approach

Throughout the paper, random variables and realizations of thereof are respectively denoted by uppercase and corresponding lowercase symbols; in addition, notations such as $f(\boldsymbol{y} \mid \boldsymbol{x})$ and $\Pr(\boldsymbol{x} \mid \boldsymbol{y})$ are employed as shorthands for $f_{\boldsymbol{Y}|\boldsymbol{X}}(\boldsymbol{y} \mid \boldsymbol{x})$ and $\Pr(\boldsymbol{X} = \boldsymbol{x} \mid \boldsymbol{Y} = \boldsymbol{y})$, whenever unambiguous.

As stated in the introduction, the image to be segmented is modeled as a hidden N-ary PRF. N-ary PRFs were studied by Pickard in a two-dimensional (2D) framework (Pickard, 1977, 1980); these fields are stationary and their joint probability is determined by a measure τ on a four-pixel elementary cell

$$\begin{pmatrix} A & B \\ C & D \end{pmatrix}$$

that must fulfill several symmetry and independence conditions (Pickard, 1980). Conversely, it is shown in Champagnat et al. (1998) that stationary MRFs on a finite rectangular lattice can be characterized by their marginal distribution on a four-pixel elementary cell, and that in some important cases (Gaussian fields, symmetric fields), the only stationary fields are PRFs. As a consequence of the stationarity of $\boldsymbol{X}$, the marginal probability of each row and column presents the structure of a stationary and reversible Markov chain whose initial and transition probabilities can be easily deduced from τ. According to Idier and Goussard (1999), most of the latter results have three-dimensional (3D) counterparts that apply to 3D PRFs.

Here, we assume that the observations $\boldsymbol{y}$ of PRF $\boldsymbol{X}$ fulfill the following properties:

$$f(\boldsymbol{y} \mid \boldsymbol{x}) = \prod_{\{i,j\}} f\left(y_{\{i,j\}} \mid x_{\{i,j\}}\right), \tag{1}$$

$$f\left(y_{\{i,j\}} \mid X_{\{i,j\}} = n\right) = \mathcal{G}_n, \tag{2}$$

where i, j and $n \in \{1, \ldots, N\}$ respectively denote the row, column and state indices, and where $\mathcal{G}_n$ represents the Gaussian density with mean u_n and variance v_n.

These assumptions correspond to the common situation of an image degraded by independent Gaussian noise and, as underlined by Devijver and Dekesel (1988) in a 2D context, they are well suited to marginal maximum a posteriori (MMAP) segmentation of $\boldsymbol{X}$ as well as to maximum likelihood (ML) estimation of the PRF and noise parameters. The key to the derivation of numerically efficient segmentation algorithms is the approximation

$$\Pr\left(x_{\{i,j\}} \mid \boldsymbol{y}\right) \approx \Pr\left(x_{\{i,j\}} \mid \boldsymbol{y}_{\{i,\cdot\}}, \boldsymbol{y}_{\{\cdot,j\}}\right), \tag{3}$$

where $\boldsymbol{y}_{\{i,\cdot\}}$ and $\boldsymbol{y}_{\{\cdot,j\}}$ respectively denote ith row and jth column of $\boldsymbol{y}$. The above approximation amounts to neglecting interactions in the diagonal directions and to rely only on interactions in the horizontal and vertical directions. This may cause a lower accuracy for segmentation of objects with diagonally-oriented boundaries. However, this effect is not severe, as shown in Devijver and Dekesel (1988) and in Section 7 of this article, and with this approximation, the marginal posterior likelihood only involves 1D restrictions of $\boldsymbol{y}$ which present Markov chain structures. In order to take advantage of this property, Bayes rule is applied to (3) and the orthogonality properties of measure τ yield the following simplified expression

$$\Pr\left(x_{\{i,j\}} \mid \boldsymbol{y}_{\{i,\cdot\}}, \boldsymbol{y}_{\{\cdot,j\}}\right) \propto f\left(\boldsymbol{y}_{\{i,\cdot\}} \mid x_{\{i,j\}}\right) f\left(\boldsymbol{y}_{\{\cdot,j\}} \mid x_{\{i,j\}}\right) \Pr\left(x_{\{i,j\}}\right). \quad (4)$$

The above expression only involves 1D quantities; this has two important consequences. First, due to the Markov chain structures of $\boldsymbol{X}_{\{i,\cdot\}}$ and $\boldsymbol{X}_{\{\cdot,j\}}$, the first two terms of the right hand side of (4) can be evaluated in an efficient manner by means of 1D forward-backward algorithms. Second, the only parameters of interest in the a priori PRF model are those which control the distribution of rows and columns $\boldsymbol{X}_{\{i,\cdot\}}$ and $\boldsymbol{X}_{\{\cdot,j\}}$, thereby simplifying the parameter estimation stage outlined below.

The PRF representation associated with assumptions (1)–(2) is also well suited to ML estimation of the model parameter vector $\boldsymbol{\theta}$. The ML estimate $\widehat{\boldsymbol{\theta}} = \arg\max_{\boldsymbol{\theta}} f(\boldsymbol{y};\boldsymbol{\theta})$ cannot be expressed in closed form. Devijver and Dekesel (1988) proposed to evaluate $\boldsymbol{\theta}$ through maximization of the following criterion:

$$J(\boldsymbol{y};\boldsymbol{\theta}) \propto \prod_i f\left(\boldsymbol{y}_{\{i,\cdot\}};\boldsymbol{\theta}\right) \prod_j f\left(\boldsymbol{y}_{\{\cdot,j\}};\boldsymbol{\theta}\right). \quad (5)$$

They showed that iterative maximization of J can be carried out by an expectation-maximization (EM) algorithm and that the quantities required for the EM iterations can be evaluated by the same forward-backward procedures as the ones used for segmentation of $\boldsymbol{X}$. Even though Devijver and Dekesel presented J as a mere approximation of the exact likelihood function, it is clear by inspection that J can be interpreted as a generalization of the pseudo-likelihood function proposed in Besag (1974). More generally, we conjecture that the above estimator can be cast within the framework of minimum contrast estimation and that its convergence and consistency properties can be investigated with techniques similar to those presented in (Guyon, 1992, pp. 157–162).

This method, in the form proposed by Devijver and Dekesel (1988), proved to provide interesting segmentation results in a non supervised framework at a reasonable computational cost. It nonetheless presents several limitations and deficiencies. First, it is limited to 2D problems; second, the distributions of $\boldsymbol{X}_{\{i,\cdot\}}$ and $\boldsymbol{X}_{\{\cdot,j\}}$ are parameterized in a standard manner by the initial and transition probabilities. Consequently, the stationarity and reversibility of each row and column of PRF $\boldsymbol{X}$ is not guaranteed. In addition, $O(N^2)$ parameters must be estimated, which requires a significant amount of computations and induces convergence difficulties, even for moderate numbers of states; third, the likelihood function used for estimation of $\boldsymbol{\theta}$ presents singular points. Intuitively, this is caused by the normal densities $f\left(\boldsymbol{y}_{\{i,\cdot\}} \mid \boldsymbol{x}_{\{i,\cdot\}};\boldsymbol{\theta}\right)$ which enter the right-hand side of (5) through decompositions of the form:

$$f\left(\boldsymbol{y}_{\{i,\cdot\}};\boldsymbol{\theta}\right) = \sum_{\boldsymbol{x}_{\{i,\cdot\}}} \prod_j f\left(y_{\{i,j\}} \mid x_{\{i,j\}};\boldsymbol{\theta}\right) \Pr\left(\boldsymbol{x}_{\{i,\cdot\}};\boldsymbol{\theta}\right), \quad (6)$$

and which degenerate when $x_{\{i,j\}} = n$, $u_n = y_{\{i,j\}}$ and $v_n \searrow 0$ for some n, j. For estimation of parameters of mixtures of Gaussian densities, this behavior is well known and well documented (Nádas, 1983, Redner and Walker, 1984). The consequence of this degeneracy is the divergence of the EM procedure if a reestimated value of $\boldsymbol{\theta}$ reaches a neighborhood of any singular point.

The main purpose of this article is to propose several extensions and refinements of the segmentation method introduced by Devijver and Dekesel (1988) in order to alleviate its main limitations. The major improvements are

1. extension of the technique to a three dimensional (3D) framework;

2. correction of the degeneracy of the likelihood function through adjunction of an appropriate penalization function, while retaining the possibility of estimating the model parameters with an EM procedure in a slightly modified form;

3. parameterization of the 1D restrictions of $\boldsymbol{X}$ with a *telegraph model* (TM) which guarantees their stationarity and reversibility while remaining compatible with the EM procedure used for model estimation. In addition, convergence of the EM procedure is improved by the reduced dimension ($O(N)$) of the TM parameter vector with respect to standard parameterization of Markov chains.

Before addressing these three points, we briefly recall the equations of the EM algorithm and derive two properties that will simplify the subsequent derivations.

3 EM Reestimation Formulas for Parameter Estimation

3.1 General EM Procedure

The EM algorithm is an iterative procedure which increases the likelihood $f(\boldsymbol{y}; \boldsymbol{\theta})$ of a parameter vector $\boldsymbol{\theta}$ given observations $\boldsymbol{y}$ at each iteration. Starting from an initial value $\boldsymbol{\theta}^0$, a series of successive estimates $\boldsymbol{\theta}^k$ is generated by alternating the following two steps:

$$\text{Expectation (E):} \qquad \text{Evaluate } Q(\boldsymbol{\theta}, \boldsymbol{\theta}^k; \boldsymbol{y}), \tag{7}$$

$$\text{Maximization (M):} \qquad \boldsymbol{\theta}^{k+1} = \arg\max_{\boldsymbol{\theta}} Q(\boldsymbol{\theta}, \boldsymbol{\theta}^k; \boldsymbol{y}), \tag{8}$$

where the function Q is defined as

$$Q(\boldsymbol{\theta}, \boldsymbol{\theta}^0; \boldsymbol{y}) \triangleq \sum_{\boldsymbol{x}} \Pr(\boldsymbol{x} \mid \boldsymbol{y}; \boldsymbol{\theta}^0) \log\big(f(\boldsymbol{y} \mid \boldsymbol{x}; \boldsymbol{\theta}) \Pr(\boldsymbol{x}; \boldsymbol{\theta})\big) \tag{9}$$

$$= E\big[\log\big(f(\boldsymbol{y} \mid \boldsymbol{X}; \boldsymbol{\theta}) \Pr(\boldsymbol{X}; \boldsymbol{\theta})\big) \mid \boldsymbol{y}; \boldsymbol{\theta}^0\big], \tag{10}$$

$\boldsymbol{X}$ being an auxiliary variable whose practical role is to make the *extended likelihood* $f(\boldsymbol{y} \mid \boldsymbol{x}; \boldsymbol{\theta}) \Pr(\boldsymbol{x}; \boldsymbol{\theta})$ easier to compute than the original likelihood $f(\boldsymbol{y}; \boldsymbol{\theta})$. The above equations are given for a continuous-valued variable $\boldsymbol{y}$ and a discrete-valued auxiliary variable $\boldsymbol{x}$, as this corresponds to our application. Transposition to a discrete-valued $\boldsymbol{y}$ and/or a continuous-valued $\boldsymbol{x}$ is straightforward. In all cases, the EM algorithm can be shown to increase the likelihood at each iteration and to converge to a critical point of the likelihood function $f(\boldsymbol{y}; \boldsymbol{\theta})$. A detailed analysis of the properties of the EM algorithm can be found in Baum et al. (1970) in the context of hidden Markov chains and in Dempster et al. (1977), Redner and Walker (1984) in a more general framework. Here, we provide the equations of a complete EM algorithm for estimation of the parameters of a 1D HMM, as this will be the base for the derivations in Sections 5 and 6. The hidden Markov chains $\boldsymbol{X}_{\{i,\cdot\}}$ and $\boldsymbol{X}_{\{\cdot,j\}}$ are discrete-valued and in accordance with assumptions (1)–(2) the observations $\boldsymbol{Y}_{\{i,\cdot\}}$ and $\boldsymbol{Y}_{\{\cdot,j\}}$ are conditionally independent and Gaussian. For a generic discrete-valued hidden Markov chain $X_t \in \{1, \ldots, N\}$, $1 \leq t \leq T$, with conditionally independent Gaussian observations y_t, $1 \leq t \leq T$, the equations of the complete EM algorithm are given in Table 2, in compliance with the compact notations defined in Table 1. It should be underlined that quantity $p^0_{t,n}$ computed by the forward-backward algorithm is precisely the marginal likelihood $\Pr(X_t = n \mid \boldsymbol{y})$ used for estimation of $\boldsymbol{X}$. This illustrates the point made in Section 2 that the forward-backward algorithm is the basic tool for both the segmentation step and the parameter estimation step.

3.2 Decoupling of the M Step

Assume that parameter vector $\boldsymbol{\theta}$ can be partitioned into two subvectors $\boldsymbol{\theta}_{Y|X}$ and $\boldsymbol{\theta}_X$ which respectively control the conditional probability function $f(\boldsymbol{y} \mid \boldsymbol{x})$ and the probability distribution $\Pr(\boldsymbol{x})$. Such a situation is commonly encountered and can be taken advantage of in order to decouple the M step of the EM algorithm into two—hopefully simpler—independent maximization problems.

Under these assumptions, the probability product which enters the definition of Q in (9) can be expressed as

$$f(\boldsymbol{y} \mid \boldsymbol{x}; \boldsymbol{\theta}) \Pr(\boldsymbol{x}; \boldsymbol{\theta}) = f(\boldsymbol{y} \mid \boldsymbol{x}; \boldsymbol{\theta}_{Y|X}) \Pr(\boldsymbol{x}; \boldsymbol{\theta}_X). \tag{11}$$

For any set value of parameter vector $\boldsymbol{\theta}^0$, define functions $Q_{Y|X}$ and Q_X as

$$Q_{Y|X}(\boldsymbol{\theta}_{Y|X}, \boldsymbol{\theta}^0; \boldsymbol{y}) \triangleq \sum_{\boldsymbol{x}} Pr(\boldsymbol{x} \mid \boldsymbol{y}; \boldsymbol{\theta}^0) \log f(\boldsymbol{y} \mid \boldsymbol{x}; \boldsymbol{\theta}_{Y|X}), \tag{12}$$

Table 1. Notations

$$y = [y_1, \dots, y_T]^t, \qquad y_s^t = [y_s, \dots, y_t]^t,$$

$$\mathcal{G}_n = f(y_t \mid X_t = n) = (2\pi v_n)^{-1/2} \exp\left[-\frac{(y-u_n)^2}{2v_n}\right],$$

$$p_n = \Pr(X_1 = n; \boldsymbol{\theta}), \qquad P_{mn} = \Pr(X_t = n \mid X_{t-1} = m; \boldsymbol{\theta}),$$

$$p_n^0 = \Pr(X_1 = n; \boldsymbol{\theta}^0), \qquad P_{mn}^0 = \Pr(X_t = n \mid X_{t-1} = m; \boldsymbol{\theta}^0),$$

$$p_{t,n}^0 = \Pr(X_t = n \mid \boldsymbol{y}; \boldsymbol{\theta}^0), \qquad p_{t,mn}^0 = \Pr(X_{t-1} = m, X_t = n \mid \boldsymbol{y}; \boldsymbol{\theta}^0),$$

$$\alpha_n^0 = \sum_{t=1}^{T} p_{t,n}^0, \qquad \beta_n^0 = \sum_{t=2}^{T-1} p_{t,n}^0, \qquad s_n^0 = \sum_{t=2}^{T} p_{t,nn}^0,$$

$$\eta_n^0 = \frac{\alpha_n^0 + \beta_n^0}{2}, \qquad \gamma_n^0 = \eta_n^0 - s_n^0,$$

$$\mathcal{F}_{t,n} = P(X_t = n \mid \boldsymbol{y}_1^t; \boldsymbol{\theta}^0), \qquad \mathcal{N}_{t,n} = f(y_t \mid \boldsymbol{y}_1^{T-1}),$$

$$\mathcal{B}_{t,n} = \frac{f(\boldsymbol{y}_{t+1}^T \mid X_t = n; \boldsymbol{\theta}^0)}{f(\boldsymbol{y}_{t+1}^T \mid \boldsymbol{y}_1^t; \boldsymbol{\theta}^0)}.$$

$$Q_X(\boldsymbol{\theta}_X, \boldsymbol{\theta}^0; \boldsymbol{y}) \triangleq \sum_{\boldsymbol{x}} \Pr(\boldsymbol{x} \mid \boldsymbol{y}; \boldsymbol{\theta}^0) \log \Pr(\boldsymbol{x}; \boldsymbol{\theta}_X)\, d\boldsymbol{x}. \tag{13}$$

It can be immediately deduced from (9) and (11) that function Q can be expressed as

$$Q(\boldsymbol{\theta}, \boldsymbol{\theta}^0; \boldsymbol{y}) = Q_{Y|X}(\boldsymbol{\theta}_{Y|X}, \boldsymbol{\theta}^0; \boldsymbol{y}) + Q_X(\boldsymbol{\theta}_X, \boldsymbol{\theta}^0; \boldsymbol{y}), \tag{14}$$

which shows that the M step of the EM algorithm can be decoupled into two operations: maximization of $Q_{Y|X}$ with respect to $\boldsymbol{\theta}_{Y|X}$ and maximization of Q_X with respect to $\boldsymbol{\theta}_X$.

3.3 Independent realizations

Another special case of interest occurs when $\boldsymbol{y}$ is made up of independent realizations $\boldsymbol{y}_i; 1 \leq i \leq I$. For instance, this corresponds to the case of the pseudo-likelihood defined in (5). As a consequence, the corresponding auxiliary processes $\boldsymbol{X}_i$ are also independent and it is not difficult to obtain

$$\Pr(\boldsymbol{x} \mid \boldsymbol{y}; \boldsymbol{\theta}^0) = \prod_{i=1}^{I} \Pr(\boldsymbol{x}_i \mid \boldsymbol{y}_i; \boldsymbol{\theta}^0), \tag{15}$$

$$\log\big(f(\boldsymbol{y} \mid \boldsymbol{x}; \boldsymbol{\theta}) \Pr(\boldsymbol{x}; \boldsymbol{\theta})\big) = \sum_{i=1}^{I} \log\big(f(\boldsymbol{y}_i \mid \boldsymbol{x}_i; \boldsymbol{\theta}) \Pr(\boldsymbol{x}_i; \boldsymbol{\theta})\big). \tag{16}$$

Table 2. Standard reestimation EM formulas that yield $\boldsymbol{\theta} = (\{p_n\}, \{P_{mn}\}, \{u_n\}, \{v_n\})$ as the maximizer of $Q(\cdot, \boldsymbol{\theta}^0, \mathbf{y})$ for a finite state homogeneous HMM with Gaussian observations. The forward-backward algorithm provided here takes the normalized form given in Devijver and Dekesel (1988).

Forward step:

- $\mathcal{N}_1 = \sum_{n=1}^N p_n^0 \mathcal{G}_n$,
- for $n = 1, \ldots, N$: $\mathcal{F}_{1,n} = p_n^0\, \mathcal{G}_n / \mathcal{N}_1$,
- for $t = 2, \ldots, T$: $\mathcal{N}_t = \sum_{n=1}^N (\sum_{m=1}^N \mathcal{F}_{t-1,m} P_{mn}^0) \mathcal{G}_n$,
- for $n = 1, \ldots, N$: $\mathcal{F}_{t,n} = (\sum_{m=1}^N \mathcal{F}_{t-1,m}\, P_{mn}^0) \mathcal{G}_n / \mathcal{N}_t$.

Backward step

- for $n = 1, \ldots, N$: $\mathcal{B}_{T,n} = 1$,
- for $t = T-1, \ldots, 1$: for $n = 1, \ldots, N$:

  $\mathcal{B}_{t,n} = \sum_{m=1}^N \mathcal{B}_{t+1,m} P_{nm}^0 \mathcal{G}_m / \mathcal{N}_{t+1}$.
- For $t = T-1, \ldots, 1$:
  - for $n = 1, \ldots, N$: $p_{t,n}^0 = \mathcal{F}_{t,n} \mathcal{B}_{t,n}$,
  - for m, $n = 1, \ldots, N$: $p_{t,mn}^0 = \mathcal{F}_{t-1,m} P_{nm}^0 \mathcal{B}_{t,n} \mathcal{G}_n / \mathcal{N}_t$.

Reestimation step:

- for $n = 1, \ldots, N$: $p_n = p_{1,n}^0$,
- for m, $n = 1, \ldots, N$: $P_{mn} = \sum_{t=2}^T p_{t,mn}^0 / \sum_{t=1}^{T-1} p_{t,m}^0$,
- for $n = 1, \ldots, N$: $u_n = \sum_{t=1}^T p_{t,n}^0\, y_t / \alpha_n^0$,

  $v_n = \sum_{t=1}^T p_{t,n}^0 (y_t - u_n)^2 / \alpha_n^0$.

Therefore, the expression of $Q(\boldsymbol{\theta}, \boldsymbol{\theta}^0; \boldsymbol{y})$ can be rewritten as

$$Q(\boldsymbol{\theta}, \boldsymbol{\theta}^0; \boldsymbol{y}) = \sum_{\boldsymbol{x}_1, \ldots \boldsymbol{x}_I} \prod_{i=1}^I \Pr(\boldsymbol{x}_i \mid \boldsymbol{y}_i; \boldsymbol{\theta}^0) \sum_{i=1}^I \log(f(\boldsymbol{y}_i \mid \boldsymbol{x}_i; \boldsymbol{\theta}) \Pr(\boldsymbol{x}_i; \boldsymbol{\theta})), \quad (17)$$

and using the fact that conditional probability distributions $\Pr(\boldsymbol{x}_i \mid \boldsymbol{y}_i; \boldsymbol{\theta}^0)$ are normalized, we finally obtain

$$Q(\boldsymbol{\theta}, \boldsymbol{\theta}^0; \boldsymbol{y}) = \sum_{i=1}^I Q^i(\boldsymbol{\theta}, \boldsymbol{\theta}^0; \boldsymbol{y}_i), \quad (18)$$

where functions Q^i are defined by

$$Q^i(\boldsymbol{\theta}, \boldsymbol{\theta}^0; \boldsymbol{y}_i) \triangleq \sum_{\boldsymbol{x}_i} \Pr(\boldsymbol{x}_i \mid \boldsymbol{y}_i; \boldsymbol{\theta}^0) \log\big(f(\boldsymbol{y}_i \mid \boldsymbol{x}_i; \boldsymbol{\theta}) \Pr(\boldsymbol{x}_i; \boldsymbol{\theta})\big), \tag{19}$$

$$= E\big[\log\big(f(\boldsymbol{y}_i \mid \boldsymbol{X}_i; \boldsymbol{\theta}) \Pr(\boldsymbol{X}_i; \boldsymbol{\theta})\big) \mid \boldsymbol{y}_i; \boldsymbol{\theta}^0\big]. \tag{20}$$

In addition, if parameter vector $\boldsymbol{\theta}$ can be partitioned into two subvectors $\boldsymbol{\theta}_{Y|X}$ and $\boldsymbol{\theta}_X$, it is straightforward to check in the same manner as in Paragraph 3.2 that each function Q_i can be decomposed as

$$Q^i(\boldsymbol{\theta}, \boldsymbol{\theta}^0; \boldsymbol{y}_i) = Q^i_{Y|X}(\boldsymbol{\theta}_{Y|X}, \boldsymbol{\theta}^0; \boldsymbol{y}_i) + Q^i_X(\boldsymbol{\theta}_X, \boldsymbol{\theta}^0; \boldsymbol{y}_i), \tag{21}$$

where the expressions of $Q^i_{Y|X}$ and Q^i_X can be deduced from (12) and (13) by substituting $\boldsymbol{y}_i$ and $\boldsymbol{x}_i$ for $\boldsymbol{y}$ and $\boldsymbol{x}$, respectively.

4 3D extension

4.1 *Segmentation of 3D PRFs*

This paragraph relies on an extension of the construction of stationnary MRFs and PRFs presented in Pickard (1977, 1980) to the 3D case. The results are available in Idier and Goussard (1999) and will not be derived here. We model $\boldsymbol{X}$ as a 3D Pickard random field and we consider MMAP estimation of a voxel $X_{\{i,j,k\}}$ of the 3D array under approximations similar to those outlined in Section 2. More specifically, the marginal likelihood $\Pr(x_{\{i,j,k\}} \mid \boldsymbol{y})$ is approximated as

$$\Pr(x_{\{i,j,k\}} \mid \boldsymbol{y}) \approx \Pr(x_{\{i,j,k\}} \mid \boldsymbol{y}_{\{i,\cdot,k\}}, \boldsymbol{y}_{\{\cdot,j,k\}}, \boldsymbol{y}_{\{i,j,\cdot\}}), \tag{22}$$

where $\boldsymbol{y}_{\{i,\cdot,k\}}$, $\boldsymbol{y}_{\{\cdot,j,k\}}$ and $\boldsymbol{y}_{\{i,j,\cdot\}}$ denote the three 1D restrictions of $\boldsymbol{y}$ which contain voxel $y_{\{i,j,k\}}$. Here again, this approximation amounts to neglecting interactions in the diagonal directions. It can be shown that (see Idier and Goussard, 1999):

$$\Pr(x_{\{i,j,k\}} \mid \boldsymbol{y}_{\{i,\cdot,k\}}, \boldsymbol{y}_{\{\cdot,j,k\}}, \boldsymbol{y}_{\{i,j,\cdot\}}) \propto \Pr(x_{\{i,j,k\}}) f(\boldsymbol{y}_{\{i,\cdot,k\}} \mid x_{\{i,j,k\}}) \\ f(\boldsymbol{y}_{\{\cdot,j,k\}} \mid x_{\{i,j,k\}}) f(\boldsymbol{y}_{\{i,j,\cdot\}} \mid x_{\{i,j,k\}}). \tag{23}$$

As in the 2D case, the terms in the right hand side of (23) only involve 1D quantities. More specifically, due to the hidden Markov chain structures of the 1D restrictions of $\boldsymbol{y}$, the conditional probabilities in the right hand side of (23) can be evaluated using the same 1D forward-backward algorithms as in the 2D case, and the only parameters of interest of the PRF prior model are those which control the behavior of 1D Markov chains $\boldsymbol{X}_{\{i,\cdot,k\}}$, $\boldsymbol{X}_{\{\cdot,j,k\}}$ and $\boldsymbol{X}_{\{i,j,\cdot\}}$.

4.2 Parameter estimation

Here again, the ML estimator of $\boldsymbol{\theta}$ cannot be expressed in closed form and an EM procedure is applied to the pseudo-likelihood obtained by taking the product of marginal likelihoods of all 1D restrictions of $\boldsymbol{Y}$. Therefore, we have:

$$f(\boldsymbol{y};\boldsymbol{\theta}) \underset{\sim}{\propto} \prod_{r\in\mathcal{R}_1} f(\boldsymbol{y}_{(r)};\boldsymbol{\theta}), \tag{24}$$

where $\{\boldsymbol{y}_{(r)}; r \in \mathcal{R}_1\}$ is a shorthand notation for $\{\boldsymbol{y}_{\{i,\cdot,k\}}; i,k\} \cup \{\boldsymbol{y}_{\{\cdot,j,k\}}; j,k\} \cup \{\boldsymbol{y}_{\{i,j,\cdot\}}; i,j\}$, the set of all 1D restrictions of $\boldsymbol{y}$. Choosing $\boldsymbol{x}$ as the auxiliary variable of the EM algorithm and applying the result of Paragraph 3.3 yields

$$Q(\boldsymbol{\theta},\boldsymbol{\theta}^0;\boldsymbol{y}) = \sum_{r\in\mathcal{R}_1} Q^{(r)}(\boldsymbol{\theta},\boldsymbol{\theta}^0;\boldsymbol{y}_{(r)}), \tag{25}$$

$$= \sum_{r\in\mathcal{R}_1} E\big[\ln(f(\boldsymbol{y}_{(r)} \mid \boldsymbol{X}_{(r)};\boldsymbol{\theta}) \Pr(\boldsymbol{X}_{(r)};\boldsymbol{\theta})) \mid \boldsymbol{y}_r;\boldsymbol{\theta}^0\big]. \tag{26}$$

The process $\boldsymbol{y}_{(r)}$ has the structure of a 1D hidden Markov model with hidden process $\boldsymbol{X}_{(r)}$, and (26) shows that functions $Q^{(r)}$ are identical to those obtained for EM estimation of the parameters of 1D hidden Markov models. In other words, the reestimation formulas essentially operate on 1D quantities, which is the key to a tractable numerical implementation. We now precisely define these quantities and derive the corresponding EM algorithm, keeping in mind that parameter vector $\boldsymbol{\theta}$ can be partitioned into $\{\boldsymbol{\theta}_{Y|X}, \boldsymbol{\theta}_X\}$ which allows decoupling of the maximization step.

5 Telegraph model

In this section, we introduce the telegraph model whose purpose is to reduce the computational cost of parameter estimation and to ensure that the necessary condition of stationarity of the 1D restrictions of PRF $\boldsymbol{X}$ are fulfilled. As indicated by (23) and (24), the prior model needs only to specify the distribution of 1D quantities. Therefore the process we consider, i.e., the telegraph model (TM), is strictly a 1D Markov chain model, the 3D nature of the problem being accounted for through the aforementioned equations. We now define the TM and its parameter vector $\boldsymbol{\theta}_X$ and then derive the corresponding EM reestimation formulas.

5.1 Telegraph model definition

The TM is a straightforward generalization of a class of discrete-valued Markov chains proposed in (Godfrey et al., 1980) for segmentation of seismic signals. The transition probability matrix $\boldsymbol{P} = (P_{mn})$ of the model is defined by

$$\boldsymbol{P} = \boldsymbol{\Lambda} + (\mathbf{1} - \boldsymbol{\lambda})\boldsymbol{\mu}^{\mathrm{t}}, \tag{27}$$

with $\boldsymbol{\lambda} \triangleq \mathrm{vect}(\lambda_n)$, $\boldsymbol{\Lambda} \triangleq \mathrm{diag}(\lambda_n)$, $\mathbf{1} = (1, \ldots, 1)^{\mathrm{t}}$.

From an intuitive ground, the telegraphic parameterization $\boldsymbol{\theta}_X = \{\boldsymbol{\mu}, \boldsymbol{\lambda}\}$ can be interpreted as follows. The transition from one state to another is the result of a two-stage sampling experiment. On the basis of the first toss, the decision of keeping the current state m is made with probability λ_m. Otherwise, a new state n is chosen with probability μ_n, independently from the previous state. Since, in the latter case, n may be equal to m with probability μ_m, the probability of keeping the current state m is actually $\lambda_m + \mu_m - \lambda_m\mu_m$. According to such values, typical trajectories of the TM are more or less "blocky". This is a one-dimensional counterpart to well-known spatial Gibbsian models available for unordered colors (Besag, 1986).

In order to ensure that the resulting Markov chain is well defined and irreducible, it is straightforward to check that the following constraints form a set of sufficient conditions:

$$\sum_{n=1}^{N} \mu_n = 1, \tag{28}$$

$$\forall n = 1, \ldots, N, \quad \mu_n > 0, \tag{29}$$

$$\forall n = 1, \ldots, N, \quad \lambda_n < 1, \tag{30}$$

$$\forall n = 1, \ldots, N, \quad \lambda_n > -\mu_n/(1 - \mu_n). \tag{31}$$

Note that λ_n is not necessarily positive, although $\lambda_n > 0$, $n = 1, \ldots, N$ was understood in the above interpretation of the TM.

The stationary probability vector of the TM is readily obtained as

$$\boldsymbol{p} = (\boldsymbol{I} - \boldsymbol{\Lambda} + \boldsymbol{\mu}\boldsymbol{\lambda}^{\mathrm{t}})^{-1}\boldsymbol{\mu}, \tag{32}$$

where $\boldsymbol{I}$ is the identity matrix. Componentwise, such a vector also reads

$$p_n = \frac{\mu_n}{1 - \lambda_n} \Big/ \sum_{m=1}^{N} \frac{\mu_m}{1 - \lambda_m}. \tag{33}$$

Moreover, it can be verified that matrix $\mathrm{diag}(\boldsymbol{p})\boldsymbol{P}$ is symmetric, so the TM is reversible in its stationary state. Therefore, as long as the initial state probability vector is equal to $\boldsymbol{p}$ and that constraints (28)–(31) are fulfilled, (27) defines a stationary and reversible Markov chain that we choose to parameterize with $\boldsymbol{\theta}_X = \{\boldsymbol{\lambda}, \boldsymbol{\mu}\}$. The resulting number of degrees of freedom is $2N - 1$, which is linear w.r.t. the number of states, as opposed to the standard HMM case, which yields $N^2 - 1$ free parameters.

5.2 *Reestimation formulas for* $\boldsymbol{\theta}_X$

One of the reasons for introducing the TM is to simplify the forward-backward algorithm used to evaluate marginal likelihood values $\Pr(X_t = n \mid \boldsymbol{y}; \boldsymbol{\theta})$. As seen in Table 2 (evaluation of quantities $\mathcal{F}_t$, $\mathcal{B}_t$ and p_t^0), each of the $T-1$ recursions of the algorithm requires matrix products involving transition matrix $\boldsymbol{P}^0$. As seen in the sequel, expressing $\boldsymbol{P}$ according to (27) allows us to bring the computational complexity of each recursion down from $O(N^2)$ to $O(N)$.

5.2.1 E-step

From the definition of Q_X (13), we have

$$\begin{aligned} Q_X(\boldsymbol{\theta}_X, \boldsymbol{\theta}^0; \boldsymbol{y}) &= \sum_{\boldsymbol{x}} \Pr(\boldsymbol{x} \mid \boldsymbol{y}; \boldsymbol{\theta}^0) \log \Pr(\boldsymbol{x}; \boldsymbol{\theta}_X) \\ &= \sum_{n=1}^{N} p_{1,n}^0 \log p_n + \sum_{m,n=1}^{N} \sum_{t=2}^{T} p_{t,mn}^0 \log p_{mn}, \end{aligned}$$

where

$$\begin{aligned} p_{t,n}^0 &\triangleq \Pr(X_t = n \mid \boldsymbol{y}; \boldsymbol{\theta}^0), \\ p_{t,mn}^0 &\triangleq \Pr(X_{t-1} = m, X_t = n \mid \boldsymbol{y}; \boldsymbol{\theta}^0). \end{aligned}$$

Then, expressions (27) and (33) allow us to express the explicit dependence of Q_X on $\boldsymbol{\lambda}, \boldsymbol{\mu}$:

$$\begin{aligned} Q_X(\boldsymbol{\theta}_X, \boldsymbol{\theta}^0; \boldsymbol{y}) = \sum_{n=1}^{N} &\alpha_n^0 \log \mu_n + \beta_n^0 \log(1-\lambda_n) \\ &+ s_n^0 \log\left(1 + \frac{\lambda_n}{\mu_n(1-\lambda_n)}\right) - \log \sum_{n=1}^{N} \frac{\mu_n}{1-\lambda_n}, \end{aligned} \quad (34)$$

with

$$\alpha_n^0 \triangleq \sum_{t=1}^{T} p_{t,n}^0, \quad \beta_n^0 \triangleq \sum_{t=2}^{T-1} p_{t,n}^0, \quad s_n^0 \triangleq \sum_{t=2}^{T} p_{t,nn}^0. \quad (35)$$

5.2.2 Approximate M-step

The major difficulty lies in the M step, which consists of maximizing Q_X under constraints (28)–(31). Because of the last term in (34), explicit maximization is intricate. On the other hand, relative simplification occurs if Q_X is approximated by

$$\begin{aligned} &\widetilde{Q}_X(\boldsymbol{\theta}_X, \boldsymbol{\theta}^0; \boldsymbol{y}) \\ &\quad \triangleq Q_X(\boldsymbol{\theta}_X, \boldsymbol{\theta}^0; \boldsymbol{y}) - E[\log P(X_1; \boldsymbol{\theta}_X) P(X_T; \boldsymbol{\theta}_X) \mid \boldsymbol{y}; \boldsymbol{\theta}^0]/2 \end{aligned}$$

$$
\begin{aligned}
&= E[\log P(X_2 \ldots, X_T \mid X_1; \boldsymbol{\theta}_X) P(X_1, \ldots, X_{T-1} \mid X_T; \boldsymbol{\theta}_X) \mid \boldsymbol{y}; \boldsymbol{\theta}^0]/2 \\
&= E[\log P(X_2, \ldots, X_T \mid X_1; \boldsymbol{\theta}_X) \mid y_1, \ldots, y_T; \boldsymbol{\theta}^0]/2 \\
&\quad + E[\log P(X_2, \ldots, X_T \mid X_1; \boldsymbol{\theta}_X) \mid y_T, \ldots, y_1; \boldsymbol{\theta}^0]/2.
\end{aligned}
$$

Apart from the fact that the difference between Q_X and $\widetilde{Q}_X$ is moderate, it is not difficult to check that $\widetilde{Q}_X$ itself is an exact auxiliary function associated to a modified likelihood function. The latter reads

$$
f_{\boldsymbol{\pi},\boldsymbol{\lambda},\boldsymbol{\mu}}(y_1, \ldots, y_T) f_{\boldsymbol{\pi},\boldsymbol{\lambda},\boldsymbol{\mu}}(y_T, \ldots, y_1), \tag{36}
$$

where $f_{\boldsymbol{\pi},\boldsymbol{\lambda},\boldsymbol{\mu}}$ is the probability density function of the data when the initial probability vector of the TM is an arbitrary vector $\boldsymbol{\pi}$, while the transition matrix is parameterized by $(\boldsymbol{\lambda}, \boldsymbol{\mu})$ according to (27). The latter property ensures that the fixed-point EM procedure based on $\widetilde{Q}_X$ does converge (towards a stationary point of (36)).

First, let us express $\widetilde{Q}_X$ as an explicit function of $\boldsymbol{\lambda}$ and $\boldsymbol{\mu}$:

$$
\widetilde{Q}_X(\boldsymbol{\theta}_X, \boldsymbol{\theta}^0; \boldsymbol{y}) = \sum_{n=1}^{N} \widetilde{Q}_n,
$$

with

$$
\widetilde{Q}_n \triangleq \eta_n^0 \log \mu_n (1 - \lambda_n) + s_n^0 \log\left(1 + \frac{\lambda_n}{\mu_n(1 - \lambda_n)}\right), \tag{37}
$$

and

$$
\eta_n^0 \triangleq (\alpha_n^0 + \beta_n^0)/2. \tag{38}
$$

It is easy to maximize $\widetilde{Q}_X$ w.r.t. $\boldsymbol{\lambda}$ when $\boldsymbol{\mu}$ is held constant, since each function $\widetilde{Q}_n$ depends on λ_n only, and its maximum is reached at a unique point

$$
\hat{\lambda}_n = \frac{s_n^0/\eta_n^0 - \mu_n}{1 - \mu_n}. \tag{39}
$$

Moreover, constraints (30), (31) are fulfilled by $\widehat{\boldsymbol{\lambda}} = (\hat{\lambda}_n)$ since $s_n^0 < \alpha_n^0$ and $s_n^0 < \beta_n^0$ according to (35), provided that $\boldsymbol{\theta}^0$ meets (28)–(31). Substituting (39) into (37) allows us to express $\widetilde{Q}_n$ as a function of μ_n to within an additive constant factor:

$$
\widetilde{Q}_n = \gamma_n^0 \log \frac{\mu_n}{1 - \mu_n}, \tag{40}
$$

with $\gamma_n^0 \triangleq \eta_n^0 - s_n^0 \geq 0$. The Lagrange multiplier technique is used for maximization of $\widetilde{Q}_X$ with respect to $\boldsymbol{\mu}$ under constraints (28) and (29). Equating the gradient of the corresponding criterion to zero yields:

$$
\forall n, \quad \nu \widehat{\mu}_n^2 - \nu \widehat{\mu}_n + \gamma_n^0 = 0, \tag{41}
$$

where ν denotes the Lagrange multiplier. When $\nu > 4\gamma_n^0$, the above equation has two distinct roots, $\mu_n^+(\nu)$ and $\mu_n^-(\nu)$, located in $(0,1)$ on either side of $1/2$:

$$\mu_n^\pm(\nu) = \frac{1}{2}(1 \pm \sqrt{1 - 4\gamma_n^0/\nu}).$$

At first glance, the set of all possible combinations of μ_n^- and μ_n^+ provides 2^N different forms for $\widehat{\boldsymbol{\mu}} = (\widehat{\mu}_n)$. However, according to (28) and (29), $\widehat{\boldsymbol{\mu}}$ may only contain one μ_n^+. This brings the number of possible combinations down to $N+1$. Furthermore, among the N combinations that include one μ_n^+, $\widetilde{Q}_X$ is maximized if and only if the corresponding state n is chosen among the maximizers of (γ_n^0): $\forall m, \gamma_m^0 \leq \gamma_n^0$. Such a result stems from the following property: let us assume that constraint (28) is fulfilled by

$$\boldsymbol{\mu}(\nu) = (\mu_1^-(\nu), \ldots, \mu_{n-1}^-(\nu), \mu_n^+(\nu), \mu_{n+1}^-(\nu), \ldots, \mu_N^-(\nu))$$

for some value of ν, and, for instance, that $\gamma_1^0 > \gamma_n^0$. Then, for the same value of ν, constraint (28) is still fulfilled after the permutation of $\mu_1^-(\nu)$ and $\mu_n^+(\nu)$ in $\boldsymbol{\mu}(\nu)$, while it is easy to check from (40) that Q_X is increased by the positive amount

$$(\gamma_1^0 - \gamma_n^0) \log \frac{\mu_n^+(\nu)(1 - \mu_1^-(\nu))}{\mu_1^-(\nu)(1 - \mu_n^+(\nu))}.$$

Only two possible forms of combination remain:

$$\boldsymbol{\mu}^-, \text{ defined by: } \quad \forall m,\ \mu_m = \mu_m^-, \tag{42}$$

$$\boldsymbol{\mu}_n^+, \text{ defined by: } \quad \begin{cases} \forall m \neq n,\ \mu_m = \mu_m^-, \\ \mu_n = \mu_n^+, \\ \forall m,\ \gamma_m^0 \leq \gamma_n^0. \end{cases} \tag{43}$$

Note that there are as many different combinations $\boldsymbol{\mu}_n^+$ as maximizers of (γ_n^0). Further analysis of the properties of the remaining combinations brings the following existence and uniqueness result: the maximum of $\widetilde{Q}_X = \sum_{n=1}^N \widetilde{Q}_n$ (where $\widetilde{Q}_n$ is given by (40)), under constraints (28) and (29), is reached by a unique vector $\widehat{\boldsymbol{\mu}}(\widehat{\nu})$, where $\widehat{\nu}$ is uniquely determined by $\sum_{n=1}^N \widehat{\mu}_n(\widehat{\nu}) = 1$, and

$$\widehat{\boldsymbol{\mu}} = \begin{cases} \boldsymbol{\mu}^- & \text{if } \sum_{n=1}^N \omega_n^0 \leq N-2, \\ \boldsymbol{\mu}^+_{\arg\max_n \gamma_n^0} & \text{otherwise}, \end{cases} \tag{44}$$

with

$$\omega_n^0 = \sqrt{1 - \gamma_n^0 / \max_n \gamma_n^0}.$$

Since $0 \leq \omega_n^0 < 1$ for all n, and $\omega_m^0 = 0$ if $\gamma_m^0 = \max_n \gamma_n^0$, it is not difficult to check that $\sum_{n=1}^N \omega_n^0 \leq N-2$ if γ_n^0 admits more than one maximizer. Hence, the (unique) maximizer $\arg\max_n \gamma_n^0$ is well defined in (44).

In practice, $\widehat{\nu}$ cannot be expressed in closed form, but tight lower and upper bounds can be easily derived and classical numerical interpolation techniques can then be employed to refine the approximation. A summary of the forward-backward algorithm and of the reestimation formulas for $\boldsymbol{\mu}$ and $\boldsymbol{\lambda}$ is given in Table 3.

6 Mixture of Gaussians

We now address the question of the degeneracy of the likelihood with respect to parameters $\boldsymbol{\theta}_{Y|X}$. Maximizing $f(\boldsymbol{y};\boldsymbol{\theta})$ with respect to

$$\boldsymbol{\theta}_{Y|X} = (\boldsymbol{u},\boldsymbol{v}) = (u_1,\ldots,u_N,v_1,\ldots,v_N) \in \Theta = \mathbb{R}^N \times \mathbb{R}_+^{*\,N}$$

is indeed a degenerate problem since $f(\boldsymbol{y};\boldsymbol{\theta})$ is not bounded above: for an arbitrary state n and an arbitrary data sample y_t, it is clear that $f(\boldsymbol{y};\boldsymbol{\theta})$ can take arbitrary large values as v_n comes close to 0, when $u_n = y_t$ and every other unknowns are held fixed to arbitrary constants. This is a well-known problem in maximum likelihood identification of some mixture models (Redner and Walker, 1984, Nádas, 1983). In order to cope with the degeneracy in the case of an independent identically distributed (i.i.d.) mixture model, Hathaway proposed to restrict the admissible domain, and he showed that an EM strategy could still be implemented to solve the resulting constrained maximization problem (Hathaway, 1985, 1986).

Here, we adopt a slightly different approach, based on the maximization on Θ of a *penalized* version of the likelihood function:

$$F(\boldsymbol{y};\boldsymbol{\theta}) = f(\boldsymbol{y};\boldsymbol{\theta})G(\boldsymbol{v})$$

where $G(\boldsymbol{v})$ is an ad hoc prior distribution for $\boldsymbol{v}$ that compensates for the degeneracy at $v_n \searrow 0, n = 1,\ldots,N$. For this purpose, the solution of choice is the i.i.d. inverted gamma model:

$$G(\boldsymbol{v}) = \prod_{n=1}^{N} g(v_n), \tag{45}$$

with

$$g(v_n) = \frac{a^{b-1}}{\Gamma(b-1)}\frac{1}{v_n^b}\exp\left\{-\frac{a}{v_n}\right\}1_{[0,+\infty)}, \tag{46}$$

which is ensured to be proper if $b > 1$ and $a > 0$. The justification is twofold:

- For small values of v, $g(v)$ vanishes fast enough to compensate for the corresponding degeneracy of $f(\boldsymbol{y};\boldsymbol{\theta})$. More precisely, it can be established that F is a bounded function on Θ, which tends to zero

Table 3. Penalized EM formulas for a telegraphic HMM with Gaussian observations.

for $n = 1, \dots, N$:

- $p_n^0 = [\mu_n^0/(1-\lambda_n^0)]/[\sum_{m=1}^N \mu_m^0/(1-\lambda_m^0)]$,
- $P_{nn}^0 = \lambda_n^0 + \mu_n^0 - \lambda_n^0\mu_n^0$.

Forward step:

- $\mathcal{N}_1 = \sum_{n=1}^N p_n^0 \mathcal{G}_n$,
- for $n = 1, \dots, N$: $\mathcal{F}_{1,n} = p_n^0 \mathcal{G}_n / \mathcal{N}_1$
- for $t = 2, \dots, T$:
  - $\mathcal{N}_t = \sum_{n=1}^N (\lambda_n^0 \mathcal{F}_{t-1,n} + (1 - \sum_{m=1}^N \lambda_m^0 \mathcal{F}_{t-1,m})\mu_n^0)\mathcal{G}_n$,
  - for $n = 1, \dots, N$:

    $$\mathcal{F}_{t,n} = (\lambda_n^0 \mathcal{F}_{t-1,n} + (1 - \sum_{m=1}^N \lambda_m^0 \mathcal{F}_{t-1,m})\mu_n^0)\mathcal{G}_n / \mathcal{N}_t,$$

Backward step:

- for $n = 1, \dots, N$: $\mathcal{B}_{T,n} = 1$,
- for $t = T-1, \dots, 1$:
  - for $n = 1, \dots, N$:

    $$\mathcal{B}_{t,n} = (\lambda_n^0 \mathcal{B}_{t+1,n}\mathcal{G}_n + (\sum_{m=1}^N (1-\lambda_m^0)\mathcal{B}_{t+1,m}\mathcal{G}_m)\mu_n)/\mathcal{N}_{t+1}.$$

- for $t = T-1, \dots, 1$:
  - for $n = 1, \dots, N$: $p_{t,n}^0 = \mathcal{F}_{t,n}\mathcal{B}_{t,n}$,

    $p_{t,nn}^0 = \mathcal{F}_{t-1,n} P_{nn}^0 \mathcal{B}_{t,n} \mathcal{G}_n / \mathcal{N}_t$.

Reestimation step:

- approximate $\widehat{\nu}$ s.t. $\sum_{n=1}^N \widehat{\mu}_n(\widehat{\nu}) = 1$ by interpolation, where, for $n = 1, \dots, N$:

  $$\widehat{\mu}_n(\nu) = \begin{cases} (1+\sqrt{1-4\gamma_n^0/\nu})/2 & \text{if } \gamma_n^0 = \max_m \gamma_m^0 \text{ and } \\ & \sum_{m=1}^N \sqrt{1-\gamma_m^0/\gamma_n^0} > N-2, \\ (1-\sqrt{1-4\gamma_n^0/\nu})/2 & \text{otherwise;} \end{cases}$$

- for $n = 1, \dots, N$: $\mu_n = \widehat{\mu}_n(\nu)$, $\lambda_n = (s_n^0/\eta_n^0 - \mu_n)/(1-\mu_n)$,
- for $n = 1, \dots, N$: $u_n = \sum_t p_{t,n}^0 y_t / \alpha_n^0$,

  $v_n = (2a + \sum_t p_{t,n}^0 (y_t - u_n)^2)/(2b + \alpha_n^0)$.

when v vanishes. Thus, the global maximum of F is finite and it is reached for strictly positive components of $\boldsymbol{v}$, whereas the degeneracy points of f are not even local maxima for F. In the case of independent Gaussian mixtures, it has been recently shown that the global maximizer of F is a strongly consistent estimator (Ciuperca et al., 2000).

- Substituting F for f allows us to maintain explicit reestimation equations in the classical EM scheme for Gaussian mixtures. The underlying reason is that the inverse gamma distribution $G(\boldsymbol{v})$ is *conjugate* for the complete-data distribution $f(\boldsymbol{y} \mid \boldsymbol{x}; \boldsymbol{u}, \boldsymbol{v}) \Pr(\boldsymbol{x})$. Contrarily to Hathaway's constrained formulation, our penalized version is as simple to derive and to implement as the original EM scheme. The resulting reestimation formula for each v_n is

$$v_n = \frac{2a + \sum_{t=1}^{T} \Pr(X_t = n \mid \boldsymbol{y}; \boldsymbol{\theta}^0)(y_t - u_n)^2}{2b + \sum_{t=1}^{T} \Pr(X_t = n \mid \boldsymbol{y}\ \boldsymbol{\theta}^0)},$$

while the other reestimation equations are unaltered.

The equations of the complete EM algorithm are given in Table 3. Note that the MMAP segmentation stage directly follows from (23) and forward-backward evaluation of quantity $p_{t,n}^0$.

7 Results

The unsupervised segmentation method described above was successfully tested on synthetic and real 2D and 3D images. In this section, we present a limited set of results in order to illustrate two points: the ability of the penalized approach to cope with the degeneracy of the likelihood and the performance of the method in real-size 3D data processing.

7.1 Likelihood degeneracy

The unsupervised segmentation method was applied to the 128×128 2D magnetic resonance image of the heart region[1] presented in Fig 1. The structures of interest are the ventricles whose approximate boundaries have been superimposed on the image. The model parameters were initialized in the following manner: the histogram of the original image was separated into N equal quantiles, and the values of $\boldsymbol{\theta}_{Y|X} = \{\boldsymbol{u}, \boldsymbol{v}\}$ were set to the empirical mean and variance of each quantile; all elements of $\boldsymbol{\mu}$ were set to $1/N$, and all elements of $\boldsymbol{\lambda}$ were set to the same value $\lambda_0 < 1/2$.

---

[1]Data courtesy of Dr. Alain Herment, INSERM U494, Hôpital de la Pitié-Salpêtrière, Paris, France

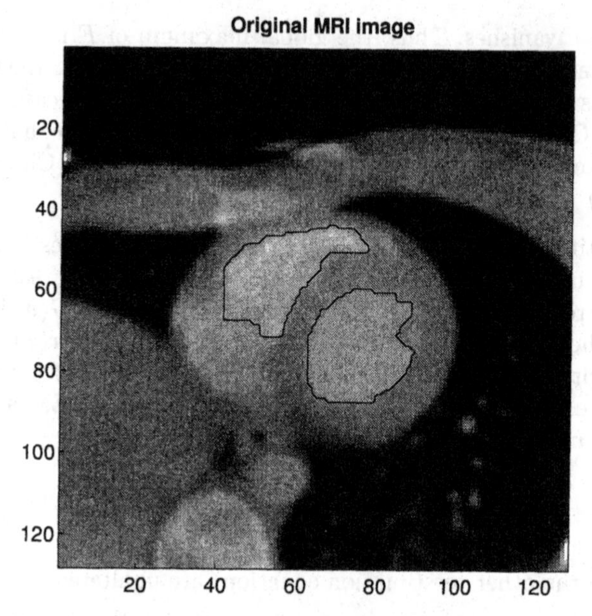

Figure 1. Original magnetic resonance image of the heart region, size 128 × 128. The structures of interest are the ventricles whose approximate boundaries have been superimposed on the image.

Without penalization of the likelihood, the method diverged after 12 iterations of the EM algorithm. The trajectories of the elements of $\boldsymbol{\theta}_X$ and $\boldsymbol{\theta}_{Y|X}$ are shown in Fig. 2. As can be observed, the divergence occurs when one of the components of the variance parameter $\boldsymbol{v}$ approaches zero. This result is consistent with the analysis of the degeneracy presented in Section 6.

The penalized method was applied to the magnetic resonance image with the same initial conditions. The parameters of the inverse gamma distribution were set to $a = 25$, $b = 1.01$. The trajectories of the estimated parameters and the resulting MMAP segmented image are shown in Figs. 3 and 4, respectively. It can be observed that convergence was reached after about 150 iterations and that even though several components of variance vector $\boldsymbol{v}$ were small, none of them approached zero closely enough for divergence to occur, thanks to the penalization term. More complete simulation results about likelihood degeneracy can be found in Ridolfi and Idier (1999).

Regarding implementation issues, it should be underlined that with $N =$ 15 labels, the TM induces a reduction of the computational cost of about one order of magnitude with respect to a standard parameterization of the Markov chains.

It should also be noted that for the particular example of Fig 1, the best results were obtained with $N = 13$ labels. In these conditions, even

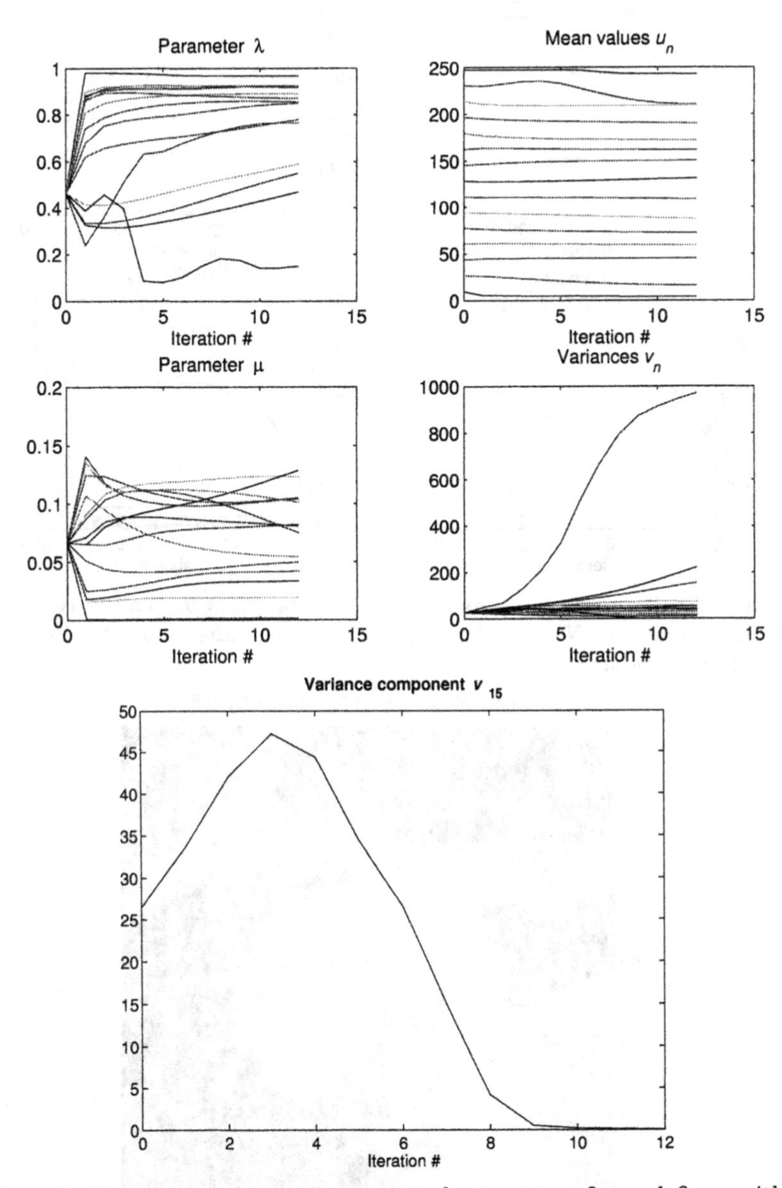

Figure 2. Trajectories of the components of parameters $\boldsymbol{\theta}_X$ and $\boldsymbol{\theta}_{Y|X}$ without penalization of the likelihood, $N = 15$. Divergence occurs after 12 iterations of the EM algorithm, as component 15 of variance vector $\boldsymbol{v}$ approaches zero.

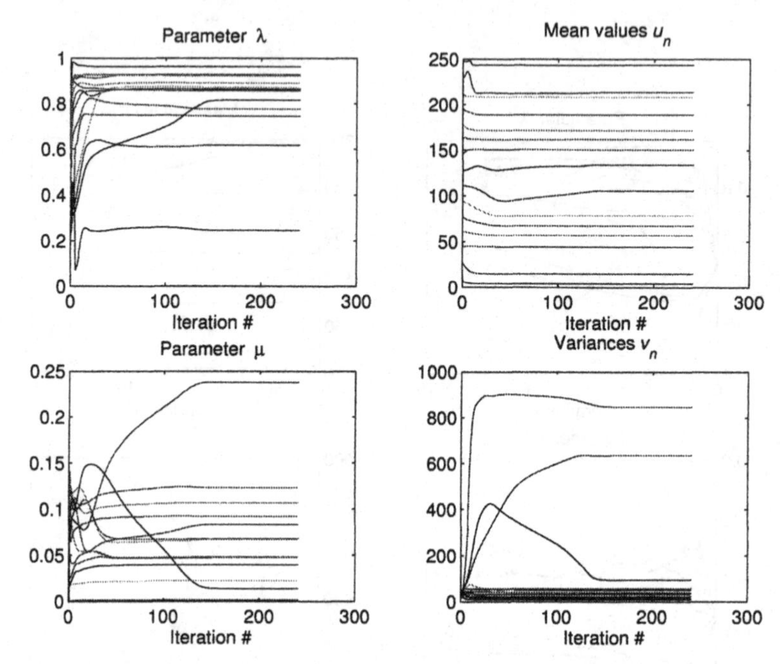

Figure 3. Trajectories of the components of parameters $\boldsymbol{\theta}_X$ and $\boldsymbol{\theta}_{Y|X}$ with a penalized likelihood, $N = 15$. Convergence takes place after about 150 iterations of the EM procedure.

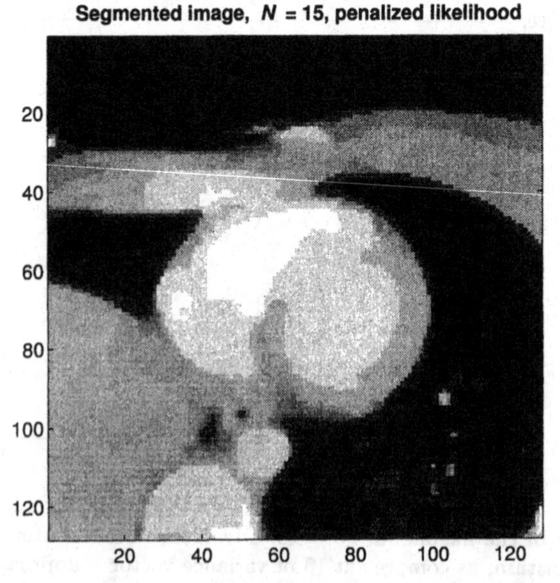

Figure 4. MMAP unsupervised segmentation result, $N = 15$. The parameters were obtained with a penalized likelihood.

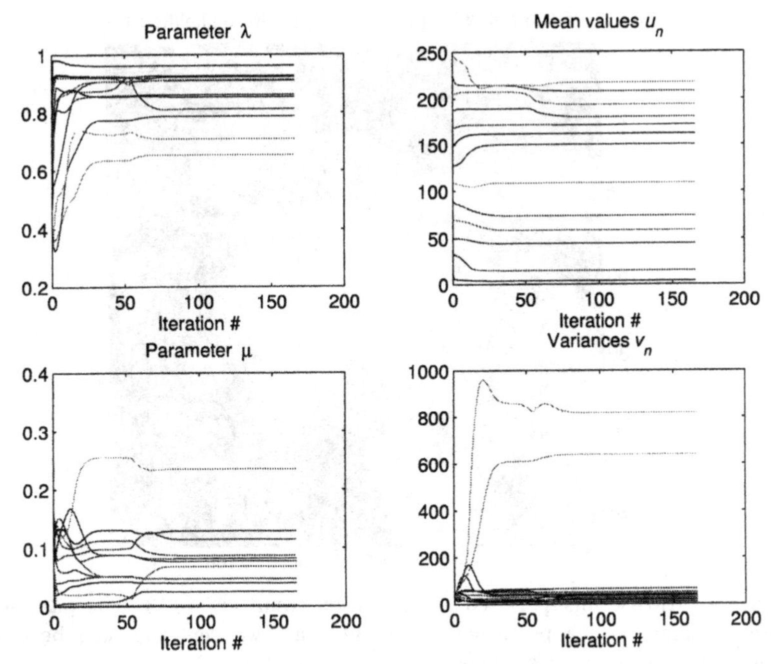

Figure 5. Trajectories of the components of parameters $\boldsymbol{\theta}_X$ and $\boldsymbol{\theta}_{Y|X}$ with a penalized likelihood, $N = 13$. Convergence takes place after less than 100 iterations of the EM procedure.

the non penalized method happens to converge. The results obtained with the penalized algorithm are presented in Figs 5 and 6. It can be observed that convergence takes place after less than 100 iterations and that in the MMAP segmented images, the two structures of interest can be clearly identified.

7.2 Segmentation of 3D data

The 3D data to be segmented were obtained with a power Doppler ultrasound echograph. This imaging modality is used for analysis of blood flow. The data set[2] presented here was collected on a synthetic blood vessel which exhibits a strongly stenosed area. It consisted of 80 frames of size 166×219, and segmentation was used to assess the dimension of the stenosis.

The number N of labels was set to four as such a number is sufficient to separate the various velocity regions present in the data, and the penalized version of the method was used.

[2]Data courtesy of Dr. Guy Cloutier, Institut de recherches cliniques de Montréal, Montreal, Quebec, Canada.

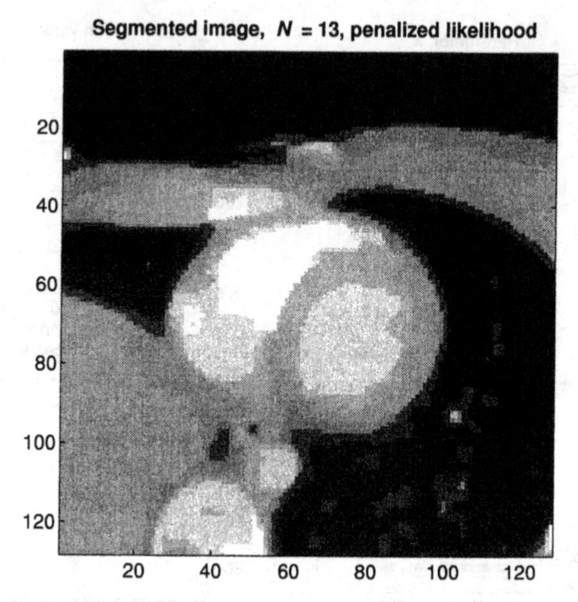

Figure 6. MMAP unsupervised segmentation result, $N = 13$. The parameters were obtained with a penalized likelihood. The two ventricles can be clearly identified in the segmented image.

Fig 7 shows a longitudinal slice of the original power Doppler data and of the segmented data. This representation is adopted because of the approximately cylindrical symmetry of the medium. The trajectories of the estimated parameters are presented in Fig 8. It can be observed that convergence occurs after a number of iterations that is much smaller than in the 2D case. This can be interpreted as the consequence of smaller number of labels and of the larger number of observations. The value of the stenosis diameter inferred from the segmented data was closer to the actual value than results provided by conventional methods of the field, thereby indicating a satisfactory behavior of our technique. It should be underlined that each iteration of the EM algorithm took approximately 100 seconds on a desktop Pentium II/300 computer, which shows that the proposed 3D method is usable even with a moderate computing power.

8 Conclusion

In this paper, we have presented a fully unsupervised method for segmenting 2D and 3D images, provided that the number N of levels is known. Following Devijver and Dekesel (1988), local dependencies were taken into account through a unilateral hidden Markov model with a conditionally Gaussian distribution of the observed pixels.

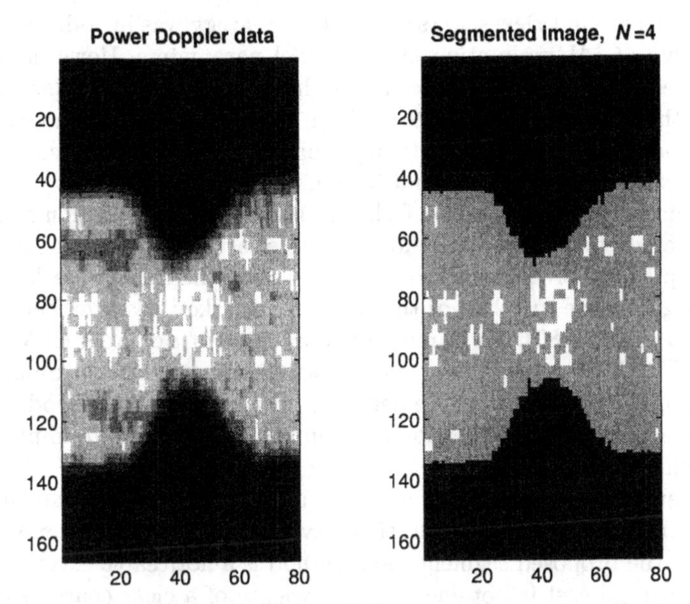

Figure 7. Longitudinal slice of 3D data. Left: original power Doppler data; right: MMAP segmented data, $N = 4$.

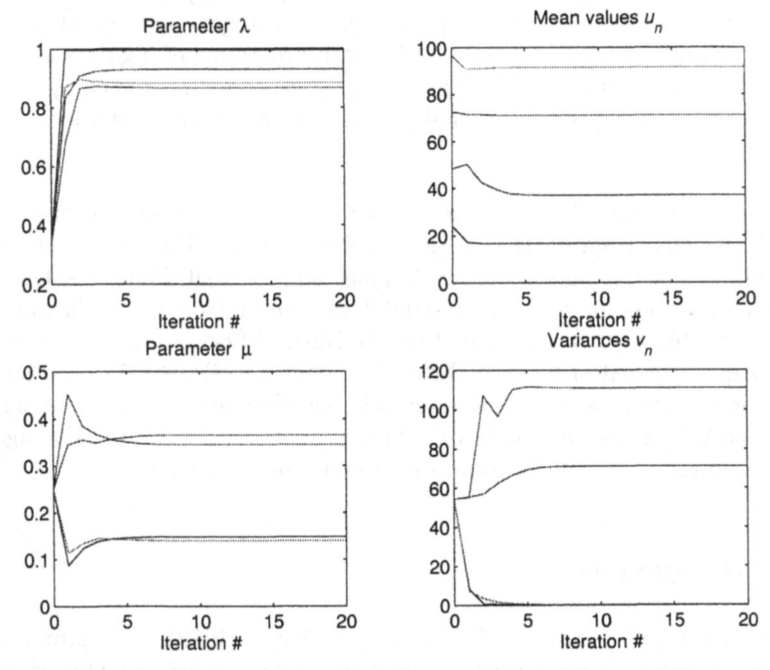

Figure 8. Trajectories of the components of parameters $\boldsymbol{\theta}_X$ and $\boldsymbol{\theta}_{Y|X}$ with a penalized likelihood, $N = 4$. Convergence takes place after less than 10 iterations of the EM procedure.

In Devijver and Dekesel (1988), an EM strategy was introduced in order to carry out ML estimation of the model parameters. However, because of its heavy computational cost and of hazardous behavior (Nádas, 1983), the authors finally preferred to perform joint estimation of model parameters and of the image, even though this technique presents controversial statistical properties (Little and Rubin, 1983).

Compared to Devijver and Dekesel (1988), our contribution makes the EM strategy truly practicable for parameter estimation. On the one hand, we adopted a more parsimonious description of the hidden Markov model. It is a generalized version of the telegraph Markov chain model found in Godfrey et al. (1980), whose number of parameters is of order $O(N)$ instead of $O(N^2)$. On the other hand, we introduced a penalized maximum likelihood approach that avoids the degeneracy of the usual likelihood function. Moreover, our penalized version is as simple to derive and to implement as the standard EM scheme.

Implementation of image processing methods based on Markov modeling usually requires heavy computations, even in supervised contexts. In this respect, the proposed segmentation method is a noticeable exception. This low numerical cost is obtained at the expense of a clear coarseness of the prior model, mostly due to its unilateral structure. It was also necessary to neglect diagonal interactions in the segmentation stage. As a consequence, the proposed method can be placed in the category of general purpose techniques best suited for automatic batch processing of big data sets. For specific types of images, more accurate segmentation results can probably be obtained using computationally more intensive Markov methods.

Acknowledgments: The authors wish to thank Dr. Alain Herment, INSERM U494, Hôpital de la Pitié-Salpêtrière, Paris, France and Dr. Guy Cloutier, Institut de recherches cliniques de Montréal, Montréal, Québec, Canada for providing the ultrasound data used in Section 7.2. Partial support for this work was provided by the Natural Sciences and Engineering Research Council of Canada (Research Grant #OGP0138417) and jointly by the ministère des Relations internationales du Québec (Cooperative Program 5.1.4, Project #7) and by the French ministère des Affaires Étrangères (Coopération scientifique franco-québécoise, projet I.1.2.1.7).

9 References

Baum, L.E., T. Petrie, G. Soules, and N. Weiss (1970). A maximization technique occuring in the statistical analysis of probabilistic functions of Markov chains. *The Annals of Mathematical Statistics 41*, 164–171.

Besag, J. (1974). Spatial interaction and the statistical analysis of lattice systems (with discussion). *Journal of the Royal Statistical Society. Series B. Methodological 36*, 192–236.

Besag, J. (1986). On the statistical analysis of dirty pictures (with discussion). *Journal of the Royal Statistical Society. Series B. Methodological 48*, 259–302.

Champagnat, F., J. Idier, and Y. Goussard (1998). Stationary Markov random fields on a finite rectangular lattice. *IEEE Transactions on Information Theory 44*, 2901–2916.

Ciuperca, G., A. Ridolfi, and J. Idier (2000). Penalized maximum likelihood estimator for normal mixtures. Technical report, Université Paris-Sud.

Dempster, A.P., N.M. Laird, and D.B. Rubin (1977). Maximum likelihood from incomplete data via the EM algorithm. *Journal of the Royal Statistical Society. Series B. Methodological 39*, 1–38.

Devijver, P.A. and M. Dekesel (1988). Champs aléatoires de Pickard et modélisation d'images digitales. *Traitement du Signal 5*, 131–150.

Godfrey, R., F. Muir, and F. Rocca (1980). Modeling seismic impedance with Markov chains. *Geophysics 45*, 1351–1372.

Guyon, X. (1992). *Champs aléatoires sur un réseau : modélisations, statistique et applications.* Traitement du Signal. Paris: Masson.

Hathaway, R. (1985). A constrained formulation of maximum-likelihood estimation for normal mixture distributions. *The Annals of Statistics 13*, 1.

Hathaway, R.J. (1986). A constrained EM algorithm for univariate normal mixtures. *Journal of Statistical Computation and Simulation 23*, 211–230.

Idier, J. and Y. Goussard (1999). Champs de Pickard 3D. Technical report, IGB/GPI-LSS.

Little, R.J.A. and D.B. Rubin (1983). On jointly estimating parameters and missing data by maximizing the complete-data likelihood. *The American Statistician 37*, 218–220.

Nádas, A. (1983). Hidden Markov chains, the forward-backward algorithm, and initial statistics. *IEEE Transactions on Acoustics Speech and Signal Processing 31*, 504–506.

Pickard, D.K. (1977). A curious binary lattice process. *Journal of Applied Probability 14*, 717–731.

Pickard, D.K. (1980). Unilateral Markov fields. *Advances in Applied Probability 12*, 655–671.

Redner, R.A. and H.F. Walker (1984). Mixture densities, maximum likelihood and the EM algorithm. *SIAM Review 26*, 195–239.

Ridolfi, A. and J. Idier (1999). Penalized maximum likelihood estimation for univariate normal mixture distributions. In *Proc. GRETSI '99*, Vannes, 1999, pp. 259–262.

7

Estimation of Motion from Sequences of Images

Verena Gelpke
Hans R. Künsch

ABSTRACT The estimation of smooth velocity fields from sequences of images is of great interest in many domains in natural sciences such as meteorology and physical oceanography. We suggest a model, which is a discretization of the continuity equation. We assume absence of divergence. This property is preserved in our discretization. Because we deal with an errors-in-variables phenomenon, we use a penalized least squares method to estimate the displacement field. The penalty term includes a difference-based estimate of noise variance.

1 Introduction

1.1 The Problem

In many scientific applications one has to deal with data that vary in space and time. In this article a special subclass of such data is considered, where the differences between two consecutive times is mainly due to motion. We then determine this motion from a sequence of images. Motion is a very general description of a phenomenon, where some objects change position. This can be a car going along an empty street or it can be the clouds in satellite images that are shown on TV to illustrate weather forecast. It should be pointed out that these two kinds of motion have different characters. In the first example, nothing except the car is moving. Hence the displacement vector field is very discontinuous on the boundary of the car. In the weather forecast images, motion is a very smooth phenomenon. In this paper we concentrate on the second kind of motion. We apply our model to successive images of measurements of total ozone on the northern hemisphere. Figure 1 shows two such images. The purpose of this work is to check whether the changes between consecutive images are due to motion and whether the motion determined from ozone measurements corresponds to the size and direction of winds.

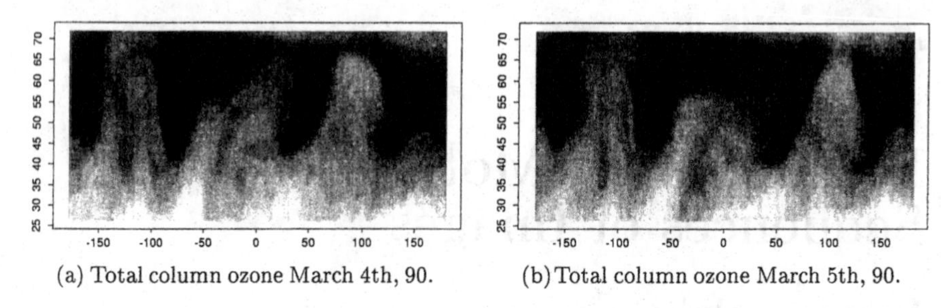

(a) Total column ozone March 4th, 90.

(b) Total column ozone March 5th, 90.

Figure 1. Ozone concentrations on two subsequent days.

1.2 Our Approach

Our model is a discretization of the continuity equation which is formulated in Section 3. In order to be able to deal with this equation, we have to make a few assumptions, namely absence of sources and sinks, pure advection and incompressibility of the medium. The discretization is discussed in Section 6. Due to the assumption of absence of divergence of the velocity field, the displacement field is supposed to be area preserving. This basic result is outlined in Section 3. This leads to an under-determined differential equation, for which a parametric class of solutions is given in Section 5. The parameter of the displacement field is estimated by a least squares method, which results in a nonlinear minimization problem. In order to obtain a consistent estimate of the displacement field, one has to add a negative penalty term. This is due to the fact that we have an errors-in-variables phenomenon, because we assume that the intensity values of the image are disturbed by an observation error. Naturally, the penalty term depends on the variance of this error process. As this noise variance is unknown, it has to be estimated first. According to ideas in nonparametric regression a difference-based estimate for the two-dimensional case is proposed. This estimate will be proposed in Section 8. In Section 9 an asymptotic theory for the estimator is outlined. We consider the behaviour of the estimate when the pixel-width is getting smaller and smaller.

1.3 Overview of the Literature

Recovery of motion in an image sequence has received considerable attention in different fields of research. In computer science, the 2-D velocities of brightness patterns in the image are often called the optical flow. There, several methods to determine optical flow have been developed. The most popular are the feature-based and the gradient-based methods. The former are related to texture segmentation, an observation window in the first image is selected and compared to windows of the second image. One decides that the pixels inside the window have moved to the place where the "textures" of source and target window are most similar. Such block matching algorithms frequently fail to produce good results, because they rely on the data only and do not model explicitly the phenomenon of motion and the laws that are associated to a motion field. Gradient-based methods determine optical flow from the change of grey values. One of the most famous gradient-based approaches was originally described in Horn and Schunck (1981). It relies on the fundamental assumption that the brightness f of a moving point is invariant between time t and $t + dt$. It can easily be shown (see, e.g., Winkler, 1995) that this assumption corresponds to the following equation

$$\frac{\partial f}{\partial x}\frac{dx}{dt} + \frac{\partial f}{\partial y}\frac{dy}{dt} = -\frac{\partial f}{\partial t} \tag{1}$$

where $(u, v) = (dx/dt, dy/dt)$ is the velocity vector. While equation (1) is postulated in these methods as an assumption, we will derive it in Section 3 from a physical law that is valid in certain circumstances. Hence, our method is appropriate only for those images where we know or believe that these circumstances are sufficiently fulfilled. Clearly, velocity discontinuities as well as intensity discontinuities (on sharp edges, or highly textured regions) are not allowed, because for such images (1) does not hold. In addition to (1) Horn and Schunck use a smoothness condition for the velocity field. Usually the square of the magnitude of the gradient of the velocity $|\nabla u|^2 + |\nabla v|^2$ is required to be small. We show in Section 3 that the above equation (1) is equivalent to a continuity equation if we assume that the field of velocities is divergence free, i.e., $\partial u/\partial x + \partial v/\partial y = 0$. Our method uses this constraint.

These first steps in feature-based and gradient-based methods are compared in, e.g., Aggarwal and Nandhakumar (1988). Some improvements have been made by Nagel and Enkelmann (1986), who use an oriented smoothness constraint which prevents smoothing of the velocity field in directions where significant variations of grey values are detected. Konrad and Dubois (1993) formulate a stochastic approach with Markov random fields to deal with discontinuities in the motion field. Heitz and Bouthemy (1993) suggested the use of several complementary constraints with the aim to preserve motion discontinuities and to process occlusions. Their theoretical and computational framework is also based on Bayesian estimation theory and Markov random fields.

The method of Horn and Schunck (1981) has been applied and used in other scientific domains. Many meteorological satellite images do not contain sharp edges or discontinuities and the method seems to be quite satisfactory. Bannehr et al. (1995, 1996) applied it on water vapour picture and on sea surface temperature images. Schnörr (1994) showed that the smoothness constraint of Horn and Schunck can be rewritten as a sum of divergence, deformation and vorticity. This fact has been used by Bannehr et al. (1997) to give different weights to these three kinematic fields . They gave the highest weight to the divergence, as we do in some sense.

In geophysical sciences there exist also two categories of methods to infer velocity fields from sequences of images. One method called *maximum cross correlation* (MCC) follows features and the other uses the conservation equation (of the temperature), which is similar to equation (1). These methods have been employed to obtain information about wind fields in the troposphere, sea surface velocities (Tokmakian et al., 1990, Kelly and Strub, 1992, e.g.,) and ice motion from satellite images. The feature-based MCC method was originally developed by Leese et al. (1971) to determine cloud motion from geosynchronous satellite pictures. They maximize the cross-correlation between blocks. The basic assumption for this method is shape invariance of the pattern under translation. Rotational motion of the pattern, however can also occur. Pattern matching techniques that also account for rotational motion are computationally very expensive. Wahl and Simpson (1990) used idealized models of uniform advection, horizontal diffusion and surface heating to show that such processes degrade the correlations between the image pairs. The second method is described in detail in Kelly (1989). It is similar to our approach in the sense that it is a discretization of a version of the differential equation (1). The velocity field is a parametric function and a penalization for the absolute value of the divergence of the velocity field is added.

A more statistical approach to estimate motion in meteorology can be found in Brillinger (1997). He suggests two methods for the velocity estimation of a moving meteorological phenomenon as it appears in the global geopotential 500mb height field, which is affected by disturbances. The first method is a Fourier-based parametric analysis, where periodicities are handled directly, but it does not give very precise estimates. The second approach is based on generalized additive models.

Further related work has been done by Berman et al. (1994). They discuss cross-covariance based and Fourier based techniques to estimate band-to-band misregistration in multi spectral imagery. This is similar to estimating a constant noninteger valued lag (usually smaller than the width of a pixel) between two time series. They show that both of these methods give biased estimators. In the first case, because of the interpolation procedures and in the second case because of the presence of aliasing in the data. They further present a method that accounts for aliasing in the data. They extend their method to two-dimensional problems and apply it to multispectral images. Another method in the domain of estimating band-to-band misregistration has been developed by Carroll et al. (1994). It has some similarities with our work. They suggest two methods for estimating the lag or amount of misregistration in two noisy signals, where the recorded data are obtained by pixel averaging. They use interpolation methods to accommodate the effect of pixel averaging. The least squares method is applied to the differences between the recorded signal in one band and an estimate of that signal from the other band which is disturbed by noise. Thus they have an errors-in-variables phenomenon which is solved by introducing a penalty term. The penalty term depends on the covariance function of the error process and must be estimated separately. The assumption of averaged images (Section 6) and the penalized least squares method with an estimated noise variance (Section 8) are used here too. Our lag however is nonconstant, which is an indispensable generalization for our applications. This will be a main part in this work. A class of nonconstant displacement functions that are physically sensible will be determined in Section 5. The second method of Carroll et al. (1994) consists in the maximization of the cross-covariance function of the observed signal and its interpolated estimate. It is computationally simpler, but the bias converges more slowly to zero than the bias of the penalized least squares.

In the domain of image restoration, the problem of estimating the dislocation of an object is also considered. Amit et al. (1991) restore degraded images that are assumed to be formed from a template by a composition of some continuous mappings plus additive noise. The restoration consists then of sampling from the posterior distribution. Their method has been successful if the location and orientation of the object was given. That is to say, they need a method that first generates an estimate of that dislocation. But typically only some objects in their images move. So this is the discontinuous type of motion where our method will not be very useful.

2 Data

In the past years there has been considerable interest in the decline in stratospheric ozone related to human activity. Substantial ozone decreases in the spring season over Antarctica are documented in Farman et al. (1985) and decreases in the mid latitudes have been reported by many people as well (e.g., Bojkov et al., 1990, Niu and Tiao, 1995). These changes in the amounts of stratospheric ozone are long term trends. But there is also substantial day to day variability believed to be caused mainly by meteorological phenomena, namely winds, in the upper troposphere that still influence the lower stratosphere.

TOMS (Total Ozone Mapping Spectrometer) ozone data are total column ozone measurements from a satellite which scans the earth in 13–14 orbits over the poles per day. The units of the TOMS column ozone measurements are Dobson Units (DU), where 1 DU $= 10^{-5}$ meters, also expressed as a milli-atmosphere-centimeter of ozone. Hence, a typical value of column ozone is of the order of 300 DU. The data are stored as averages on a grid. This means that we have a grid on the earth and in each cell of the grid the satellite makes many measurements, which then are averaged to a single datapoint for that particular cell. The grid is constructed such that the area of each cell is roughly constant. The earth is divided into 1 degree latitude zones, each of which is subdivided into a number of longitude cells. The number of cells is allowed to vary with latitude. Because the constant separation in degrees between orbits corresponds to a smaller distance in kilometers closer to the poles, a particular cell may be viewed from several adjacent orbits in the higher latitudes (at approx. 100 minutes time interval). In the available data files, higher latitude measurements are repeated such that there is a constant number of measurements per latitude zone and the grid has rectangular shape. For simplicity we ignore that we have repeated measurements.

3 The Continuity Equation

As mentioned in Section 1 we concentrate on motions associated with continuous displacement fields. Let us denote the concentration or temperature of a substance by $\rho(\mathbf{x}, t)$, the flow of the substance by $\mathbf{q}(\mathbf{x}, t)$ and the production rate by $k(\mathbf{x}, t)$. We assume that the space variable $\mathbf{x}$ is in a subset $\mathbb{P}$ of the plane $\mathbb{R}^2$. The transition from the sphere to the plane will be discussed in Section 4. The temporal change of the concentration is given by a general form of the *continuity equation* (see, e.g., Holton, 1979)

$$\frac{\partial \rho}{\partial t} = -\operatorname{div}(\mathbf{q}(\mathbf{x}, t)) + k(\mathbf{x}, t) \tag{2}$$

This equation is very general in the sense that $k(\mathbf{x}, t)$ and the vector field $\mathbf{q}(\mathbf{x}, t)$ are arbitrary functions. In order to make progress, we have to make a number of additional assumptions. Although, as mentioned in Section 2, ozone is continuously built and destroyed by natural processes and human activity, we set the production rate equal to zero,

$$k(\mathbf{x}, t) = 0. \tag{3}$$

This is justified because the day to day variability is dominated by transport. Next we assume that the space-time variation is induced by pure advection. The flow can then be written as the product of the velocity $\mathbf{v}(\mathbf{x}, t)$ and the concentration $\rho(\mathbf{x}, t)$,

$$\mathbf{q}(\mathbf{x}, t) = \mathbf{v}(\mathbf{x}, t)\rho(\mathbf{x}, t). \tag{4}$$

Finally we assume incompressibility of the medium, that is

$$\operatorname{div}(\mathbf{v}(\mathbf{x}, t)) = 0. \tag{5}$$

A motivation for this assumption can be found in our application to the ozone. As mentioned in Section 2, we assume that the wind is at least partially responsible for the transport of the ozone. The geostrophic wind (see, e.g., Holton, 1979, p. 37), which approximates the actual wind field for large-scale motion is almost free of divergence. Under assumptions (3)–(5), the continuity equation (2) becomes

$$\frac{\partial \rho}{\partial t} = -\sum_{i=1}^{2} v_i \frac{\partial \rho}{\partial x_i}. \tag{6}$$

The displacement $\mathbf{h}$ induced by the velocity $\mathbf{v}$ is obtained from the ordinary differential equation

$$\begin{aligned} \frac{d}{dt}\mathbf{h}(\mathbf{x},t) &= \mathbf{v}(\mathbf{x}+\mathbf{h}(\mathbf{x},t),t) \\ \mathbf{h}(\mathbf{x},0) &= \mathbf{0}. \end{aligned} \tag{7}$$

We then can show the following theorem.

Theorem 1. (a) *The solution of equation* (6) *with initial condition* $\rho(\mathbf{x},0)$ *is given by*

$$\rho(\mathbf{x}+\mathbf{h}(\mathbf{x},t),t) = \rho(\mathbf{x},0). \tag{8}$$

(b) *Assumption* (5) *is equivalent to*

$$\det \frac{\partial}{\partial \mathbf{x}}(\mathbf{x}+\mathbf{h}(\mathbf{x},t)) = 1. \tag{9}$$

The first part of the theorem states that mass that was at $\mathbf{x}$ at time 0 will be at $\mathbf{x}+\mathbf{h}(\mathbf{x},t)$ at time t, and the second part of the theorem shows that divergence free motion is equivalent to an area preserving transformation. This equation will be used for the parametrization in Section 5.

Proof. (a) follows by differentiating both sides of (8) with respect to t.
(b) is proved as the theorem of Liouville in classical mechanics (Volume preservation of phase flow); see, e.g., Arnold (1989, p. 69). □

4 Projection from the Sphere to the Plane

The measurements in our application are taken on the earth, we assume a sphere. But the data are usually presented in the plane, an unfolded

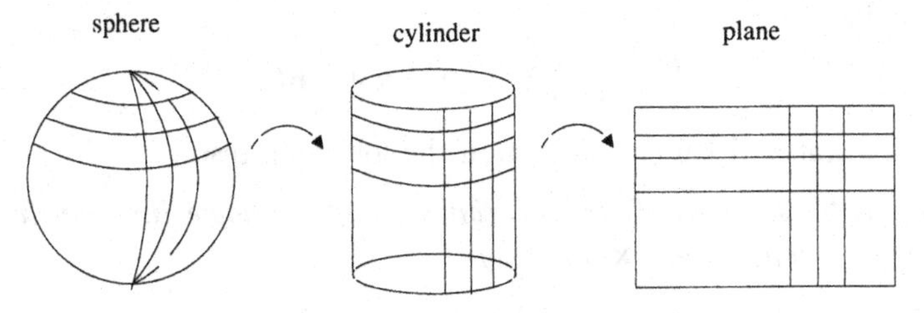

Figure 2. Projection from the sphere to the plane.

cylinder, see Figure 2. As we want to do the computations in the plane, we have to find a projection from the sphere to the cylinder that preserves the continuity equation (2) and the absence of divergence (5). Denote the coordinates on the sphere and on the cylinder by $\boldsymbol{\alpha} = (\alpha_1, \alpha_2) =$ (longitude, latitude) and $\mathbf{x} = (x_1, x_2) =$ (const $\cdot$ longitude, height) respectively. On the sphere call the concentration $\rho(\boldsymbol{\alpha})$ and the velocity field $\mathbf{u} = (u, v) = \mathbf{u}(\boldsymbol{\alpha})$, whereas $\tilde{\rho}(\mathbf{x})$ and $\tilde{\mathbf{u}} = (\tilde{u}, \tilde{v}) = \tilde{\mathbf{u}}(\mathbf{x})$ denote concentration and velocity on the cylinder. Any coordinate transformation $\mathbf{x} = \mathbf{x}(\alpha)$ of the form $x_1 = x_1(\alpha_1)$ and $x_2 = x_2(\alpha_2)$ determines the relation between ρ and $\tilde{\rho}$, $\mathbf{u}$ and $\tilde{\mathbf{u}}$, respectively, by invariance of mass and by invariance of motion. It can be shown that the continuity equation is automatically invariant. But the absence of divergence does not transfer automatically from $\mathbf{u}$ to $\tilde{\mathbf{u}}$. It does if

$$x_1 = R\alpha_1, \quad x_2 = R\sin(\alpha_2), \tag{10}$$

where R denotes the radius of the sphere, but does *not* if for example

$$x_1 = R\alpha_1, \quad x_2 = R\alpha_2.$$

5 Parametrization of the Displacement Field

The area preserving equation (9) is an underdetermined differential equation. It has infinitely many solutions. In order to estimate the displacement field $\mathbf{h}(\mathbf{x}, t)$ from data, we look for a parametric family of displacement fields satisfying (9). This family should be flexible. Ideally, we would have several families whose parameter dimension varies, and we would select in a final step the best fitting family. Furthermore, computation should be fast. The general form of displacements $\mathbf{h}(\mathbf{x}, t)$ satisfying (9) is given in Courant and Hilbert (1968, p. 49). Without loss of generality we set $t = 1$ and thus we suppress this variable in this section. For convenience we summarize the result of Courant and Hilbert (1968). Let

$$\mathbf{y} = \mathbf{g}(\mathbf{x}) = \mathbf{x} + \mathbf{h}(\mathbf{x}).$$

Here $\mathbf{y}$ is the position of a particle at time 1 which starts at $\mathbf{x}$ and moves in the velocity field $\mathbf{v}$. Courant and Hilbert use the coordinate transformation which associates to each $\mathbf{x}$ the mean of the starting point $\mathbf{x}$ and the end point $\mathbf{y}$, i.e.

$$\begin{pmatrix}\alpha\\ \beta\end{pmatrix} = \frac{1}{2}(\mathbf{x} + \mathbf{y}) = \mathbf{x} + \frac{1}{2}\mathbf{h}(\mathbf{x}).$$

From Courant and Hilbert we obtain the following result.

Theorem 2. *Let $\omega(\alpha, \beta)$ be an arbitrary differentiable function from $\mathbb{R}^2$ to $\mathbb{R}$. If the map $\binom{\alpha}{\beta} \longrightarrow \mathbf{x}$ given by*

$$\mathbf{x} = \begin{pmatrix}\alpha\\ \beta\end{pmatrix} + \begin{pmatrix}\partial\omega/\partial\beta\\ -\partial\omega/\partial\alpha\end{pmatrix} \tag{11}$$

has an inverse, then $\mathbf{h}(\mathbf{x}) = 2\big(\binom{\alpha}{\beta} - \mathbf{x}\big)$ *satisfies* $\det \partial(\mathbf{x} + \mathbf{h}(\mathbf{x}))/\partial \mathbf{x} = 1$. *Conversely, any* $\mathbf{h}(\mathbf{x})$ *can be written in this form.*

A constant translation in space, i.e., $\mathbf{h}(\mathbf{x}) \equiv (h_1, h_2)$, is obtained by choosing $\partial\omega/\partial\alpha = \frac{1}{2}h_2$, $\partial\omega/\partial\beta = -\frac{1}{2}h_1$. This means

$$\omega = \tfrac{1}{2}(\alpha h_2 - \beta h_1).$$

This example is too simple for our application, because in practice motion is not constant in space. It is however important, because it describes a smooth motion. In order to allow more complex displacements, we modify this function ω and consider

$$\omega(\alpha, \beta) = \tfrac{1}{2}(\alpha h_2 - \beta(h_1 + d(\alpha)) + c\beta^2). \tag{12}$$

The corresponding displacement field has the form

$$\mathbf{h}(\mathbf{x}) = \begin{pmatrix} h_1 + d(\alpha(\mathbf{x})) - 2c\beta(\mathbf{x}) \\ h_2 - \beta(\mathbf{x})d'(\alpha(\mathbf{x})) \end{pmatrix}. \tag{13}$$

The inversion of (11) must be done numerically, but it is quite simple, because it can be reduced to a one dimensional problem. If we use ω as in (12), (11) becomes

$$x_1 = \alpha - \frac{h_1}{2} - \frac{d(\alpha)}{2} + c\beta$$
$$x_2 = \beta - \frac{h_2}{2} + \frac{\beta}{2}d'(\alpha).$$

By eliminating β from these equations, we obtain the following equation for $\alpha(\mathbf{x})$

$$\left(1 + \frac{d'(\alpha)}{2}\right)\left(x_1 - \alpha + \frac{h_1}{2} + \frac{d(\alpha)}{2}\right) = c\left(\frac{h_2}{2} + x_2\right).$$

It can be solved by discretizing the axis and then applying a few steps of the Newton–Raphson algorithm (see, e.g., Press et al., 1992, p. 254).

In Figure 3 three examples of motion are shown. The accompanying displacement vectors $\mathbf{h}(\mathbf{x})$ are of the form (13), where $d(\alpha)$ is a linear combination of trigonometric functions,

$$d(\alpha) = \sum_{j=1}^{m} a_j \sin\left(j\frac{\alpha}{R}\right) + \sum_{j=1}^{m} b_j \cos\left(j\frac{\alpha}{R}\right) \tag{14}$$

where R is the radius of the earth.

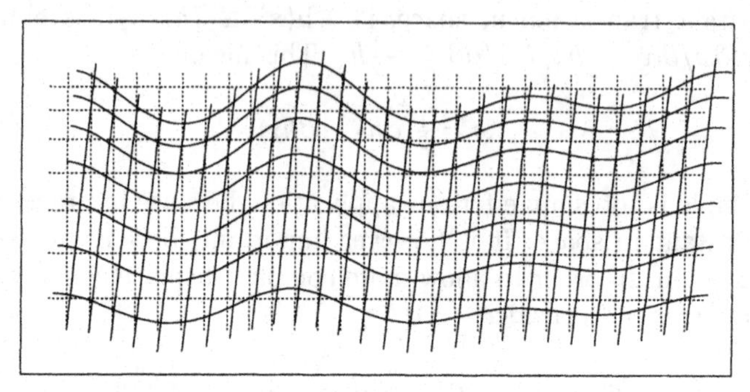

(a) Estimated motion between two ozone images

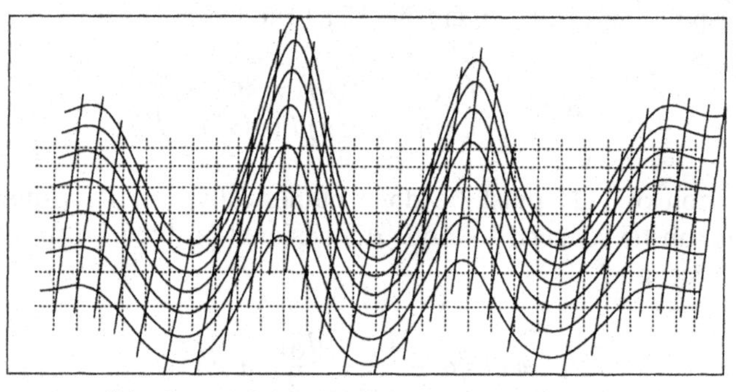

(b) Motion of the grid for an artificial example

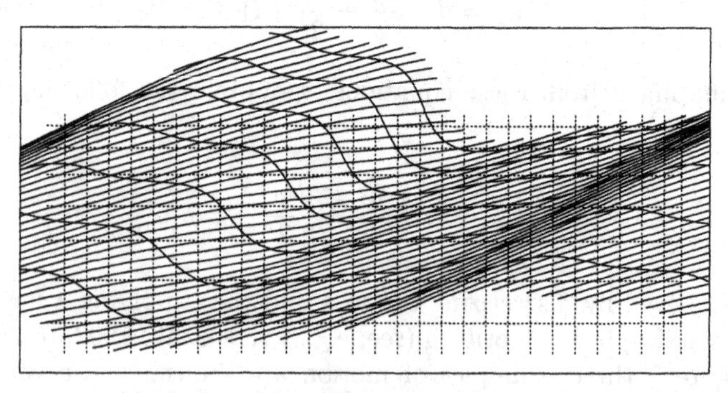

(c) Motion of the grid for an artificial example

Figure 3. Several examples of divergence free motion. The figures show the shape of the originally rectangular grid after the motion.

6 Discretization of the Continuity Equation

To get an appropriate model on which we can base the estimation of the motion, we have to find a good discretization in space of equation (8). Let us assume a grid on which the measurements are taken. Denote by $B_{\mathbf{l}}$ the pixel of the grid with centre $\mathbf{x_l}$ and volume $|B_{\mathbf{l}}|$. Here $\mathbf{x_l} \in \mathbb{P} \subset \mathbb{R}^2$, where $\mathbf{l} \in \mathbb{E} \subset \mathbb{Z}^2$, $\mathbf{l} = (l_1, l_2)$. We do not have measurements of the available concentration $\rho(\mathbf{x_k}, t)$ but mean concentrations over the whole pixel,

$$Z(\mathbf{x_k}, t) = \frac{1}{|B_{\mathbf{k}}|} \int_{B_{\mathbf{k}}} \rho(\mathbf{x}, t)\, d\mathbf{x}. \tag{15}$$

In some application it might make sense to model a weighted mean.

Because $\rho(\mathbf{x}, t)$ in (8) is linear in $\rho(\mathbf{x}, 0)$, the natural discretization of (8) has the form

$$Z(\mathbf{x_k}, t) = \sum_{\mathbf{l}} a(\mathbf{x_k}, \mathbf{x_l}, \mathbf{h}(\cdot, t)) Z(\mathbf{x_l}, 0) \tag{16}$$

where $|B_{\mathbf{k}}| a(\mathbf{x_k}, \mathbf{x_l}, \mathbf{h}(\cdot, t))/|B_{\mathbf{l}}|$ is the part of substance in pixel $B_{\mathbf{l}}$, that moves to pixel $B_{\mathbf{k}}$. To complete the model, we add to (16) a small error $\varepsilon(\mathbf{x_k}, t)$ due to physical processes, like diffusion, that do not follow (8),

$$Z(\mathbf{x_k}, t) = \sum_{\mathbf{l}} a(\mathbf{x_k}, \mathbf{x_l}, \mathbf{h}(\cdot, t)) Z(\mathbf{x_l}, 0) + \varepsilon(\mathbf{x_k}, t). \tag{17}$$

We assume the $\varepsilon(\mathbf{x_k}, t)$ to be i.i.d. The aim of this section is to determine the properties the coefficients $a(\mathbf{x_k}, \mathbf{x_l}, \mathbf{h}(\cdot, t))$ should satisfy and then propose a form for $a(\mathbf{x_k}, \mathbf{x_l}, \mathbf{h}(\cdot, t))$. One can easily show that equation (8) implies conservation of mass. It further follows that a constant field remains constant. These two properties should be preserved in the discretization (17).

Proposition 1. (a) *The total mass* $\sum_{\mathbf{k}} |B_{\mathbf{k}}| Z(\mathbf{x_k}, t)$ *is preserved by* (16) *iff*

$$|B_{\mathbf{l}}| = \sum_{\mathbf{k}} a(\mathbf{x_k}, \mathbf{x_l}, \mathbf{h}(\cdot, t)) |B_{\mathbf{k}}| \quad \forall \mathbf{l}. \tag{18}$$

(b) *Equation* (16) *holds for* $\rho(\mathbf{x}, 0) = \text{const}$ *iff*

$$1 = \sum_{\mathbf{l}} a(\mathbf{x_k}, \mathbf{x_l}, \mathbf{h}(\cdot, t)) \quad \forall \mathbf{k}. \tag{19}$$

Proof. (a) Let $\rho(\mathbf{x}, 0) = I_{B_{\mathbf{l}}}(\mathbf{x})$ for a fixed $\mathbf{l}$. Then we obtain $Z(\mathbf{x_k}, t) = a(\mathbf{x_k}, \mathbf{x_l}, \mathbf{h}(\cdot, t))$ and with (15) the total mass $\int \rho(\mathbf{x}, t) d\mathbf{x}$ is equivalent to $\sum_{\mathbf{k}} |B_{\mathbf{k}}| Z(\mathbf{x_k}, t) = \sum_{\mathbf{k}} |B_{\mathbf{k}}| a(\mathbf{x_k}, \mathbf{x_l}, \mathbf{h}(\cdot, t))$. This should be equal to $\int \rho(\mathbf{x}, 0) d\mathbf{x} = |B_{\mathbf{l}}|$. Conversely, the total mass at time t is

$$\begin{aligned} \sum_{\mathbf{k}} |B_{\mathbf{k}}| Z(\mathbf{x_k}, t) &= \sum_{\mathbf{k}} |B_{\mathbf{k}}| \sum_{\mathbf{l}} a(\mathbf{x_k}, \mathbf{x_l}, \mathbf{h}(\cdot, t)) Z(\mathbf{x_l}, 0) \\ &= \sum_{\mathbf{l}} \sum_{\mathbf{k}} |B_{\mathbf{k}}| a(\mathbf{x_k}, \mathbf{x_l}, \mathbf{h}(\cdot, t)) Z(\mathbf{x_l}, 0). \end{aligned}$$

Applying (18) one obtains $\sum_{\mathbf{l}} |B_{\mathbf{l}}| Z(\mathbf{x}_{\mathbf{l}}, 0)$. This is the total mass at time $t = 0$.

(b) Because of (8) it follows from $\rho(\mathbf{x}, 0) = \text{const}$ that $\rho(\mathbf{x}, t) = \text{const}$ $\forall t$, i.e., $Z(\mathbf{x}_{\mathbf{k}}, t) = \text{const}$ $\forall \mathbf{k}$, $\forall t$. Equation (16) leads then to the result. □

For the determination of $a(\mathbf{x}_{\mathbf{k}}, \mathbf{x}_{\mathbf{l}}, \mathbf{h}(\cdot, t))$ recall equation (9), which states that $\mathbf{g}(\mathbf{x}, t) = \mathbf{x} + \mathbf{h}(\mathbf{x}, t)$ is an area preserving transformation. Thus, $\forall B \subset \mathbb{R}^2$

$$|g(B, t)| = |B|,$$

where $g(B, t) = \{\mathbf{y} | \mathbf{y} = \mathbf{x} + \mathbf{h}(\mathbf{x}, t) \text{ for some } \mathbf{x} \in B\}$. Therefore,

$$|B_{\mathbf{l}}| = |g(B_{\mathbf{l}}, t)| = \sum_{\mathbf{k}} |g(B_{\mathbf{l}}, t) \cap B_{\mathbf{k}}|. \tag{20}$$

Consequently, if we choose

$$a(\mathbf{x}_{\mathbf{k}}, \mathbf{x}_{\mathbf{l}}, \mathbf{h}(\cdot, t)) = \frac{|g(B_{\mathbf{l}}, t) \cap B_{\mathbf{k}}}{|B_{\mathbf{k}}|} \tag{21}$$

(18) and (19) are satisfied; $a(\mathbf{x}_{\mathbf{k}}, \mathbf{x}_{\mathbf{l}}, \mathbf{h}(\cdot, t))$ represents the portion of area of pixel $B_{\mathbf{k}}$ that is covered by pixel $B_{\mathbf{l}}$ after the motion.

The coefficients $a(\mathbf{x}_{\mathbf{k}}, \mathbf{x}_{\mathbf{l}}, \mathbf{h}(\cdot, t))$ in the model equation (17) are defined in (21) as the area of the intersection of pixel $B_{\mathbf{l}}$, displaced by $\mathbf{h}$, with pixel $B_{\mathbf{k}}$, devided by $|B_{\mathbf{k}}|$. Because the displaced pixel does not necessarily have rectangular shape, it is not easy to calculate this area. We approximate the shifted pixel by a quadrangle, i.e. we displace the vertices of the original pixel $B_{\mathbf{l}}$ and connect them to a new quadrangle. We then collect the vertices of the polygon of intersection, by "walking" counter clockwise on the boundary of the two polygons. The algorithm is based on O'Rourke (1998) and is described in Gelpke (1999). In spite of the approximation by a quadrangle condition (19) is exactly and condition (18) is up to order $O(\mathbf{a}(\mathbf{l}))$ fulfilled.

Proposition 2. *The area of a displaced pixel* $B_{\mathbf{l}}$ *approximated by a polygon is* $B_{\mathbf{l}} \cdot (1 + O(|\mathbf{a}(\mathbf{l})|))$, *where* $\mathbf{a}(\mathbf{l}) = (a_1(\mathbf{l}), a_2(\mathbf{l}))$ *are the side lengths of the pixel* $B_{\mathbf{l}}$.

Proof. See Gelpke (1999). □

7 Penalized Least Squares Estimation

In the following we consider the discretization (17) with a displacement field $\mathbf{h}(\mathbf{x}, t) = \mathbf{h}_{\boldsymbol{\theta}}(\mathbf{x}, t)$ obtained from (13) where d is given in (14). The unknown parameters are $\boldsymbol{\theta} = (h_1, h_2, c, a_1, \ldots, a_m, b_1, \ldots, b_m)^T$. If we can observe $Z(\mathbf{x_k}, t)$ and $Z(\mathbf{x_l}, 0)$ it is natural to choose the unknown parameter $\boldsymbol{\theta}$ by minimizing the residual sum of squares

$$\widehat{\boldsymbol{\theta}} = \arg\min_{\boldsymbol{\theta}} \frac{1}{N} \sum_{\mathbf{k}} \{Z(\mathbf{x_k}, t) - \sum_{\mathbf{l}} a(\mathbf{x_k}, \mathbf{x_l}, \mathbf{h}_{\boldsymbol{\theta}}(\cdot, t)) Z(\mathbf{x_l}, 0)\}^2 \tag{22}$$

where N denotes the number of pixels in the grid. But in view of the complicated measurement process we should assume that we do not observe Z, but only a quantity perturbed by noise,

$$U(\mathbf{x_k}, t) = Z(\mathbf{x_k}, t) + \eta(\mathbf{x_k}, t). \tag{23}$$

In order to keep things manageable, we assume the observation errors η are i.i.d. Because we only observe U we have an errors-in-variables phenomenon. To take this problem into account we add a negative penalty term. To motivate and to choose this penalty term we observe that

$$\begin{aligned} U(\mathbf{x_k}, t) &- \sum_{\mathbf{l}} a(\mathbf{x_k}, \mathbf{x_l}, \mathbf{h}_{\boldsymbol{\theta}}(\cdot, t)) U(\mathbf{x_l}, 0) \\ &= \Big(Z(\mathbf{x_k}, t) - \sum_{\mathbf{l}} a(\mathbf{x_k}, \mathbf{x_l}, \mathbf{h}_{\boldsymbol{\theta}}(\cdot, t)) Z(\mathbf{x_l}, 0) \\ &\qquad + (\eta(\mathbf{x_k}, t) - \sum_{\mathbf{l}} a(\mathbf{x_k}, \mathbf{x_l}, \mathbf{h}_{\boldsymbol{\theta}}(\cdot, t)) \eta(\mathbf{x_l}, 0) \Big). \end{aligned}$$

The variance of the second term on the right hand side of this last expression is

$$\sigma_\eta^2 \Big(1 + \sum_{\mathbf{l}} a(\mathbf{x_k}, \mathbf{x_l}, \mathbf{h}_{\boldsymbol{\theta}}(\cdot, t))^2 \Big). \tag{24}$$

We thus estimate $\boldsymbol{\theta}$ by

$$\begin{aligned} \widehat{\boldsymbol{\theta}} = \arg\min_{\boldsymbol{\theta}} \frac{1}{N} \sum_{\mathbf{k}} \Big\{ \Big(U(\mathbf{x_k}, t) - \sum_{\mathbf{l}} a(\mathbf{x_k}, \mathbf{x_l}, \mathbf{h}_{\boldsymbol{\theta}}(\cdot, t)) U(\mathbf{x_l}, 0) \Big)^2 \\ - \sigma_\eta^2 \Big(1 + \sum_{\mathbf{l}} a(\mathbf{x_k}, \mathbf{x_l}, \mathbf{h}_{\boldsymbol{\theta}}(\cdot, t))^2 \Big) \Big\}. \end{aligned} \tag{25}$$

Since σ_η^2 is unknown, we replace it by an estimated $\hat{\sigma}_\eta^2$ (see Section 8).

8 Estimation of the Variance of the Observation Error

Several difference-based estimators of the noise variance in nonparametric regression have been proposed in the past few years. Gasser et al. (1986) suggested a simple estimator based on local linear fitting. Hall et al. (1990)

made a proposal which minimizes asymptotic variance. In Seifert et al. (1993) these and other methods are compared for finite sample size, moreover they give a more general form of the estimator suggested by Gasser et al. (1986). Hall et al. (1991) give some suggestions for the estimation of noise variance in two-dimensional signal processing. The main difference between the one- and higher-dimensional case is the very rich variety of configurations available to take differences. Also Herrmann et al. (1995) give an example for an estimator for the two-dimensional case. These difference-based estimators have better properties than the naive estimators, obtained by subtracting an appropriately smoothed curve from the observations. Some of the disadvantages of the naive estimator are mentioned in Hall et al. (1990). Our proposition is somehow a generalization of the estimator proposed by Gasser et al. (1986) to the two-dimensional case.

Let $U_\mathbf{j}$, $Z_\mathbf{j}$ and $\eta_\mathbf{j}$ be short notations for $U(\mathbf{x_j},\cdot)$, $Z(\mathbf{x_j},\cdot)$ and $\eta(\mathbf{x_j},\cdot)$ respectively, with $\mathbf{j} \in \mathbb{E} \subset \mathbb{Z}^2$. Then $U_\mathbf{j} = Z_\mathbf{j} + \eta_\mathbf{j}$, $\eta_\mathbf{j}$ i.i.d and $E(\eta_\mathbf{j}) = 0$. For given coefficients $d_\mathbf{ij}$ define the **i**-th pseudo-residual as

$$e_\mathbf{i} = \sum_\mathbf{j} d_\mathbf{ij} U_\mathbf{j} \tag{26}$$

The difference-based estimator for σ_η^2 is then given by

$$\hat{\sigma}_\eta^2 = \frac{1}{n} \sum_\mathbf{i} e_\mathbf{i}^2 \tag{27}$$

where n is the number of pseudo-residuals that have been constructed. Regarding the choice of the $d_\mathbf{ij}$ we consider the following,

$$e_\mathbf{i} = \sum_\mathbf{j} d_\mathbf{ij} \eta_\mathbf{j} + \sum_\mathbf{j} d_\mathbf{ij} Z_\mathbf{j}, \tag{28}$$

so if we could choose the $d_\mathbf{ij}$ such that

$$\sum_\mathbf{j} d_\mathbf{ij} Z_\mathbf{j} = 0 \tag{29}$$

and

$$\sum_\mathbf{j} d_\mathbf{ij}^2 = 1, \tag{30}$$

the pseudo-residuals would have the desirable properties

$$E(e_\mathbf{i}) = 0, \quad E(e_\mathbf{i}^2) = \sigma_\eta^2$$

which justifies the estimator (27). Because $Z_\mathbf{j}$ is assumed to be smooth, we choose $d_\mathbf{ij}$ such that (29) holds for low order polynomials,

$$\sum_\mathbf{j} d_\mathbf{ij} \mathbf{x_j}^\gamma = 0 \quad \forall \gamma \text{ with } |\gamma| \leq p, \tag{31}$$

where $\gamma = (\gamma_1, \gamma_2)$, $\mathbf{x}^\gamma = x_1^{\gamma_1} x_2^{\gamma_2}$, $|\gamma| = \gamma_1 + \gamma_2$ and p is the order of the polynomial. Now, defining $\mathcal{J}(\mathbf{i}) := \{\mathbf{j} \mid d_{\mathbf{ij}} \neq 0\}$ we observe that if $|\mathcal{J}(\mathbf{i})| = (p+1)(p+2)/2+1$ then the $d_{\mathbf{ij}}$ are up to sign uniquely determined by (31) and (30). If we choose $|\mathcal{J}(\mathbf{i})|$ larger than $(p+1)(p+2)/2+1$, we can use the remaining degrees of freedom to minimize the variance of the estimator. In this regard we can prove the following; see Gelpke (1999).

Proposition 3. *Set* $b_{\mathbf{jk}} = \sum_{\mathbf{i}} d_{\mathbf{ij}} d_{\mathbf{ik}} = b_{\mathbf{kj}}$. *If* (29) *is fulfilled*

$$\operatorname{Var}\left(\sum_{\mathbf{i}} e_{\mathbf{i}}^2\right) = 2\sum_{\mathbf{j},\mathbf{k}} b_{\mathbf{jk}}^2 \sigma_\eta^4 + \sum_{\mathbf{j}} b_{\mathbf{jj}}^2 (\operatorname{Var}(\eta_{\mathbf{j}}^2) - 2\sigma_\eta^4). \tag{32}$$

Example with $p = 1$. Assume Z to be a polynomial of order 1, so that (31) and (30) have four equations. Let us take the simplest configuration, namely $\mathbf{i}$ and its four nearest neighbours: $\mathcal{J}_{\mathbf{i}} = \{\mathbf{i}, \mathbf{i} \pm \mathbf{e}_1, \mathbf{i} \pm \mathbf{e}_2\}$ where $\mathbf{e}_1 = (1,0)'$ and $\mathbf{e}_2 = (0,1)'$ and let $\mathcal{D}(\mathbf{i}) = \{d_{\mathbf{i}0} = d_{\mathbf{ii}}, d_{\mathbf{i}1} = d_{\mathbf{i}(\mathbf{i}+\mathbf{e}_1)}, d_{\mathbf{i}2} = d_{\mathbf{i}(\mathbf{i}+\mathbf{e}_2)}, d_{\mathbf{i}3} = d_{\mathbf{i}(\mathbf{i}-\mathbf{e}_1)}, d_{\mathbf{i}4} = d_{\mathbf{i}(\mathbf{i}-\mathbf{e}_2)}\}$ the corresponding coefficients. If the distances between horizontal and vertical neighbours respectively are the same, that is $|\mathbf{x}_{(\mathbf{i}+\mathbf{e}_1)} - \mathbf{x}_{\mathbf{i}}| = |\mathbf{x}_{(\mathbf{i}-\mathbf{e}_1)} - \mathbf{x}_{\mathbf{i}}|$ and $|\mathbf{x}_{(\mathbf{i}+\mathbf{e}_2)} - \mathbf{x}_{\mathbf{i}}| = |\mathbf{x}_{(\mathbf{i}-\mathbf{e}_2)} - \mathbf{x}_{\mathbf{i}}|$, then the conditions above lead to the following class of solutions:

$$\begin{aligned} -2/\sqrt{5} \leq d_{\mathbf{i}0} &\leq 2/\sqrt{5} \\ d_{\mathbf{i}1} = d_{\mathbf{i}3} &= -\frac{d_{\mathbf{i}0}}{4} \mp \frac{\sqrt{4 - 5d_{\mathbf{i}0}^2}}{4} \\ d_{\mathbf{i}2} = d_{\mathbf{i}4} &= -\frac{d_{\mathbf{i}0}}{4} \pm \frac{\sqrt{4 - 5d_{\mathbf{i}0}^2}}{4}. \end{aligned}$$

Minimizing (32) is complicated, because of the unknown σ_η^2 and $Var(\eta^2)$. A simple approach is to assume the error process to be Gaussian. Then, the minimization problem for this example reduces to finding a set of $\{d_{\mathbf{i}0}\}$ that minimizes $\sum_{\mathbf{j},\mathbf{k}} b_{\mathbf{jk}}^2$. We assume that we have a periodic continuation at the boundary, i.e., we do not have boundary effects. Further, without loss of generality, we assume that the $d_{\mathbf{i}0}$ are all positive. To minimize $\sum_{\mathbf{j},\mathbf{k}} b_{\mathbf{jk}}^2$ is rather complicated for a general set of $\{d_{\mathbf{i}k}\}$. Thus we just look at some plausible examples. One simple possibility is to choose $d_{\mathbf{i}r} =: \delta_r$ independent of $\mathbf{i}$ Then, the $b_{\mathbf{jk}}$ depend only on the difference $\mathbf{j} - \mathbf{k}$. With some calculation one can show that

$$\sum_{\mathbf{j}} \sum_{\mathbf{k}} b_{\mathbf{jk}}^2 = n \frac{1}{16}(-29\delta_0^4 + 12\delta_0^2 + 36).$$

This is minimal, if $\delta_0 = \sqrt{\frac{6}{29}}$.

A second simple case is the one where $d_{\mathbf{i}0} =: \delta_0$ is independent of $\mathbf{i}$, but the signs of $d_{\mathbf{i}1}$ and $d_{\mathbf{i}2}$ are distributed according to a chessboard, i.e. if

$i_1 + i_2$ even then

$$d_{\mathbf{i}1} = d_{\mathbf{i}3} = -\frac{d_{\mathbf{i}0}}{4} - \frac{\sqrt{4 - 5d_{\mathbf{i}0}^2}}{4} =: \delta_1$$
$$d_{\mathbf{i}2} = d_{\mathbf{i}4} = -\frac{d_{\mathbf{i}0}}{4} + \frac{\sqrt{4 - 5d_{\mathbf{i}0}^2}}{4} =: \delta_2$$

and if $i_1 + i_2$ odd, then

$$d_{\mathbf{i}1} = d_{\mathbf{i}3} = \delta_2$$
$$d_{\mathbf{i}2} = d_{\mathbf{i}4} = \delta_1.$$

Then

$$\sum_{\mathbf{l}} \sum_{\mathbf{k}} b_{\mathbf{jk}}^2 = n\frac{1}{16}(51\delta_0^4 - 52\delta_0^2 + 36). \tag{33}$$

This is minimal for $\delta_0 = \sqrt{\frac{26}{51}}$. The value of $\sum_{\mathbf{l}} \sum_{\mathbf{k}} b_{\mathbf{jk}}^2$ is $n \cdot 2.33$ in the first case and $n \cdot 1.42$ in the second example. Thus the second choice of $d_{\mathbf{i}r}$ is preferable.

We applied this estimator in the ozone example. In Hall et al. (1991) the same configuration as ours with nonzero weights at a pixel and its 4 nearest neighbours is considered. But as in their work (31) is not required, they obtain weights that are not symmetric about the centre, but the largest weight is assigned to one of the extremities of the cross.

9 Asymptotics

We want to examine the asymptotic properties of our estimation of motion. As a first step we study the asymptotic behaviour of the criterion which is minimized by the estimator $\widehat{\boldsymbol{\theta}}$. This result indicates the intuitive condition under which the estimator will be well determined and consistent. The proof of consistency itself will not be given here, because it is rather theoretical in the sense that one has to justify the interchange of taking the argmin and letting n tend to infinity in the minimization criterion. The details of this proof can be found in Gelpke (1999). As a last step one could investigate the rate of convergence. This is also outlined in Gelpke (1999).

First we have to choose a sensible type of asymptotics. To keep notation simple, assume the observations are on a regular rectangular grid. One can either consider enlarging the image by adding new pixels on the boundary, or one can keep the frame of the image fixed and let the size of the pixels decrease (and the number of pixels increase), compare Cressie (1993, Section 5.8). The second kind seems more appropriate, because the process we model is continuous and can be expressed via a differential equation. If we let the pixel size decrease, we get closer and closer to that continuous process. If we add new pixels, we may get a better estimate of the motion, but we are always far away from a continuous process.

9.1 Notation and Assumptions

Let $\mathbb{P}$ be a fixed rectangle in $\mathbb{R}^2$. We consider a sequence of partitions of $\mathbb{P}$ into a finer and finer grid of pixels as follows. The centers of pixels for the n-th partition are $\mathbf{x}_{n,\mathbf{k}}$, $\mathbf{k} \in \{1, \ldots, N_1(n)\} \times \{1, \ldots, N_2(n)\} = \mathbb{E}_n$. The pixels itself are denoted by $B_{n,\mathbf{k}}$ and the side length of pixels are $\mathbf{a}_n = (a_{1n}, a_{2n})$. We assume that $\lim_{n\to\infty} \frac{N_2(n)}{N_1(n)}$ exists and is in $(0, \infty)$ (e.g., $N_i(n) = 2^n N_i(0)$). The total number of pixels is $N_n = N_1(n) \cdot N_2(n)$. The initial distribution $\rho(\mathbf{x}, 0)$ is assumed to be continuously differentiable. We have observations at two fixed times which are taken to be 0 and 1 without loss of generality, so

$$U_{n,\mathbf{k}}(t) = Z_{n,\mathbf{k}}(t) + \eta_{n,\mathbf{k}}(t) \qquad (t = 0, 1) \tag{34}$$

$$Z_{n,\mathbf{k}}(0) = \frac{1}{|B_{n,\mathbf{k}}|} \int_{B_{n,\mathbf{k}}} \rho(\mathbf{x}, 0)\, d\mathbf{x} \tag{35}$$

$$Z_{n,\mathbf{k}}(1) = \frac{1}{|B_{n,\mathbf{k}}|} \int_{B_{n,\mathbf{k}}} \rho(\mathbf{x}, 1)\, d\mathbf{x} + \varepsilon_{n,\mathbf{k}} \tag{36}$$

$$\rho(\mathbf{x}, 1) = \rho(g_{\boldsymbol{\theta}_0}^{-1}(\mathbf{x}), 0). \tag{37}$$

We assume that for each n, $(\varepsilon_{n,\mathbf{k}})$, $(\eta_{n,\mathbf{k}}(0))$ and $(\eta_{n,\mathbf{k}}(1))$ are three independent vectors with independent components and

$$E(\varepsilon_{n,\mathbf{k}}) = E(\eta_{n,\mathbf{k}}(0)) = E(\eta_{n,\mathbf{k}}(1)) = 0,$$
$$E(\varepsilon_{n,\mathbf{k}}^2) = \sigma_\varepsilon^2, \quad E(\eta_{n,\mathbf{k}}(0)^2) = E(\eta_{n,\mathbf{k}}(1)^2) = \sigma_\eta^2,$$
$$\sup_{n,\mathbf{k}} E(\varepsilon_{n,\mathbf{k}}^4) < \infty, \quad \sup_{n,\mathbf{k}} E(\eta_{n,\mathbf{k}}(0)^4) < \infty, \quad \sup_{n,\mathbf{k}} E(\eta_{n,\mathbf{k}}(1)^4) < \infty.$$

Finally, for $\mathbf{g}_{\boldsymbol{\theta}}$ we assume that for each $\boldsymbol{\theta} \in \boldsymbol{\Theta} \subset \mathbb{R}^p$, $\mathbf{g}_{\boldsymbol{\theta}}$ is a differentiable, area preserving and invertible map: $\mathbb{P} \to \mathbb{P}$. Moreover $\mathbf{g}_{\boldsymbol{\theta}}(\mathbf{x})$ should be Lipschitz continuous in $\boldsymbol{\theta}$, uniformly in $\mathbf{x}$. With the notation used before $\mathbf{g}_{\boldsymbol{\theta}}(\mathbf{x}) = \mathbf{x} + \mathbf{h}_{\boldsymbol{\theta}}(x)$.

As mentioned before, we estimate $\boldsymbol{\theta}$ by minimizing

$$S_n(\boldsymbol{\theta}) = \frac{1}{N_n}\Bigg\{\sum_{\mathbf{k}}\Big(U_{n,\mathbf{k}}(1) - \sum_{\mathbf{l}} b_n(\mathbf{k}, \mathbf{l}, \boldsymbol{\theta}) U_{n,\mathbf{l}}(0)\Big)^2 - \sigma_\eta^2 \sum_{\mathbf{k}}\Big(1 + \sum_{l} b_n(\mathbf{k}, \mathbf{l}, \boldsymbol{\theta})^2\Big)\Bigg\}.$$

Here $b_n(\mathbf{k}, \mathbf{l}, \boldsymbol{\theta})$ stands for $a(\mathbf{x_k}, \mathbf{x_l}, \mathbf{h}_{\boldsymbol{\theta}}(\cdot, t)) = |g_{\boldsymbol{\theta}}(B_{n,\mathbf{l}}) \cap B_{n,\mathbf{k}}|/|B_{n,\mathbf{k}}|$.

9.2 Main Result

The estimator $\widehat{\boldsymbol{\theta}}$ is determined by minimizing $S_n(\boldsymbol{\theta})$. Thus it is important to analyse the asymptotic behaviour of $S_n(\boldsymbol{\theta})$. To get a sensible estimator for $\boldsymbol{\theta}$, asymptotically the minimum of $S_n(\boldsymbol{\theta})$ should exist and should be equal to the value of $S_n(\boldsymbol{\theta})$ at the true value $\boldsymbol{\theta}_0$. In Theorem 3 one can see that this is fulfilled, if there is enough variation in the image.

Theorem 3. *Under the above conditions and if $\hat{\sigma}_\eta^2$ is a consistent estimator of σ_η^2, then for any fixed $\boldsymbol{\theta}$, $S_n(\boldsymbol{\theta}) - S_n(\boldsymbol{\theta}_0)$ converges in probability to*

$$\frac{1}{|\mathbb{P}|}\int_{\mathbb{P}} (\rho(\mathbf{g}_{\boldsymbol{\theta}_0}^{-1}(\mathbf{x}),0) - \rho(\mathbf{g}_{\boldsymbol{\theta}}^{-1}(\mathbf{x}),0))^2 \, d\mathbf{x}.$$

The integrand is greater or equal to zero. From this, we see that consistency of $\widehat{\boldsymbol{\theta}}$ can hold only if $\rho(\mathbf{x},0)$ varies sufficiently so that for all $\boldsymbol{\theta} \neq \boldsymbol{\theta}_0$, $\rho(\mathbf{g}_{\boldsymbol{\theta}}^{-1}(\mathbf{x}),1) \neq \rho(\mathbf{g}_{\boldsymbol{\theta}_0}^{-1}(\mathbf{x}),1)$ for some $\mathbf{x} \in \mathbb{P}$. From equation (8) one can see that a constant field remains constant, independent of the displacement field $\mathbf{h}(\mathbf{x},t)$. Thus it is clear that for a constant field an estimation of $\boldsymbol{\theta}$ is not possible, because the integrand in Theorem 3 is zero and thus $S_n(\boldsymbol{\theta})$ has asymptotically no minimum.

9.3 Proofs

Before we prove Theorem 3, we show some auxiliary results. First we investigate the deterministic error which is induced by the discretization of the continuity equation. The following proposition gives an upper bound for this systematic component of the error.

Proposition 4. *For any $\boldsymbol{\theta}$*

$$\left| \frac{1}{|B_{n,\mathbf{k}}|}\int_{B_{n,\mathbf{k}}} \rho(\mathbf{g}_{\boldsymbol{\theta}}^{-1}(\mathbf{x}),0)\, d\mathbf{x} - \sum_{\mathbf{l}} b_n(\mathbf{k},\mathbf{l},\boldsymbol{\theta}) \frac{1}{|B_{n,\mathbf{l}}|}\int_{B_{n,\mathbf{l}}} \rho(\mathbf{x},0)\, d\mathbf{x} \right| = O(|\mathbf{a}_n|)$$

Proof. By the mean value theorem $Z_{n,\mathbf{k}}(0) = \rho(\xi_{\mathbf{n},\mathbf{k}},0)$ for some value $\xi_{n,\mathbf{k}} \in B_{n,\mathbf{k}}$. On the other hand, because $g_{\boldsymbol{\theta}}$ is area preserving, we obtain by substitution

$$\begin{aligned} Z_{n,\mathbf{k}}(1) - \varepsilon_{n,\mathbf{k}} &= \frac{1}{|B_{n,\mathbf{k}}|}\int_{\mathbb{P}} \rho(\mathbf{g}_{\boldsymbol{\theta}}^{-1}(\mathbf{x}),0) I_{B_{n,\mathbf{k}}}(\mathbf{x})\, d\mathbf{x} \\ &= \frac{1}{|B_{n,\mathbf{k}}|}\int_{\mathbb{P}} \rho(\mathbf{y},0) I_{B_{n,\mathbf{k}}}(g_{\boldsymbol{\theta}}(\mathbf{y}))\, d\mathbf{y} \\ &= \sum_{\mathbf{l}} \frac{1}{|B_{n,\mathbf{k}}|}\int_{B_{n,\mathbf{l}}} \rho(\mathbf{y},0) I_{B_{n,\mathbf{k}}}(g_{\boldsymbol{\theta}}(\mathbf{y}))\, d\mathbf{y} \end{aligned}$$

$$= \sum_{\mathbf{l}} \frac{1}{|B_{n,\mathbf{k}}|} \int_{B_{n,\mathbf{l}}} I_{B_{n,\mathbf{k}}}(g_{\boldsymbol{\theta}}(\mathbf{y}))\, d\mathbf{y}\ Z_{n,\mathbf{l}}(0)$$
$$+ \sum_{\mathbf{l}} \frac{1}{|B_{n,\mathbf{k}}|} \int_{B_{n,\mathbf{l}}} (\rho(\mathbf{y},0) - \rho(\xi_{\mathbf{n},\mathbf{l}},0)) I_{B_{n,\mathbf{k}}}(g_{\boldsymbol{\theta}}(\mathbf{y}))\, d\mathbf{y}.$$

The first term on the right in the last expression is $\sum_{\mathbf{l}} b_n(\mathbf{k},\mathbf{l},\boldsymbol{\theta}) Z_{n,\mathbf{l}}(0)$ and the second term is bounded by

$$\sum_{\mathbf{l}} b_n(\mathbf{k},\mathbf{l},\boldsymbol{\theta}) \Big(\sup_{\mathbf{x}\in B_{n,\mathbf{k}}} \Big\{ \Big| \frac{\partial}{\partial x_1} \rho(\mathbf{x},0) \Big| \cdot a_{1n} \Big\} + \sup_{\mathbf{x}\in B_{n,\mathbf{k}}} \Big\{ \Big| \frac{\partial}{\partial x_2} \rho(\mathbf{x},0) \Big| \cdot a_{2n} \Big\} \Big)$$
$$\leq \mathrm{const}(a_{1n} + a_{n2}). \quad \square$$

Second we show that also $g_{\boldsymbol{\theta}}^{-1}$ is Lipschitz continous in $\boldsymbol{\theta}$.

Proposition 5. *$g_{\boldsymbol{\theta}}^{-1}(\mathbf{y})$ is Lipschitz continuous in $\boldsymbol{\theta}$.*

Proof. Let $\mathbf{y} = g_{\boldsymbol{\theta}}(\mathbf{x})$. Then

$$\begin{aligned} \|g_{\boldsymbol{\theta}}^{-1}(\mathbf{y}) - g_{\boldsymbol{\theta}'}^{-1}(\mathbf{y})\| &= \|\mathbf{x} - g_{\boldsymbol{\theta}'}^{-1}(g_{\boldsymbol{\theta}}(\mathbf{x}))\| \\ &= \|\mathbf{x} - g_{\boldsymbol{\theta}'}^{-1}(g_{\boldsymbol{\theta}}(\mathbf{x})) + g_{\boldsymbol{\theta}'}^{-1}(g_{\boldsymbol{\theta}'}(\mathbf{x})) - g_{\boldsymbol{\theta}'}^{-1}(g_{\boldsymbol{\theta}'}(\mathbf{x}))\| \\ &= \|g_{\boldsymbol{\theta}'}^{-1}(g_{\boldsymbol{\theta}'}(\mathbf{x})) - g_{\boldsymbol{\theta}'}^{-1}(g_{\boldsymbol{\theta}}(\mathbf{x}))\| < K\|g_{\boldsymbol{\theta}'}(\mathbf{x}) - g_{\boldsymbol{\theta}}(\mathbf{x})\| \\ &\leq KC\|\boldsymbol{\theta} - \boldsymbol{\theta}'\|, \end{aligned}$$

because $g_{\boldsymbol{\theta}}^{-1}(\mathbf{y})$ is Lipschitz continuous in $\mathbf{y}$ due to the inverse function theorem and because $g_{\boldsymbol{\theta}}(\mathbf{y})$ is Lipschitz continuous in $\boldsymbol{\theta}$. $\square$

Next we prove that an upper bound exists for the number of pixels that can put mass into a specific pixel after the displacement. This upper bound does not depend on n, the partition of the grid.

Proposition 6. *There exists a constant $M(\boldsymbol{\theta})$ such that for all $\mathbf{k}$ and n, the set $\{\mathbf{l} \in \mathbb{E}_n \mid b_n(\mathbf{k},\mathbf{l},\boldsymbol{\theta}) \neq 0\}$ contains at most $M(\boldsymbol{\theta})$ elements.*

Proof. If $b_n(\mathbf{k},\mathbf{l},\boldsymbol{\theta}) \neq 0$ and $b_n(\mathbf{k},\mathbf{l}',\boldsymbol{\theta}) \neq 0$ then, there exist $\mathbf{x}$, $\mathbf{x}' \in B_{n,\mathbf{k}}$ such that $g_{\boldsymbol{\theta}}^{-1}(\mathbf{x}) \in B_{n,\mathbf{l}}$ and $g_{\boldsymbol{\theta}}^{-1}(\mathbf{x}') \in B_{n,\mathbf{l}'}$. By the inverse function theorem, $g_{\boldsymbol{\theta}}^{-1}$ is differentiable and thus there is a constant $K(\boldsymbol{\theta})$ such that $\|g_{\boldsymbol{\theta}}^{-1}(\mathbf{x}) - g_{\boldsymbol{\theta}}^{-1}(\mathbf{x}')\| \leq K(\boldsymbol{\theta})\|\mathbf{a}_n\|$. Because $0 < c_1 \leq a_{1,n}/a_{2,n} \leq c_2 < \infty$, this implies the desired result. $\square$

Conversely to the last proposition, after the displacement a pixel can only intersect with a fixed number of pixels . Again, this number is independent of n.

Proposition 7. *There exists a constant $L = L(\boldsymbol{\theta})$ such that for all $\mathbf{l}$ and n, the set $\{\mathbf{k} \in \mathbb{E}_n \mid b_n(\mathbf{k},\mathbf{l},\boldsymbol{\theta}) \neq 0\}$ contains at most $L(\boldsymbol{\theta})$ elements.*

Proof. Analogous to the proof of Proposition 6, but $g_{\boldsymbol{\theta}}(\mathbf{x})$ is used instead of $g_{\boldsymbol{\theta}}^{-1}(\mathbf{x})$. □

Proof of Theorem 3. Since $0 \le b_n(\mathbf{k}, \mathbf{l}, \boldsymbol{\theta}) \le 1$, we have $\sum_{\mathbf{l}} b_n(\mathbf{k}, \mathbf{l}, \boldsymbol{\theta})^2 \le \sum_{\mathbf{l}} b_n(\mathbf{k}, \mathbf{l}, \boldsymbol{\theta}) = 1$. Therefore, $(\hat{\sigma}_\eta^2 - \sigma_\eta^2) \cdot \sum_{\mathbf{k}} (1 + \sum_{\mathbf{l}} b_n(\mathbf{k}, \mathbf{l}, \boldsymbol{\theta})^2)/N_n \xrightarrow{P} 0$ and we may replace $\hat{\sigma}_\eta^2$ by σ_η^2 in the proof. Next we compute $E_{\boldsymbol{\theta}_0}(S_n(\boldsymbol{\theta}) - S_n(\boldsymbol{\theta}_0))$ and $\mathrm{Var}_{\boldsymbol{\theta}_0}(S_n(\boldsymbol{\theta}))$ and then use the Chebyshev inequality to prove the result. We have

$$S_n(\boldsymbol{\theta}) = \frac{1}{N_n} \sum_{\mathbf{k}} \Big\{ \varepsilon_{n,\mathbf{k}} + \eta_{n,\mathbf{k}}(1) - \sum_{\mathbf{l}} b_n(\mathbf{k}, \mathbf{l}, \boldsymbol{\theta}) \eta_{n,\mathbf{l}}(0) + \frac{1}{|B_{n,\mathbf{k}}|} \int_{B_{n,\mathbf{k}}} \rho(\mathbf{x}, 1)\, d\mathbf{x} - \sum_{\mathbf{l}} b_n(\mathbf{k}, \mathbf{l}, \boldsymbol{\theta}) \frac{1}{|B_{n,\mathbf{l}}|} \int_{B_{n,\mathbf{l}}} \rho(\mathbf{x}, 0) d\mathbf{x} \Big\}^2 - \sigma_\eta^2 \frac{1}{N_n} \sum_{\mathbf{k}} \Big(1 + \sum_{\mathbf{l}} b_n(\mathbf{k}, \mathbf{l}, \boldsymbol{\theta})^2 \Big). \quad (38)$$

We use Proposition 4 and set

$$\sum_{\mathbf{l}} b_n(\mathbf{k}, \mathbf{l}, \boldsymbol{\theta}) \frac{1}{|B_{n,\mathbf{l}}|} \int_{B_{n,\mathbf{l}}} \rho(\mathbf{x}, 0)\, d\mathbf{x} = \frac{1}{|B_{n,\mathbf{k}}|} \int_{B_{n,\mathbf{k}}} \rho(\mathbf{g}_{\boldsymbol{\theta}}^{-1}(\mathbf{x}), 0)\, d\mathbf{x} + R_{n,\mathbf{k}} \quad (39)$$

where $R_{n,\mathbf{k}} = O(a_{1,n} + a_{2,n})$. Furthermore using $\rho(\mathbf{x}, 1) = \rho(\mathbf{g}_{\boldsymbol{\theta}_0}^{-1}(\mathbf{x}), 0)$, we have

$$E_{\boldsymbol{\theta}_0}(S_n(\boldsymbol{\theta})) = \sigma_\varepsilon^2 + \frac{1}{N_n} \sum_{\mathbf{k}} \Big\{ \frac{1}{|B_{n,\mathbf{k}}|} \Big(\int_{B_{n,\mathbf{k}}} \rho(\mathbf{g}_{\boldsymbol{\theta}_0}^{-1}(\mathbf{x}), 0)\, d\mathbf{x} - \int_{B_{n,\mathbf{k}}} \rho(\mathbf{g}_{\boldsymbol{\theta}}^{-1}(\mathbf{x}), 0)\, d\mathbf{x} \Big) + R_{n,\mathbf{k}} \Big\}^2.$$

Now, because ρ and $g_{\boldsymbol{\theta}}^{-1}$ are differentiable there exist constants L_1 and $L_2(\boldsymbol{\theta})$ for which the following inequalities hold

$$\begin{aligned} \Big| \frac{1}{|B_{n,\mathbf{k}}|} \int_{B_{n,\mathbf{k}}} \rho(\mathbf{g}_{\boldsymbol{\theta}}^{-1}(\mathbf{x}), 0)\, d\mathbf{x} - \rho(\mathbf{g}_{\boldsymbol{\theta}}^{-1}(\mathbf{x}_{\mathbf{k}}), 0) \Big| &\le \frac{1}{|B_{n,\mathbf{k}}|} \int_{B_{n,\mathbf{k}}} L_1 \| g_{\boldsymbol{\theta}}^{-1}(\mathbf{x}) - g_{\boldsymbol{\theta}}^{-1}(\mathbf{x}_{\mathbf{k}}) \|\, d\mathbf{x} \\ &\le \frac{1}{|B_{n,\mathbf{k}}|} \int_{B_{n,\mathbf{k}}} L_2(\boldsymbol{\theta}) \| \mathbf{x} - \mathbf{x}_{\mathbf{k}} \|\, d\mathbf{x} \\ &\le \mathrm{const}(a_{1n} + a_{2n}). \end{aligned}$$

Hence,

$$E_{\boldsymbol{\theta}_0}(S_n(\boldsymbol{\theta})) = \sigma_\varepsilon^2 + \frac{1}{N_n} \sum_{\mathbf{k}} \big(\rho(\mathbf{g}_{\boldsymbol{\theta}_0}^{-1}(\mathbf{x}_{\mathbf{k}}), 0) - \rho(\mathbf{g}_{\boldsymbol{\theta}}^{-1}(\mathbf{x}_{\mathbf{k}}), 0) + O(a_{1n} + a_{2n}) \big)^2.$$

The sum can be approximated by the Riemann integral. Therefore

$$E_{\boldsymbol{\theta}_0}(S_n(\boldsymbol{\theta}))-\sigma_\varepsilon^2 = \frac{1}{|\mathbb{P}|}\int_{\mathbb{P}}\{\rho(\mathbf{g}_{\boldsymbol{\theta}_0}^{-1}(\mathbf{y}),0)-\rho(\mathbf{g}_{\boldsymbol{\theta}}^{-1}(\mathbf{y}),0)\}^2\,d\mathbf{y}+O(|\mathbf{a_n}|). \quad (40)$$

Finally we get

$$E_{\boldsymbol{\theta}_0}(S_n(\boldsymbol{\theta}) - S_n(\boldsymbol{\theta}_0)) = \frac{1}{|\mathbb{P}|}\int_{\mathbb{P}}\{\rho(\mathbf{g}_{\boldsymbol{\theta}_0}^{-1}(\mathbf{y}),0) - \rho(\mathbf{g}_{\boldsymbol{\theta}}^{-1}(\mathbf{y}),0)\}^2\,d\mathbf{y} + O(|\mathbf{a_n}|). \quad (41)$$

The variance of $S_n(\theta)$ is

$$\mathrm{Var}_{\boldsymbol{\theta}_0}\left(\frac{1}{N_n}\sum_{\mathbf{k}}\{U_{n,\mathbf{k}}(1) - \sum_{\mathbf{l}} b_n(\mathbf{k},\mathbf{l},\boldsymbol{\theta})U_{n,\mathbf{l}}(0)\}^2\right) = \frac{1}{N_n^2}\sum_{\mathbf{k},\mathbf{k}'}\mathrm{Cov}(A_{\mathbf{k}}, A'_{\mathbf{k}}),$$

where $A_{\mathbf{k}} = \{U_{n,\mathbf{k}}(1) - \sum_{\mathbf{l}} b_n(\mathbf{k},\mathbf{l},\boldsymbol{\theta})U_{n,\mathbf{l}}(0)\}^2$; $A_{\mathbf{k}}$ contains $\varepsilon_{n,\mathbf{k}}$, $\eta_{n,\mathbf{k}}(1)$ and $\eta_{n,\mathbf{l}}(0)$ for $\mathbf{l}$ with $b_n(\mathbf{k},\mathbf{l},\boldsymbol{\theta}) \neq 0$. Hence $\mathrm{Cov}(A_{\mathbf{k}}, A_{\mathbf{k}'}) \neq 0$ only if $A_{\mathbf{k}}$ and $A_{\mathbf{k}'}$ have terms in common. So, either $\mathbf{k} = \mathbf{k}'$ or $\exists\mathbf{l}$ with $b_n(\mathbf{k},\mathbf{l},\boldsymbol{\theta}) \neq 0$ and $b_n(\mathbf{k}',\mathbf{l},\boldsymbol{\theta}) \neq 0$. From Proposition 7 we conclude that for any n and $\mathbf{k} \in \mathbb{E}_n$ there are at most a fixed number of indices $\mathbf{k}' \in \mathbb{E}_n$ such that $A_{\mathbf{k}}$ and $A_{\mathbf{k}'}$ are not independent. Thus it is sufficient to calculate the variance of $A_{\mathbf{k}}$ and use the Schwarz inequality to obtain an upper bound for $\mathrm{Cov}(A_{\mathbf{k}}, A_{\mathbf{k}'})$.

$$\begin{aligned}
\mathrm{Var}(A_{\mathbf{k}}) &= \mathrm{Var}\left(\left\{\varepsilon_{n,\mathbf{k}} + \eta_{n,\mathbf{k}}(1) - \sum_{\mathbf{l}} b_n(\mathbf{k},\mathbf{l},\boldsymbol{\theta})\eta_{n,\mathbf{l}}(0)\right\}^2\right)\\
&= \mathrm{Var}(\eta_{n,\mathbf{k}}(1)^2) + \mathrm{Var}(\varepsilon_{n,\mathbf{k}}^2) + \sum_{\mathbf{l}} b_n(\mathbf{k},\mathbf{l},\boldsymbol{\theta})^4\,\mathrm{Var}(\eta_{n,\mathbf{k}}(0)^2)\\
&\quad + 4\sigma_\varepsilon^2\sigma_\eta^2 + 4(\sigma_\varepsilon^2 + \sigma_\eta^2)\sum_{\mathbf{l}} b_n(\mathbf{k},\mathbf{l},\boldsymbol{\theta})^2\sigma_\eta^2\\
&\quad + \sum_{\mathbf{l}\neq\mathbf{l}'} b_n(\mathbf{k},\mathbf{l},\boldsymbol{\theta})^2 b_n(\mathbf{k},\mathbf{l}',\boldsymbol{\theta})^2\sigma_\eta^4.
\end{aligned}$$

With Proposition 6 and the fact that $0 \leq b_n(\mathbf{k},\mathbf{l},\boldsymbol{\theta}) \leq 1\ \forall n,\mathbf{l},\mathbf{k},\boldsymbol{\theta}$ this is bounded by a constant B, because $\varepsilon_{n,\mathbf{k}}$ and $\eta_{n,\mathbf{k}}(i)$ $i = 0,1$ are assumed to have uniformly bounded fourth moments. So

$$\frac{1}{N_n^2}\sum_{\mathbf{k},\mathbf{k}'}\mathrm{Cov}(A_{\mathbf{k}}, A_{\mathbf{k}'}) \leq \frac{1}{N_n^2}\sum_{\mathbf{k},\mathbf{k}'}\sqrt{Var(A_{\mathbf{k}})\,\mathrm{Var}(A_{\mathbf{k}'})} \leq \frac{1}{N_n^2}N_nL(\boldsymbol{\theta})B.$$

The theorem now follows with the Chebyshev inequality. □

10 Application to Ozone Images

The method has been applied on successive images (separated by 24 hours) of total ozone measurements. In Figure 4(a) one can see the thickness of

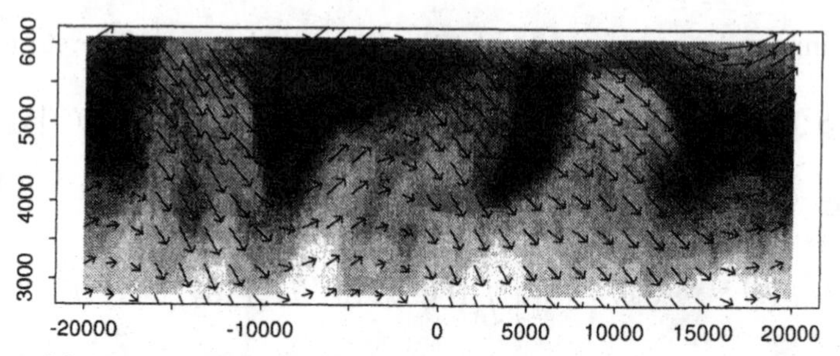

(a) Ozone measurements March 4th, 1990, estimated displacement field superimposed.

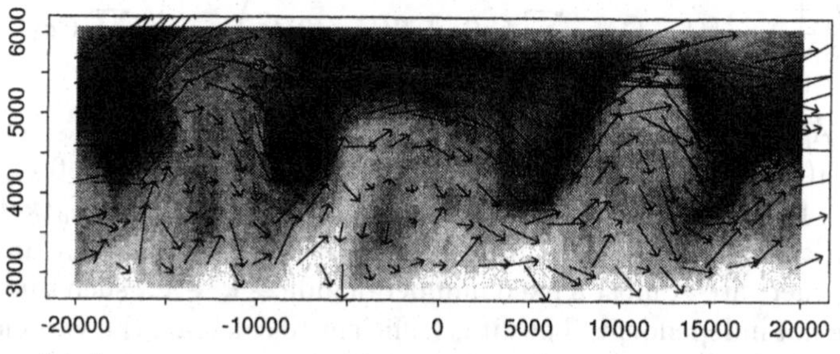

(b) Ozone measurements March 5th, 1990 with true wind field.

Figure 4. Estimated displacement field and true wind field.

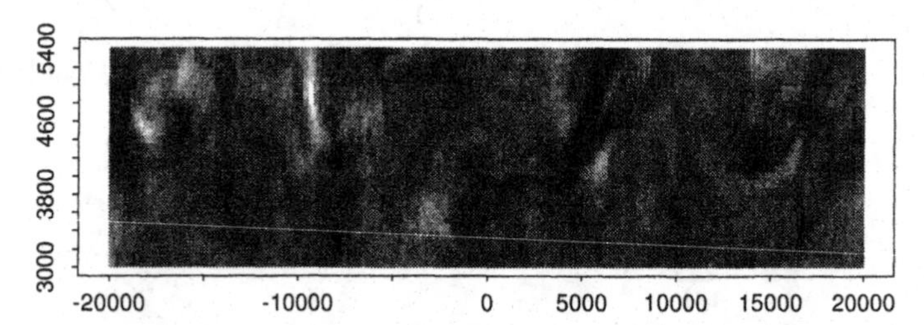

Figure 5. Residuals for the estimation of the fit in Figure 4(a).

the ozone layer on March 4th, 1990. The estimate of the displacement field is superimposed. The displacement field was parameterized according to (12) and (14) with $m = 3$. The estimate of σ_η^2 was calculated according to the second case of the example in Section 8 with $\delta_0 = \sqrt{26/51}$. For the March 4th $\hat{\sigma}_\eta^2 = 27.6$. In Figure 4(b) we show the thickness of the ozone layer of March 5th and superimpose the "true" wind field on a height of 70hPa, i.e., the field calculated from the big meteorological model of the European centre for medium-term weather forecast (ECMWF).

In Figure 5 one can see the residuals that correspond to the fit in Figure 4(a). There is still quite a few structure in the residuals. In Figure 6(a) and 6(b) one can see that the spatial correlation of the residuals disappears

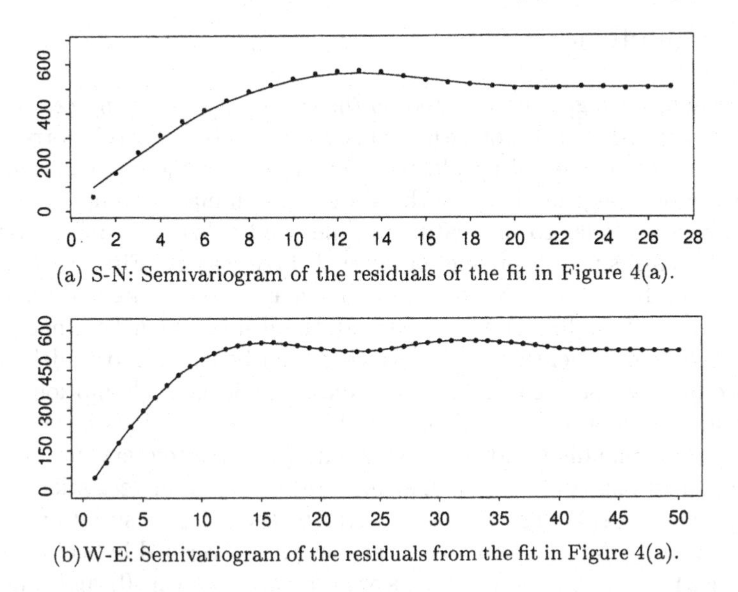

(a) S-N: Semivariogram of the residuals of the fit in Figure 4(a).

(b) W-E: Semivariogram of the residuals from the fit in Figure 4(a).

Figure 6. Semivariogram of the residuals in distances of pixel-units.

at a lag of about 12 pixels. Some spatial structure in the residuals should be expected, as can be seen in Gelpke (1999), but typically this correlation that is induced by the observation error η should die out at a lag of about 2–4 pixel units. It seems that there is more structure left.

We considered two ways to get a starting value for the nonlinear least squares problem. First, we used the wind fields. With least squares we fitted a displacement field directly to the wind fields and used the resulting parameter estimates for starting values. The other way was first to fit only a constant lag and setting the other parameters to zero. This second choice is more useful as will be explained below. In the south and in the north of the image the conservation of mass may be hurt, because some mass crosses the border of our image. One does not know where these places are where we do not have enough data, before starting with minimizing $S_n(\boldsymbol{\theta})$. In each step of the minimization algorithm such places can appear somewhere else. If one simply omits these terms from $S_n(\boldsymbol{\theta})$ the minimum may not be found, because $S_n(\boldsymbol{\theta})$ has not the same meaning anymore in each step. Hence one has to omit them from the beginning. Consequently we just took the middle zone of the images to calculate $S_n(\boldsymbol{\theta})$. After determining $\widehat{\boldsymbol{\theta}}$ for this zone we checked how much larger the zone could be for this $\widehat{\boldsymbol{\theta}}$ and restarted the estimation procedure with this new zone and with starting value $\widehat{\boldsymbol{\theta}}$. When starting with a constant lag one can usually use much broader zones at the beginning, because the displacement is very smooth. When having a starting value from the wind, which can be very rough, then only a very small zone can be used which may not contain enough structure to find a unique solution.

11 Conclusion

The model we suggested is based on the continuity equation and thus can only be applied to situations where this law is fulfilled. But this corresponds to a quite large class of problems, as mentioned in the introduction. The estimates give a general idea of the motion and in many examples the main features are similar to the features in the wind fields. The details can not be seen. This is not surprising, because of three reasons. First, we assumed absence of divergence. This assumption is maybe too strong and we locally have other effects like diffusion. Second, the displacement field had only 9 parameters to model the motion, which cannot be enough to model details. Hence our displacement field shows (only) the large scale motion. As we estimate the motion by nonlinear methods the calculations become more and more demanding with increasing number of parameters. In future investigations one could try to find other flexible parameterizations of the displacement field, that may suit better to a particular situation. Third, an ozone image is not taken instantaneously, as described in Section 2, but within 24 hours. So, we do not have observations $U(\mathbf{x_k}, 0)$ and $U(\mathbf{x_k}, 1)$, but $U(\mathbf{x_k}, t(\mathbf{x_k}))$ with $t(\mathbf{x_k}) \in [0, 2]$.

The computations have been done on an Ultra-SPARC with 256MB memory and a 200 MHz processor. It took about 1 hour to estimate the motion of two successive images with 144×23 pixels, but already 5 to 7 hours with 288×46 pixels. This is not prohibitive, but it does not allow to use the method on a routine basis. The large computing time is due to several reasons. The minimization of $S_n(\boldsymbol{\theta})$ is a nonlinear problem, the determination of $\mathbf{h}_{\boldsymbol{\theta}}(\mathbf{x}, t)$ is not straightforward but requires root finding and the calculation of the intersection of the two quadrangles is computationally very expensive. We do not know how much could be gained by optimizing the code.

The main advantage of our method compared to other methods is the possibility of interpolation of images. Most methods like the block matching algorithms produce only the displacement fields or the velocity fields, but they do not explicitly construct the image that appears, if the first image has been displaced with the estimated field of motion.

12 References

Aggarwal, J.K. and N. Nandhakumar (1988). On the computation from sequences of images—a review. *Proceedings of the IEEE 76*, 917–935.

Amit, Y., U. Grenander, and M. Piccioni (1991). Structural image restoration through deformable templates. *Journal of the American Statistical Association 86*, 376–387.

Arnold, V.I. (1989). *Mathematical Methods of Classical Mechanics* (2nd ed.), Volume 60 of *Graduate Texts in Mathematics*. New York: Springer.

Bannehr, L., M. Rohn, and C. Schnörr (1997). Ein Variationsprinzip zur Ableitung von Vektorfelden aus Satellitenbildfolgen in der Meteorologie. *Annalen der Meteorologie 31.*

Bannehr, L., M. Rohn, and G. Warnecke (1995). Determination of displacement vector fields from satellite image sequences. *Advances in Space Research 16*, (10)103–(10)106.

Bannehr, L., M. Rohn, and G. Warnecke (1996). A functional analytic method to derive displacement vector fields from satellite image sequences. *International Journal of Remote Sensing 17*, 383–392.

Berman, M., L.M. Bischof, S.J. Davies, A.A. Green, and M. Craig (1994). Estimating band-to-band misregistrations in aliased imagery. *Graphical Models and Image Processing 56*, 479–493.

Bojkov, R.D., L. Bishop, W.J. Hill, G.C. Reinsel, and G.C. Tiao (1990). A statistical trend analysis of revised Dobson total ozone data over the Northern Hemisphere. *Journal of Geophysical Research 95*, 9795–9807.

Brillinger, D.D. (1997). An application of statistics to meteorology: estimation of motion. In D. Pollard, E. Torgersen, and G. Yang (Eds.), *Festschrift for Lucien Le Cam*, pp. 93–105. New York: Springer.

Carroll, R.J., P. Hall, and D. Ruppert (1994). Estimation of lag in misregistered, averaged images. *Journal of the American Statistical Association 89*, 219–229.

Courant, R. and D. Hilbert (1968). *Methoden der Mathematischen Physik. II*, Volume 31 of *Heidelberger Taschenbücher.* Berlin: Springer.

Cressie, N.A.C. (1993). *Statistics for Spatial Data* (revised ed.). Wiley Series in Probability and Mathematical Statistics: Applied Probability and Statistics. New York: Wiley.

Farman, J.C., B.G. Gardiner, and J.D. Shanklin (1985). Large losses of total ozone in Antarctica reveal seasonal Cl0x/NOx interaction. *Nature 315*, 207–210.

Gasser, T., L. Sroka, and C. Jennen-Steinmetz (1986). Residual variance and residual pattern in nonlinear regression. *Biometrika 73*, 625–633.

Gelpke, V.M. (1999). *Estimation of Motion in Consecutive Images.* Ph.D. thesis, ETH, Zürich.

Hall, P., J.W. Kay, and D.M. Titterington (1990). Asymptotically optimal difference-based estimation of variance in nonparametric regression. *Biometrika 77*, 521–528.

Hall, P., J.W. Kay, and D.M. Titterington (1991). On estimation of noise variance in two-dimensional signal processing. *Advances in Applied Probability 23*, 476–495.

Heitz, F. and P. Bouthemy (1993). Multimodal estimation of discontinuous optical flow using Markov random fields. *IEEE Transactions on Pattern Analysis and Machine Intelligence 15*, 1217–1232.

Herrmann, E., M.P. Wand, J. Engel, and T. Gasser (1995). A bandwidth selector for bivariate kernel regression. *Journal of the Royal Statistical Society. Series B. Methodological 57*, 171–180.

Holton, J.R. (1979). *An Introduction to Dynamic Meteorology.* New York: Academic Press.

Horn, B.K.P. and B.G. Schunck (1981). Determining optical flow. *Artificial Intelligence 17*, 185–203.

Kelly, K.A. (1989). An inverse model for near-surface velocity from infrared images. *Journal of Physical Oceanography 19*, 1845–1864.

Kelly, K.A. and P.T. Strub (1992). Comparison of velocity estimates from advanced very high resolution radiometer in the coastal transition zone. *Journal of Geophysical Research 97*, 9653–9668.

Konrad, J. and E. Dubois (1993). Bayesian estimation of motion vector fields. *IEEE Transactions on Pattern Analysis and Machine Intelligence 14*, 910–927.

Leese, J.A., C.S. Novak, and B.B. Clark (1971). An automated technique for obtaining cloud motion from geosynchronous satellite data using cross correlation. *Journal of Applied Meteorology 10*, 118–132.

Nagel, H.H. and W. Enkelmann (1986). An investigation of smoothness constraints for the estimation of displacement vector fields from image sequences. *IEEE Transactions on Pattern Analysis and Machine Intelligence 8*, 565–593.

Niu, X. and G.C. Tiao (1995). Modeling satellite ozone data. *Journal of the American Statistical Association 90*, 969–983.

O'Rourke, J. (1998). *Computational Geometry in C* (2nd ed.). Cambridge: Cambridge University Press.

Press, W.H., S.A. Teukolsky, W.T. Vetterling, and B. Flannery (1992). *Numerical Recipes: The Art of Scientific Computing in C* (2nd ed.). Cambridge: Cambridge University Press.

Schnörr, C. (1994). Segmentation of visual motion by minimizing convex non-quadratic functionals. In *Proceedings of the 12th International Conference on Pattern Recognition*, Jerusalem, 1994, pp. 661–663. The International Association for Pattern Recognition.

Seifert, B., T. Gasser, and A. Wolf (1993). Nonparametric estimation of residual variance revisited. *Biometrika 80*, 373–383.

Tokmakian, R., P.T. Strub, and J. McClean-Padman (1990). Evaluation of the maximum cross-correlation method of estimating sea surface velocities from sequential satellite images. *Journal of Atmospheric and Oceanic Technology 7*, 852–865.

Wahl, D.D. and J.J. Simpson (1990). Physical processes affecting the objective determination of near-surface velocity from satellite data. *Journal of Geophysical Research 95*, 13511–13528.

Winkler, G. (1995). *Image Analysis, Random Fields and Dynamic Monte Carlo Methods. A Mathematical Introduction*, Volume 27 of *Applications of Mathematics.* Berlin: Springer.

8

Applications of Random Fields in Human Brain Mapping

J. Cao
K.J. Worsley

ABSTRACT The goal of this short article is to summarize how random field theory has been used to test for activations in brain mapping applications. It is intended to be a general discussion and hence it is not very specific to individual applications. Tables of most widely used random fields, examples of their applications, as well as references to distributions of some of their relevant statistics are provided.

1 Introduction

In a lot of situations that arise in the analysis of PET and fMRI data, we often seek relationships between a set of dependent variables Y and a set of predictor variables X. For example, we want to study how brain functional activity (Y) changes with the experimental condition of the stimulus (X), how brain structure (Y) changes with the status of the subject (X), e.g., normal or diseased, female or male, etc. The predictor variables X contain both variables whose effects are of interest, and variables whose effects are not of interest but have predicting power on Y, e.g., confounding variables.

The most widely used method for assessing such a relationship is the *general linear model.* For a detailed account of this in brain mapping, one should refer to Friston et al. (1995) and Worsley and Friston (1995). The model can be expressed as follows:

$$Y(t) = X\beta(t) + \sigma(t)\epsilon(t), \quad t \text{ in } C$$

where t represents a voxel in the brain region C, $Y(t)$ is a vector of brain activity measures, X is a design matrix whose columns are the predictor variables, $\beta(t)$, the vector of *regression coefficients*, represents on average, how $Y(t)$ changes with a unit change in X, $\sigma(t)$ is the (scalar) standard deviation of $Y(t)$ and the vector $\epsilon(t)$, the noise process, represents random changes in $Y(t)$ that are not captured by the linear relation $X\beta(t)$. We assume that the components of $\epsilon(t)$ are independent and identically distributed isotropic *Gaussian random fields* (GRF) with zero mean and unit variance.

Note X is usually external and hence voxel independent. In the following discussion, we shall refer to the brain region C where linear relations between Y and X are assessed as the *search region.* To assess an effect of interest such as a linear combination of the components of $\beta(t)$, we perform the following hypothesis test:

$$\text{Null model:} \quad c'\beta(t) = 0 \quad \text{for all } t \text{ in } C \tag{1}$$

$$\text{Alternative:} \quad c'\beta(t) \neq 0 \quad \text{for some } t \text{ in } C, \tag{2}$$

where c is a vector of *contrasts* that define the linear combination of the parameters that we wish to test. For future convenience, we shall refer to the areas for which such an effect exists as *activated regions*, and the value of $c'\beta$ at activations as *activations* or *signals.*

2 Test Statistics

To assess the effect $c'\beta(t)$ at each voxel, a statistic $T(t)$ is calculated at each voxel t. This gives rise to a *statistical map* or SPM (Statistical Parametric Map). To test for distributed activations, the sum of squares of the statistic at all voxels has been proposed (Worsley et al., 1995). For localized and intense signals, the maximum of the statistical image $T_{\max}$ has been proposed as a test statistic (Friston et al., 1991, Worsley et al., 1992). This is especially powerful for detecting signals whose shape matches the correlation function of the noise process provided that the noise is stationary Gaussian (Siegmund and Worsley, 1995). In the light of this, some spatial smoothing of Y is usually performed before applying this test statistic. For signals with different extent, different amounts of smoothing should be applied to optimally detect them. This leads to the scale space approach first suggested by Poline and Mazoyer (1994a,b) and later developed by Siegmund and Worsley (1995), Worsley et al. (1996). Shafie (1998), Shafie et al. (1998) extend this to rotated as well as scaled filters. The spatial extent $S_{\max}$ of the largest set of contiguous voxels above a threshold has also been proposed as the test statistic Friston et al. (1994), which favors the detection of diffuse and broad signals. A combination based on both spatial extent and intensity of the signal has also been proposed by (Poline et al., 1997).

3 Distribution of the Test Statistics Based on Random Field Theory

The distribution of these test statistics under the null model can be obtained based on *random field* theory. The classical book on random fields

Table 1. Types of random fields

Random field $T(t)$	$\sigma(t)$	number of contrasts	number of components of $Y(t)$
Gaussian	known	1	1
χ^2	known	≥ 1	1
t	unknown	1	1
F	unknown	≥ 1	1
Hotelling's T^2	unknown	1	≥ 1
Wilk's Λ	unknown	≥ 1	≥ 1

most relevant to our discussion is the book by Adler (1981) on the geometry of random fields. Once the distribution of the test statistic is found, the hypothesis testing in the previous sections can be completed. For example, if $T_{\max}$ is used as the test statistic, a threshold based on the upper quantiles of its distribution under the null model would be chosen and areas where the statistical image $T(t)$ exceeds that threshold are declared to be statistically significant activations. Under the assumption that the errors form a smooth isotropic Gaussian random field, the statistical image $T(t)$ is a random field of the appropriate kind depending on how the statistic is calculated. The standard deviation $\sigma(t)$ may be known, or unknown and estimated from the error sum of squares; we may wish to test several contrasts simultaneously, one for each column of the matrix c; we may have multivariate observations Y at every voxel. Table 1 gives some of the random fields $T(t)$ for each of these situations.

In the following, we give some basic principles of how distributions of test statistics, $T_{\max}$ and $S_{\max}$, are derived using random field theory.

3.1 The Maximum of the Random Field, $T_{\max}$

For a high threshold z, the probability that the maximum statistic $T_{\max}$ exceeds z can be accurately approximated by the average *Euler characteristic* (EC) of the *excursion set* of the random field T above the threshold z (Hasofer, 1978, Adler, 2000). Here the excursion set is defined as the set of voxels where the random field T exceeds z (see Figure 1).

Moreover, exact calculation of the average Euler characteristic is usually possible for smooth stationary random fields. This is due to a closed form expression for the Euler characteristic derived by Adler (1981) for any smooth random field. Worsley (1995) added a correction for when the excursion set touches the boundary of the search region and the random field is isotropic (non-isotropic but stationary random fields can be handled by a simple linear transformation of t). This is important especially for brain mapping applications, since activations often appear on the cerebral cortex which is also part of the brain boundary. An excellent review can be found in Adler (2000).

Let D be the dimension of the domain of field T. Define $\mu_i(C)$ to be proportional to the i-dimensional Minkowski functional of C, as follows. Let $a_i = 2\pi^{i/2}/\Gamma(i/2)$ be the surface area of a unit $(i-1)$-sphere in $\Re^i$. Let M be the inside curvature matrix of ∂C at a point t, and let $\mathrm{detr}_j(M)$ be the sum of the determinants of all $j \times j$ principal minors of M. For $i = 0, \ldots, d-1$

$$\mu_i(C) = \frac{1}{a_{d-i}} \int_{\partial C} \mathrm{detr}_{d-1-i}(M)\, dt,$$

and define $\mu_d(C)$ to be the Lebesgue measure of C. Minkowski functionals for some simple shapes are given in Table 2.

The i-dimensional EC intensity of $T(t)$ is defined as:

$$\rho_i(z) = \mathrm{E}\{(T \geq z)\det(-\ddot{T}_i) \mid \dot{T}_i = 0\}\, \mathrm{P}\{\dot{T}_i = 0\},$$

where dot notation with subscript i means differentiation with respect to the first i components of t. Then

$$\mathrm{P}\{T_{\max} \geq z\} \approx \mathrm{E}\{\chi(A_z(T, C))\} = \sum_{i=0}^{d} \mu_i(C)\rho_i(z),$$

where $\chi(A_z(T,C))$ is the excursion set of T above threshold z inside C.

Using the EC intensities for the field T, we can also find the p-values for the global minimum $T_{\min}$, i.e., $\mathrm{P}(T_{\min} \leq z)$. This is due to a simple relationship between $\mathrm{E}(\chi(A_z(T,C)))$ and $\mathrm{E}(\chi(A_{-z}(-T,C)))$ when the field T is homogeneous:

$$\mathrm{E}(\chi(A_z(T,C)) = (-1)^{d-1}\, \mathrm{E}(\chi(A_{-z}(-T,C))).$$

Therefore,

$$\mathrm{P}(T_{\min} \leq z) \approx \mathrm{E}\{\chi(A_{-z}(-T,C))\} = (-1)^{d-1}\, \mathrm{E}\{\chi(A_z(T,C))\}.$$

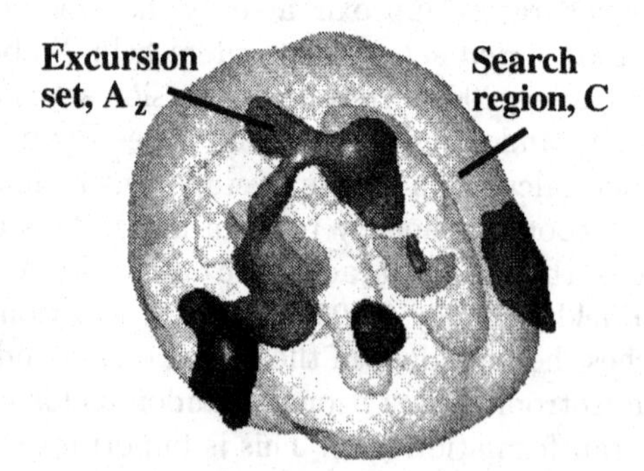

Figure 1. The excursion set of a Gaussian random field above $z = 3$ for testing for a difference in PET CBF between a hot and warm stimulus.

3.2 The Maximum Spatial Extent, $S_{\max}$

Distributions of $S_{\max}$ are derived asymptotically as the threshold z goes to infinity, based on the Poisson clumping heuristic (Aldous, 1989). The essence of this approach is that connected regions in the excursion set can be viewed as clumps that are centered at points of a Poisson process. Hence the distribution of the size of the largest connected region in the excursion set can be derived from the distribution of the size of an individual connected component. By approximating the underlying random process locally by a simpler, known process, explicit calculations are possible for the distribution of the size of an individual connected component. We now give some details of this approach.

Let S be the size of one connected component in the excursion set $A_z(T, C)$, and L be the total number of such connected components. By the Poisson clumping heuristic (Aldous, 1989), we can express the distribution of $S_{\max}$ in terms of the distribution of S and $\mathrm{E}(L)$ by:

$$\mathrm{P}(S_{\max} \leq s \mid L \geq 1) \approx \frac{\exp\{-\mathrm{E}(L)\,\mathrm{P}(S \geq s)\} - \exp\{-\mathrm{E}(L)\}}{1 - \exp\{-\mathrm{E}(L)\}}.$$

At high thresholds, $\mathrm{E}(L)$ can be approximated accurately by the average Euler Characteristic (Hasofer, 1978, Adler, 1981):

$$\mathrm{E}(L) \approx \mathrm{E}(\chi(A_z(T, C))) = \sum_{i=0}^{d} \mu_i(C)\rho_i(z). \tag{3}$$

To find the distribution of S, we study the conditional field

$$\tilde{T}_z(t) = T(t) \mid\mid \mathcal{E}_z,$$

where $\mathcal{E}_z$ denotes the event that $T(t)$ has a local maximum of height z at $t = 0$. By using horizontal window (or ergodic) conditioning and Slepian model process (Kac and Slepian, 1959), we can approximate this field locally by a simpler field and hence derive the distribution of S as the threshold z

Table 2. Minkowski functionals for some simple shapes

C	$\mu_0(C)$	$\mu_1(C)$	$\mu_2(C)$	$\mu_3(C)$
Sphere, radius r	1	$4r$	$2\pi r^2$	$\frac{4}{3}\pi r^3$
Hemisphere, radius r	1	$(2+\pi/2)r$	$\frac{3}{2}\pi r^2$	$\frac{2}{3}\pi r^3$
Disk, radius r	1	πr	πr^2	0
Sphere surface, radius r	2	0	$4\pi r^2$	0
Hemisphere surface, radius r	1	πr	$2\pi r^2$	0
Box, $a \times b \times c$	1	$a+b+c$	$ab+bc+ac$	abc
Rectangle, $a \times b$	1	$a+b$	ab	0
Line, length a	1	a	0	0

goes to infinity. For any practical use of the asymptotic distribution of S, a mean correction is always recommended to improve the approximation, based on the following identity (Aldous, 1989, Friston et al., 1994):

$$\mathrm{E}(L)\,\mathrm{E}(S) = \mathrm{E}(|A_z(T,C)|) = |C|F_T(-z), \tag{4}$$

where $F_T(\cdot)$ is the cumulative density function for the marginal distribution of T, and $|\cdot|$ is Lebesgue measure.

4 Results

Table 3 summarizes references for the distributions of the two test statistics for random fields in Table 1, except for the Wilk's Λ field, for which results are still unknown. The distribution $S_{\max}$ for the Hotelling's T^2 field is also not yet derived. References to other types of random field for which results are known are also added.

In the following subsections, we shall list all of the known results for different random fields from the above references for which there are explicit algebraic expressions. We give explicit formulae of the EC intensities for $d \leq 3$ dimensions, the most common case in practical applications. General formulae for any number of dimensions can be found in the above references. The EC intensities depend on a single parameter λ, the *roughness* of the random field, defined as the variance of the derivative of any component of $\epsilon(t)$.

We also provide the asymptotic distribution of S for any dimension d. To be complete, besides the distribution of S, we shall also provide the distribution of the size of one connected region in the excursion set $A_{-z}(-T,C)$. To distinguish these two cases, we shall add a superscript to S and denote them by S^+ and S^- respectively. When S^+ and S^- have the same distribution, we omit the superscript and denote both of them by S. For simplicity

Table 3. Distribution of $T_{\max}$ and $S_{\max}$ for various random fields.

Random field	$T_{\max}$	$S_{\max}$
Gaussian	Adler (1981)	Nosko (1969)
χ^2, t, F	Worsley (1994)	Aronowich and Adler (1986, 1988), Cao (1999)
Hotelling's T^2	Cao and Worsley (1999a)	?
Wilk's Λ	?	?
Correlation	Cao and Worsley (1999b)	Cao and Worsley (1999b)
Gaussian scale space	Siegmund and Worsley (1995)	?
χ^2 scale space	Worsley (1999)	?
Gaussian rotation space	Shafie (1998), Shafie et al. (1998)	?

and uniformity, we express the distribution of S in the form of $\alpha\ S_0$, where α is a constant and S_0 is a random variable. We give the expectation μ_0 of the random variable S_0. By the mean correction formula (4), we have

$$\alpha = |C| F_T(-z) / (\mu_0 \, \mathrm{E}(L)),$$

where $\mathrm{E}(L)$ can be derived from the EC intensities using (3).

We shall use the following notation to represent distributions. Let χ^2_ν denote the χ^2 distribution with ν degrees of freedom, $\mathrm{Exp}(\mu)$ denote the exponential distribution with expectation μ, $\mathrm{Beta}(\alpha, \beta)$ denote the beta distribution with parameters α and β, and $\mathrm{Wishart}_d(\Sigma, \nu)$ denote Wishart distribution of a $d \times d$ matrix with expectation $\nu\Sigma$ and ν degrees of freedom. Finally, let I denote the identity matrix.

4.1 Gaussian Field

$$\rho_0(z) = \int_z^\infty \frac{1}{(2\pi)^{1/2}} e^{-u^2/2}\, du \qquad \rho_1(z) = \frac{\lambda^{1/2}}{2\pi} e^{-z^2/2}$$

$$\rho_2(z) = \frac{\lambda}{(2\pi)^{3/2}} e^{-z^2/2} z \qquad \rho_3(z) = \frac{\lambda^{3/2}}{(2\pi)^2} e^{-z^2/2}(z^2 - 1)$$

$$S \sim \alpha\, \mathrm{Exp}(1), \quad \mu_0 = 1$$

4.2 χ^2 Field with ν Degrees of Freedom

$$\rho_0(z) = \int_z^\infty \frac{u^{1/2(\nu-2)} e^{-1/2u}}{2^{\nu/2}\Gamma(\nu/2)}\, du, \quad \rho_1(z) = \frac{\lambda^{1/2}}{(2\pi)^{1/2}} \frac{z^{(\nu-1)/2} e^{-z/2}}{2^{(\nu-2)/2}\Gamma(\nu/2)}$$

$$\rho_2(z) = \frac{\lambda}{(2\pi)} \frac{z^{(\nu-2)/2} e^{-z/2}}{2^{(\nu-2)/2}\Gamma(\nu/2)} [z - (\nu - 1)]$$

$$\rho_3(z) = \frac{\lambda^{3/2}}{(2\pi)^{3/2}} \frac{z^{(\nu-3)/2} e^{-z/2}}{2^{(\nu-2)/2}\Gamma(\nu/2)} [z^2 - (2\nu - 1)z + (\nu - 1)(\nu - 2)]$$

$$S^+ \sim \alpha\, \mathrm{Exp}(1), \qquad \mu_0 = 1$$

$$S^- \sim \alpha\ B^{d/2} \det(Q)^{-\frac{1}{2}}, \qquad \mu_0 = \frac{2^{d/2}\Gamma(d/2+1)(\nu - d)!}{\nu!}$$

where $B \sim \mathrm{Beta}(1, (\nu - d)/2)$ and $Q \sim \mathrm{Wishart}_d(I, \nu + 1)$ independently.

4.3 t Field with ν Degrees of Freedom, $\nu \geq d$

$$\rho_0(z) = \int_z^\infty \frac{\Gamma([\nu+1]/2)}{(\nu\pi)^{1/2}\Gamma(\nu/2)} \left(1 + \frac{u^2}{\nu}\right)^{-(\nu+1)/2} du$$

$$\rho_1(z) = \frac{\lambda^{1/2}}{2\pi}\left(1+\frac{z^2}{\nu}\right)^{-(\nu-1)/2}$$

$$\rho_2(z) = \frac{\lambda}{(2\pi)^{3/2}}\frac{\Gamma([\nu+1]/2)}{(\nu/2)^{1/2}\Gamma(\nu/2)}\left(1+\frac{z^2}{\nu}\right)^{-(\nu-1)/2} z$$

$$\rho_3(z) = \frac{\lambda^{3/2}}{(2\pi)^2}\left(1+\frac{z^2}{\nu}\right)^{-(\nu-1)/2}\left(\frac{\nu-1}{\nu}z^2-1\right)$$

$$S \sim \alpha\, B^{d/2}U^{d/2}\det(Q)^{-1/2}, \quad \mu_0 = \frac{\Gamma(d/2+1)\Gamma([\nu-d]/2+1)}{\Gamma(\nu/2+1)}$$

where $B \sim \mathrm{Beta}(1, [\nu-d]/2)$, $U \sim \chi^2_{\nu+1-d}$ and $Q \sim \mathrm{Wishart}_d(I, \nu+1)$, all independently.

4.4 F Field with k and ν Degrees of Freedom, $k+\nu > d$

$$\rho_0(z) = \int_z^\infty \frac{\Gamma([\nu+k-2]/2)}{\Gamma(\nu/2)\Gamma(k/2)}\frac{k}{\nu}\left(\frac{ku}{\nu}\right)^{(k-2)/2}\left(1+\frac{ku}{\nu}\right)^{-(\nu+k)/2} du$$

$$\rho_1(z) = \frac{\lambda^{1/2}}{(2\pi)^{1/2}}\frac{\Gamma([\nu+k-1]/2)2^{1/2}}{\Gamma(\nu/2)\Gamma(k/2)}\left(\frac{kz}{\nu}\right)^{(k-1)/2}\left(1+\frac{kz}{\nu}\right)^{-(\nu+k-2)/2}$$

$$\rho_2(z) = \frac{\lambda}{2\pi}\frac{\Gamma([\nu+k-2]/2)}{\Gamma(\nu/2)\Gamma(k/2)}\left(\frac{kz}{\nu}\right)^{(k-2)/2}\left(1+\frac{kz}{\nu}\right)^{-(\nu+k-2)/2} \times\left[(\nu-1)\frac{kz}{\nu}-(k-1)\right]$$

$$\rho_3(z) = \frac{\lambda^{3/2}}{(2\pi)^{3/2}}\frac{\Gamma([\nu+k-3]/2)2^{-1/2}}{\Gamma(\nu/2)\Gamma(k/2)}\left(\frac{kz}{\nu}\right)^{(k-3)/2}\left(1+\frac{kz}{\nu}\right)^{-(\nu+k-2)/2} \times\left[(\nu-1)(\nu-2)\left(\frac{kz}{\nu}\right)^2-(2\nu k-\nu-k-1)\left(\frac{kz}{\nu}\right)+(-1)(k-2)\right]$$

$$S^+ \sim \alpha B^{d/2}U^{d/2}\det(Q)^{-1/2}, \quad \mu_0 = \frac{2^d(\nu-d)!\,\Gamma(d/2+1)\Gamma([\nu+k]/2)}{\nu!\,\Gamma([\nu+k-d]/2)},$$

$S^- \sim S^+$ for F field with ν and k degrees of freedom,

where $B \sim \mathrm{Beta}(1, [\nu-d]/2)$, $U \sim \chi^2_{\nu+k-d}$ and $Q \sim \mathrm{Wishart}_d(I, \nu+1)$, all independently.

4.5 Hotelling's T^2 Field with k Components and ν Degrees of Freedom, $\nu > k+d$

$$\rho_0(z) = \int_z^\infty \frac{\Gamma([\nu+1]/2)}{\Gamma(k/2)\Gamma([\nu-k+1]/2)}\left(1+\frac{u}{\nu}\right)^{-(\nu+1)/2}\frac{u^{(k-2)/2}}{\nu^{k/2}}\,du$$

$$\rho_1(z) = \frac{\lambda^{1/2}\pi^{-1/2}\Gamma([\nu+1]/2)}{\Gamma(k/2)\Gamma([\nu-k+2]/2)}\left(1+\frac{z}{\nu}\right)^{-(\nu-1)/2}\left(\frac{z}{\nu}\right)^{(k-1)/2}$$

$$\rho_2(z) = \frac{\lambda\pi^{-1}\Gamma([\nu+1]/2)}{\Gamma(k/2)\Gamma([\nu-k+1]/2)}\left(1+\frac{z}{\nu}\right)^{-(\nu-1)/2}\left(\frac{z}{\nu}\right)^{(k-2)/2} \times\left[\left(\frac{z}{\nu}\right)-\frac{k-1}{\nu-k+1}\right]$$

$$\rho_3(z) = \frac{\lambda^{3/2}\pi^{-3/2}\Gamma([\nu+1]/2)}{\Gamma(k/2)\Gamma([\nu-k]/2)}\left(1+\frac{z}{\nu}\right)^{-(\nu-1)/2}\left(\frac{z}{\nu}\right)^{(k-3)/2} \times\left[\left(\frac{z}{\nu}\right)^2-\frac{2k-1}{\nu-k}\left(\frac{z}{\nu}\right)+\frac{(k-1)(k-2)}{(\nu-k+2)(\nu-k)}\right]$$

4.6 *Homologous Correlation Field with ν Degrees of Freedom, $\nu > d$*

The homologous correlation field is defined as

$$T(t) = \frac{X(t)'Y(t)}{\sqrt{X(t)'X(t)\,Y(t)'Y(t)}},$$

where the ν components of $X(t)$ are i.i.d. isotropic Gaussian random fields with roughness λ_x, and the ν components of $Y(t)$ are i.i.d. isotropic Gaussian random fields with roughness λ_y. For $\lambda_x = \lambda_y$, $\nu > d$, $d \le 3$ the EC intensities are shown below. Results for $\lambda_x \neq \lambda_y$ involve one non-explicit integral, so for simplicity we have omitted them, though they are given in Cao and Worsley (1999b).

$$\rho_0(z) = \int_z^\infty \frac{\Gamma(\nu/2)}{\pi^{1/2}\Gamma([\nu-1]/2)}(1-u^2)^{(\nu-3)/2}du$$

$$\rho_1(z) = \frac{\lambda^{1/2}\Gamma(\nu-1/2)}{2^{\nu-1}\pi^{1/2}\Gamma(\nu/2)^2}(1-z^2)^{[\nu-2]/2}$$

$$\rho_2(z) = \frac{\lambda\Gamma(\nu/2)}{\pi^{3/2}\Gamma([\nu-1]/2)}(1-z^2)^{\frac{(\nu-3)/2}{z}}$$

$$\rho_3(z) = \frac{\lambda^{3/2}\Gamma(\nu-3/2)}{2^{\nu+1}\pi^{3/2}\Gamma(\nu/2)^2}(1-z^2)^{(\nu-4)/2}[(4\nu^2-12\nu+11)z^2-(4\nu-5)]$$

$$S \sim \alpha[q(1-q)]^{d/2}B^{d/2}U^{d/2}\det(Q)^{-1/2},$$

$$\mu_0 = \frac{(\nu-d-1)!\,\Gamma(d/2+1)\Gamma(\nu/2)\Gamma([\nu-d+1]/2)}{\Gamma(\nu-d/2)\Gamma([\nu+1]/2)\Gamma([\nu-d]/2)}$$

where $q \sim \text{Beta}([\nu-d]/2, [nu-d]/2)$, $B \sim \text{Beta}(1, [\nu-d-1]/2)$, $U \sim \chi^2_{2\nu-d}$ and $Q \sim \text{Wishart}_d(I, \nu)$, all independently.

4.7 Cross Correlation Field with ν Degrees of Freedom, $\nu > d$

The cross correlation field is defined as

$$T(s,t) = \frac{X(s)'Y(t)}{sqrtX(s)'X(s)\,Y(t)'Y(t)},$$

where the ν components of $X(s)$, $s \in C_x$ are i.i.d. isotropic Gaussian random fields with roughness λ_x, and the ν components of $Y(t)$, $t \in C_y$ are i.i.d. isotropic Gaussian random fields with roughness λ_y. In this case,

$$\mathrm{E}\{\chi((C_x \oplus C_y) \cap A_z)\} = \sum_{i=0}^{d_x} \sum_{j=0}^{d_y} \mu_i(C_x)\mu_j(C_y)\rho_{ij}(z).$$

Let $d = d_x + d_y$ be the dimension of the random field. For $\nu > d$, $d_x \leq d_y \leq 3$,

$$\rho_{0,0}(z) = \int_z^\infty \frac{\Gamma(\nu/2)}{\pi^{1/2}\Gamma([\nu-1]/2)}(1-u^2)^{(\nu-3)/2}\,du$$
$$\rho_{0,1}(z) = \lambda_y^{1/2}(2\pi)^{-1}(1-z^2)^{(\nu-2)/2}$$
$$\rho_{0,2}(z) = \lambda_y \frac{\Gamma(\nu/2)}{2\pi^{3/2}\Gamma([\nu-1]/2)} z(1-z^2)^{(\nu-3)/2}$$
$$\rho_{0,3}(z) = \lambda_y^{3/2}(2\pi)^{-2}(1-z^2)^{(\nu-4)/2}[(\nu-1)z^2-1]$$
$$\rho_{1,1}(z) = \lambda_x^{1/2}\lambda_y^{1/2}\frac{\Gamma([\nu-1]/2)}{2\pi^{3/2}\Gamma([\nu-2]/2)} z(1-z^2)^{(\nu-3)/2}$$
$$\rho_{1,2}(z) = \lambda_x^{1/2}\lambda_y(2\pi)^{-2}(1-z^2)^{(\nu-4)/2}[(\nu-2)z^2-1]$$
$$\rho_{1,3}(z) = \lambda_x^{1/2}\lambda_y^{3/2}\frac{\Gamma([\nu-1]/2)}{2^2\pi^{5/2}\Gamma([\nu-2]/2)} z(1-z^2)^{[\nu-5]/2}[(\nu-1)z^2-3]$$
$$\rho_{2,2}(z) = \lambda_x\lambda_y\frac{\Gamma([\nu-2]/2)}{2^3\pi^{5/2}\Gamma([\nu-1]/2)} z(1-z^2)^{(\nu-5)/2}[(\nu-2)^2z^2-(3\nu-8)]$$
$$\rho_{2,3}(z) = \lambda_x\lambda_y^{1/2}(2\pi)^{-3}(1-z^2)^{(\nu-6)/2}[(\nu-1)(\nu-2)z^4-3(2\nu-5)z^2+3]$$
$$\rho_{3,3}(z) = \lambda_x^{3/2}\lambda_y^{3/2}\frac{\Gamma([\nu-3]/2)}{2^4\pi^{7/2}\Gamma([\nu-2]/2)} z(1-z^2)^{(\nu-7)/2}$$
$$\times\,[(\nu-1)^2(\nu-3)z^4-2(\nu-3)(5\nu-11)z^2+3(5\nu-17)]$$

$$S \sim \alpha B^{d/2}U^{d_x/2}V^{d_y/2}\det(Q)^{-1/2},$$

$$\mu_0 = \frac{2^d(\nu-1-d)!\,\Gamma(d/2+1)\Gamma(\nu/2)^2}{(\nu-1)!\,\Gamma([\nu-d_x]/2)\Gamma([\nu-d_y]/2)},$$

where $B \sim \mathrm{Beta}(1,[nu-d-1]/2)$, $U \sim \chi^2_{\nu-d_x}$, $V \sim \chi^2_{\nu-d_y}$ and $Q \sim \mathrm{Wishart}_d(I,\nu)$ independently.

4.8 Gaussian Scale Space Field

The Gaussian scale space field is defined as

$$T(t, w) = \int w^{-D/2} f[(t-s)/w]\, dZ(s)$$

where $Z(s)$ is a Brownian sheet and f is an isotropic smoothing kernel normalised so that $\int f^2 = 1$. Let

$$\kappa = \int [t'\dot{f} + (D/2)f]^2\, dt \quad \text{and} \quad \lambda = \int \dot{f}'\dot{f}\, dt/(Dw^2);$$

for an isotropic Gaussian shaped kernel, $\kappa = D/2$ and $\lambda = 1/(2w^2)$. Then for searching in $t \in C$ and w in an interval such that $\lambda \in [\lambda_1, \lambda_2]$, the EC intensities are:

$$\rho_0(z) = \int_z^\infty \frac{1}{(2\pi)^{1/2}} e^{-u^2/2}\, du + \frac{e^{-z^2/2}}{(2\pi)^{1/2}} \log\left(\frac{\lambda_2^{1/2}}{\lambda_1^{1/2}}\right)\sqrt{\frac{\kappa}{2\pi}}$$

$$\rho_1(z) = \frac{e^{-z^2/2}}{2\pi}\left\{\frac{\lambda_1^{1/2} + \lambda_2^{1/2}}{2} + (\lambda_2^{1/2} - \lambda_1^{1/2})\sqrt{\frac{\kappa}{2\pi}} z\right\}$$

$$\rho_2(z) = \frac{e^{-z^2/2}}{(2\pi)^{3/2}}\left\{\frac{\lambda_1 + \lambda_2}{2} z + \frac{\lambda_2 - \lambda_1}{2}\sqrt{\frac{\kappa}{2\pi}}\left[z^2 - 1 + \frac{1}{\kappa}\right]\right\}$$

$$\rho_3(z) = \frac{e^{-z^2/2}}{(2\pi)^2}\left\{\frac{\lambda_1^{3/2} + \lambda_2^{3/2}}{2}[z^2 - 1] + \frac{\lambda_2^{3/2} - \lambda_1^{3/2}}{3}\sqrt{\frac{\kappa}{2\pi}}\left[z^3 - 3z + \frac{3z}{\kappa}\right]\right\}$$

4.9 χ^2 Scale Space Field

The χ^2 scale space field with ν degrees of freedom is defined as the sum of squares of ν i.i.d. Gaussian scale space fields, and its EC intensities are:

$$\rho_0(z) = \int_z^\infty \frac{u^{\nu/2-1} e^{-u/2}}{2^{\nu/2}\Gamma(\nu/2)}\, du + \frac{z^{\nu/2} e^{-z/2}}{2^{(\nu-2)/2}\Gamma(\nu/2)} \log\left(\frac{\lambda_2^{1/2}}{\lambda_1^{1/2}}\right)\sqrt{\frac{\kappa}{2\pi z}}$$

$$\rho_1(z) = \frac{z^{(\nu-1)/2} e^{-z/2}}{(2\pi)^{1/2} 2^{(\nu-2)/2}\Gamma(\nu/2)}\left\{\frac{\lambda_1^{1/2} + \lambda_2^{1/2}}{2} + (\lambda_2^{1/2} - \lambda_1^{1/2})\sqrt{\frac{\kappa}{2\pi z}}[z - (\nu - 1)]\right\}$$

$$\rho_2(z) = \frac{z^{(\nu-2)/2} e^{-z/2}}{(2\pi) 2^{(\nu-2)/2}\Gamma(\nu/2)}\left\{\frac{\lambda_1 + \lambda_2}{2}[z - (\nu - 1)] + \frac{\lambda_2 - \lambda_1}{2}\sqrt{\frac{\kappa}{2\pi z}}\left[z^2 - (2\nu - 1)z + (\nu - 1)(\nu - 2) + \frac{z}{\kappa}\right]\right\}$$

$$\rho_3(z) = \frac{z^{(\nu-3)/2} e^{-z/2}}{(2\pi)^{3/2} 2^{(\nu-2)/2}\Gamma(\nu/2)}\left\{\frac{\lambda_1^{3/2} + \lambda_2^{3/2}}{2}[z^2 - (2\nu - 1)z + (\nu - 1)(\nu - 2)]\right.$$

$$+\frac{\lambda_2^{3/2}-\lambda_1^{3/2}}{3}\sqrt{\frac{\kappa}{2\pi z}}\left[z^3-3\nu z^2+3(\nu-1)^2 z-(\nu-1)(\nu-2)(\nu-3)\right.$$
$$\left.\left.+3\frac{z}{\kappa}[z-(\nu-1)]\right]\right\}$$

4.10 Gaussian Rotation Space Field

The Gaussian rotation space field is defined as

$$T(t,W)=\int \det(W)^{-1/2} f[W^{-1}(t-s)]\, dZ(s),$$

where W is a symmetric matrix with eigenvalues in a fixed range. Shafie (1998), Shafie et al. (1998) report EC intensities for the case $D=2$ but no simple closed form expressions are available.

5 Conclusion

This article summarizes how random field theory can be applied to to test for activations in brain mapping applications. Brain mapping has initiated a lot of recent research in random fields and it will continue to stimulate further methodological developments. Non-stationary random fields, random fields on manifolds, etc. are some of the future research directions pointed out by Adler (2000). We have a lot to look forward to in the future.

6 References

Adler, R.J. (1981). *The Geometry of Random Fields.* New York: Wiley.

Adler, R.J. (2000). On excursion sets, tube formulae, and maxima of random fields. *The Annals of Applied Probability 10*, 1–74.

Aldous, D. (1989). *Probability Approximations via the Poisson Clumping Heuristic*, Volume 77 of *Applied Mathematical Sciences.* New York: Springer.

Aronowich, M. and R.J. Adler (1986). Extrema and level crossings of χ^2 processes. *Advances in Applied Probability 18*, 901–920.

Aronowich, M. and R.J. Adler (1988). Sample path behavior of χ^2 surfaces at extrema. *Advances in Applied Probability 20*, 719–738.

Cao, J. (1999). The size of the connected components of excursion sets of χ^2, t and f fields. *Advances in Applied Probability 31*, 579–595.

Cao, J. and K.J. Worsley (1999a). The detection of local shape changes via the geometry of hotelling's t^2 fields. *The Annals of Statistics 27*, 925–942.

Cao, J. and K.J. Worsley (1999b). The geometry of correlation fields, with an application to functional connectivity of the brain. *The Annals of Applied Probability 9*, 1021–1057.

Friston, K.J., C.D. Frith, P.F. Liddle, and R.S.J. Frackowiak (1991). Comparing functional (PET) images: the assessment of significant change. *Journal of Cerebral Blood Flow and Metabolism 11*, 690–699.

Friston, K.J., A.P. Holmes, J.-B. Poline, K.J. Worsley, C.D. Frith, and R.S.J. Frackowiak (1995). Statistical parametric maps in functional imaging: a general linear approach. *Human Brain Mapping 2*, 189–210.

Friston, K.J., K.J. Worsley, R.S.J. Frackowiak, J.C. Mazziotta, and A.C. Evans (1994). Assessing the significance of focal activations using their spatial extent. *Human Brain Mapping 1*, 214–220.

Hasofer, A.M. (1978). Upcrossings of random fields. *Advances in Applied Probability 10, suppl.*, 14–21.

Kac, M. and D. Slepian (1959). Large excursions of Gaussian process. *The Annals of Mathematical Statistics 30*, 1215–1228.

Nosko, V.P. (1969). Local structure of Gaussian random fields in the vicinity of high-level light sources. *Soviet Mathematics. Doklady 10*, 1481–1484.

Poline, J, B. and B.M. Mazoyer (1994a). Enhanced detection in activation maps using a multifiltering approach. *Journal of Cerebral Blood Flow and Metabolism 14*, 690–699.

Poline, J.-B. and B.M. Mazoyer (1994b). Analysis of individual brain activation maps using hierarchical description and multiscale detection. *IEEE Transactions on Medical Imaging 13*, 702–710.

Poline, J.-B., K.J. Worsley, A.C. Evans, and K.J. Friston (1997). Combining spatial extent and peak intensity to test for activations in functional imaging. *NeuroImage 5*, 83–96.

Shafie, Kh. (1998). *The Geometry of Gaussian Rotation Space Random Fields.* Ph.D. thesis, Department of Mathematics and Statistics, McGill University.

Shafie, Kh., K.J. Worsley, M. Wolforth, and A.C. Evans (1998). Rotation space: detecting functional activation by searching over rotated and scaled filters. *NeuroImage 7*, S755.

Siegmund, D.O. and K.J. Worsley (1995). Testing for a signal with unknown location and scale in a stationary Gaussian random field. *The Annals of Statistics 23*, 608–639.

Worsley, K.J. (1994). Local maxima and the expected Euler characteristic of excursion sets of χ^2, F and t fields. *Advances in Applied Probability 26*, 13–42.

Worsley, K.J. (1995). Boundary corrections for the expected Euler characteristic of excursion sets of random fields, with an application to astrophysics. *Advances in Applied Probability 27*, 943–959.

Worsley, K.J. (1999). Testing for signals with unknown location and scale in a χ^2 random field, with an application to fMRI. *Advances in Applied Probability*. accepted.

Worsley, K.J., A.C. Evans, S. Marrett, and P. Neelin (1992). A three dimensional statistical analysis for CBF activation studies in human brain. *Journal of Cerebral Blood Flow and Metabolism 12*, 900–918.

Worsley, K.J. and K.J. Friston (1995). Analysis of fMRI time-series revisited—again. *NeuroImage 2*, 173–181.

Worsley, K.J., S. Marrett, P. Neelin, A.C. Vandal, K.J. Friston, and A.C. Evans (1996). A unified statistical approach for determining significant signals in images of cerebral activation. *Human Brain Mapping 4*, 58–73.

Worsley, K.J., J-B. Poline, A.C. Vandal, and K.J. Friston (1995). Test for distributed, non-focal brain activations. *NeuroImage 2*, 183–194.

9

In Praise of Tedious Permutation

Edward Bullmore, John Suckling, and Michael Brammer

ABSTRACT Tests of the null hypothesis by repeated permutation of the observed data are brutally simple, valid on the basis of only a few assumptions, and adaptable to any test statistic of interest. The history of permutation tests is briefly reviewed, from the pioneering work of R.A. Fisher in the 1930s, to contemporary applications to human brain mapping by several independent groups. Permutation tests for a difference between two sets of brain images at voxel and cluster levels are described, and compared wherever possible to the corresponding theoretical tests. It is shown that nominal Type I error control by permutation is generally excellent; and that some interesting statistics, such as the "mass" of a suprathreshold voxel cluster, are more accessible to testing by permutation than theory. Some limitations of permutation tests are discussed. On balance, the case for permutation tests of the null hypothesis in brain mapping is strong.

1 R.A. Fisher (1890–1962), the Anthropologists and Darwin's Peas

In many accounts of the history of permutation tests, the earliest cited reference is to a paper published by Fisher in the *Journal of the Royal Anthropological Society* in 1936. Fisher was then already established as one of the leading British statisticians of his generation, famous for elucidating the theoretical distribution of Pearson's correlation coefficient and for his work on factorial design and analysis of experiments.

In his communication to the anthropologists, Fisher is clearly determined to be understood on a question of data analysis raised by a readership one suspects he regarded as mathematically challenged. The specific analytic problem arises once anatomical measurements have been made on two geographically or culturally separate samples of human subjects. On the basis of these data, an anthropologist might wonder, what can we say about the probability that the samples are drawn from the same racial population? Fisher states that this question can be addressed by a test of significance, which he somewhat condescendingly introduces as a technical term that "no one need fail to understand, for it can be made plain in very simple terms."

Fisher imagines that the data consist simply of measurements of height on a random sample of 100 English men and a random sample of 100 French men. The observed difference in mean height between the two samples is one inch.

> *The simplest way of understanding quite rigorously, yet without mathematics, what the calculations of the test of significance amount to, is to consider what would happen if our two hundred actual measurements were written on cards, shuffled without regard to nationality, and divided at random into two new groups of a hundred each. This division could be dône in an enormous number of ways, but though the number is enormous it is finite and calculable. We may suppose that for each of these ways the difference between the two average statures is calculated. Sometimes it will be less than an inch, sometimes greater. If it is very seldom greater than an inch, in only one hundredth, for example, of the ways in which the subdivision can possibly be made, the statistician will have been right in saying the samples differed significantly. For if, in fact, the two populations were homogeneous, there would be nothing to distinguish the particular [observed] subdivision in which the Frenchmen are separated from the Englishmen from among the aggregate of the other possible separations which might have been made.* (Fisher, 1936)

This simple thought experiment distils the essence of any permutation test. The main advantage of the procedure, apart from its conceptual accessibility, is that it is valid regardless of the population distribution of the test statistic. This is in contrast to Student's test, for example, which is only valid assuming the statistic has a t distribution in the sampled population. Its main disadvantage, in Fisher's opinion, is just that it would be intolerably boring to do.

> *Actually, the statistician does not carry out this very simple and very tedious process, but his conclusions have no justification beyond the fact that they agree with those that could have been arrived at by this elementary method.* (Fisher, 1936)

Here he expresses an interesting ambivalence about permutation testing: it is crude, compared to the mathematical elegance of distributional theory, and yet the gold standard by which the validity of theory must be judged. If there exists a discrepancy between the results of a theoretical test of significance and a test by permutation it seems we should prefer to trust the latter, however unrefined or pedestrian or tedious we may consider the permutation test to be as an intellectual exercise.

This is not to say that permutation tests are unconditionally valid. In this example, validity of the permutation test, like its parametric alternative,

Table 1. Charles Darwin's pea data

Pair	1	2	3	4	5	6	7	8	9	10	11	12	13	14	15
Cross-fertilised	92	0	72	80	57	76	81	67	50	77	90	72	81	88	0
Self-fertilised	43	67	64	64	51	53	53	26	36	48	34	48	6	28	48
Difference	49	−67	8	16	6	23	28	41	14	29	56	24	75	60	−48

is conditional on random sampling of English and French men from their respective populations. Random sampling is one guarantee of the generally necessary condition for validity of a permutation test that the observations should be exchangeable under the null hypothesis.

Another example of a permutation test provided by Fisher is his treatment of some experimental observations originally made by Charles Darwin (Fisher, 1935). Darwin wanted to know whether cross-fertilisation (sexual reproduction) resulted in more robust growth than self-fertilisation. He sampled 15 pairs of pea plants, the members of each pair being of the same age and raised under identical conditions from the same parents. In each pair, one plant was fertilised by itself, the other was fertilised by pollen from another plant. The treatments, self- or cross-fertilisation, were randomly assigned within each pair. Seeds were eventually harvested from all plants and grown in randomly assigned plots. After a period of time, the heights of each pair of offspring were measured to the nearest eighth of an inch, and the difference in heights was estimated for each pair; see Table 1.

Under the null hypothesis that there is no difference between growth of self- and cross-fertilised plants, or that the mean difference is zero, the sign of any observed difference in each sampled pair is as likely to be positive as negative. Fisher computed the $2^{15} = 32{,}768$ possible sums of differences obtained by attaching a positive or negative sign to each of the 15 differences, and found that the absolute observed difference was exceeded by a positive or negative difference in 5.2% of the possible permutations of the data. He concluded that the experimental evidence was "scarcely sufficient" to refute the null hypothesis.

In this example, Fisher's use of permutation is justified both by invocation of random sampling (of parental plants from a population of peas) and by Darwin's exemplary experimental technique of randomly assigning treatments to units within each pair of parent plants, and within each pair of seeds. Experimental designs like this, in which the experimenter has the power randomly to assign a treatment of interest to observational units, guarantee exchangeability of the observations under the null hypothesis and thus the validity of a permutation test of that hypothesis.

2 Permutation Tests in Human Brain Mapping

Sixty years after Fisher's pioneering studies, the availability of powerful

Table 2. Some recent applications of permutation testing to human brain mapping

Year	Modality	Authors	Application
1994	EEG	(Blair and Karniski, 1994)	"omnibus" tests and multiple comparisons
1996	PET	(Holmes et al., 1996)	multiple comparisons procedures
1996	PET	(Arndt et al., 1996)	"omnibus" tests compared to theory
1996	fMRI	(Bullmore et al., 1996)	time series regression
1997	MRI	(Lange et al., 1997)	general additive modelling
1997	PET	(McIntosh et al., 1996)	partial least squares
1997	fMRI	(Brammer et al., 1997)	generic brain activation mapping

microprocessors has largely circumvented his argument against permutation tests on the grounds of tedium. And in the last 5 years, several groups have applied permutation tests to various problems in human brain mapping; see Table 2. These studies collectively illustrate some of the traditional strengths of permutation testing in a contemporary area of application.

2.1 *Versatility*

> *One of the most appealing properties of ... permutation tests lies in their adaptability to the research situation. Test statistics used in conjunction with classic parametric procedures must be expressed in forms that have known, or derivable, sampling distributions. This generally limits the domain of such statistics thereby excluding from consideration many otherwise attractive tests.* (Blair and Karniski, 1994)

Almost all human brain mapping studies using permutation tests have done so in the context of testing an innovative statistic, which might have a theoretically unknown or intractable distribution. Innovative statistics can be appealing even in relatively straightforward situations, such as a voxel-wise comparison of mean response between two sets of functional brain images; but they seem particularly attractive in the context of multivariate analysis, modern regression, spatial statistics or robust statistics.

For example, Blair and Karniski (1994) and Arndt et al. (1996) have both used permutation to test alternatives to Hotelling's T^2 statistic for an overall or omnibus difference between two sets of images. Holmes et al. (1996) estimated and tested by permutation a "pseudo-T" statistic, with locally pooled error variance in the denominator, as an alternative to the T statistic conventionally tested against a t distribution. McIntosh et al. (1996) proposed permutation to test the proportion of variance R^2 in a set of latent variable scores accounted for by a set of predictor variables, or to test the voxel-wise saliences of latent variable expression. Lange et al.

(1997) tested the ratios of non-Gaussian errors obtained by separately fitting a general additive model (GAM) to two sets of structural magnetic resonance (MR) images. Poline et al. (1997) suggested using permutation to test the "mass" of spatially contiguous voxel clusters generated by arbitrarily thresholding a smooth statistic image (an idea which is explored in further detail below). Finally, Brammer et al. (1997) extended methods initially developed by Bullmore et al. (1996) for analysis of a single functional MR image to analysis of a group of such images. Since sample sizes in fMRI are generally small, the test statistic for group activation tested by Brammer et al. (1997) is the median, rather than mean, power of experimentally determined signal change at each voxel.

Even if one is not interested in devising a novel test statistic, permutation tests have an important advantage for brain mapping in that permutation distributions can be conditioned on the correlational structure of the observed images.

> *[A permutation] test "automatically" adapts critical values to data dependencies without the necessity of estimating quantified representations of these dependencies. This has the important property of avoiding threats to test validity that often accompany such ... procedures.* (Blair and Karniski, 1994)

It is probably too strong to say that permutation *automatically* conditions the null distribution on the correlational structure of the observed data; but it is generally possible to design a permutation test to achieve this effect (Rabe-Hesketh et al., 1997).

Thus permutation tests have been shown to be versatile both in terms of the range of statistics that can be tested, and in terms of the correlational structure of those statistics estimated in an image of the brain. A third way in which permutation tests can be considered versatile is with respect to sampling. Parametric significance tests are only valid on the assumption that the data in hand represent a random sample from a population. We contend that this assumption of random sampling is almost never justified in human brain mapping.

At the First International Conference on Functional Mapping of the Human Brain (HBM 95), at Paris in 1995, a total of 410 studies were presented as abstracts. Half of these studies were based on data from 8 or fewer subjects; in about a third, the number of subjects studied was less than or equal to the number of authors (Vitouch and Gluck, 1997).

> *Small samples (often consisting of those rare people that volunteer for long and boring or stressful [brain mapping] experiments) are not representative of any population of interest.* (Vitouch and Gluck, 1997)

There are several possible reasons for this preponderance of small and unrepresentative (dodgy) samples, including high per capita costs of functional imaging, and the pressure on laboratories and individuals to maintain a high volume of published output regardless of sample size. But even in a more perfect world, where scanning was free and everybody had performance-unrelated tenure, would it be realistic to expect large, random samples of the population of interest?

> *It must be stressed that violation of the random sampling assumption invalidates parametric statistical tests not just for the occasional experiment but for virtually all experiments ... because experiments of a basic nature are not designed to find out something about a particular finite existing population ... the intention is to draw inferences applicable to individuals already dead and individuals not yet born, as well as those who are alive at the present time. If we were concerned only with an existing population, we would have extremely transitory biological laws because every minute some individuals are born and some die, producing a continual change in the existing population. Thus the population of interest in most experimentation ... cannot be sampled randomly.* (Edgington, 1980)

Edgington (1980) also draws attention to a number of other experimental situations, such as study of a single "interesting" subject or study of a sample that *is* the existing population, where assumptions of random sampling from a population are hard to justify.

Whether we believe that dodgy sampling in brain mapping is potentially remediable, by affording researchers more money and time, or an irreducible consequence of our Newtonian ambitions to discover universal laws of human brain function, it is clearly a problem for the validity of parametric analysis here and now. It is less of a problem for permutation testing provided we can convince ourselves in some other way that the data are exchangeable.

2.2 Type I Error Control

> *The importance of an exact test cannot be overestimated, particularly a test that is exact regardless of the underlying distribution. If a test that is nominally at level α is actually at level χ, we may be in trouble before we start. If $\chi > \alpha$, the risk of a Type I error is greater than we are willing to bear. If $\chi < \alpha$, then our test is suboptimal and we can improve on it by enlarging its rejection region.* (Good, 1994)

A significance test of size α can be called *exact* if, under the null hypothesis, the expected number of positive tests over a search volume of N voxels $= \alpha N$. A test is *conservative*, under the same circumstances, if the number of positive tests $\leq \alpha N$. An exact test is conservative; a conservative test need not be exact, it could be suboptimal. It can be shown that exchangeability of the data is a sufficient condition for a permutation test to be exact; see Lehmann (1986) for proof.

The exactness of the test can also be verified, without recourse to exchangeability, by calibrating it against simulated or experimental data for which the null hypothesis holds true. This is sometimes called demonstration of nominal Type I error control, and several groups have used this empirical approach to assess exactness of permutation tests in brain mapping. For example, Blair and Karniski (1994) used simulated EEG data to demonstrate that their permutation test of an alternative to Hotelling's T^2 statistic was exact. Bullmore et al. (1996) and Brammer et al. (1997) used "resting" fMRI datasets to demonstrate nominal Type I error control by permutation tests of standard power at an experimentally designed frequency in single subjects and groups. Arndt et al. (1996) demonstrated that a permutation test of $t_{\max}$ was exact, using simulated PET data, and that a theoretical test of the same data was over-conservative.

It is also possible to design permutation tests that achieve strong Type I error control. A test with strong control declares nonsignificant voxels to be significant with probability of overall error at most α, regardless of any truly significant voxels elsewhere in the image, and therefore has localising power.

Blair and Karniski (1994) describe sequentially rejective permutation tests of the maximal test statistic $t_{\max}$ which maintain strong Type I error control. Holmes et al. (1996) describes and applies computationally efficient algorithms for "step-down-in-jumps" permutation tests of $t_{\max}$ in PET data. These algorithms have recently been applied to functional MRI data analysis by Raz et al. (1998). Strong Type I error control by these methods requires ascertainment of the permutation distribution of a maximum statistic and is correspondingly time-consuming to compute:

> *Even so, considering the vast amounts of time and money spent on a typical functional mapping experiment, a day of computer time seems a small price to pay for an analysis whose validity is guaranteed.* (Holmes et al., 1996)

2.3 Power

For small samples, a parametric test may be more powerful than the comparable permutation test IF the assumptions of the parametric test are justified. Parametric tests in brain mapping make many assumptions and approximations.

> *In theory, assumptions can be checked. This is not routinely done .... In addition, the small data sets result in tests for*

departures from the assumptions with very low power. (Holmes et al., 1996)

There have been few direct comparisons of the power correctly to reject the null hypothesis by parametric and permutation tests of the same test statistic estimated in the same (necessarily simulated) data. However, Arndt et al. (1996) showed that a permutation test of $T_{\max}$ in "activated" PET data, simulated with a variety of effect sizes, more frequently rejected the null hypothesis than theoretical tests. These authors also make the point that the most powerful test will generally be of the statistic that best characterises the effect of interest, and the versatility of permutation tests allows a greater range of potential statistics to be considered.

3 Permutation Tests of Spatial Statistics

Suppose we have acquired brain images from two groups of subjects, and wish to test the (broad) null hypothesis that there is no difference between these groups of data. There are clearly several levels of spatial resolution at which the analysis could be conducted. We could estimate some *global* property of each image, and test the between-group difference in this. Or we could estimate and test a measure of between-group difference at each *voxel.* Global tests are likely to be relatively insensitive unless the difference between groups is diffuse; voxel tests are more sensitive to focal differences but entail a very large number of tests and neglect the spatially coordinated nature of the data. Poline and Mazoyer (1993) therefore advocated a third level of analysis, based on testing the 2-dimensional area of spatially contiguous voxel *clusters* generated by arbitrarily thresholding a voxel statistic image. This approach has the advantage of requiring fewer tests than analysis at voxel level, and has been shown to be more sensitive to effects in simulated PET data than either global (Poline and Mazoyer, 1993) or voxel tests (Friston et al., 1994).

The only major problem associated with use of cluster statistics seems to relate to their distribution under the null hypothesis. Several groups have used Monte Carlo procedures to sample the distribution of cluster size in images simulated under the null hypothesis (Poline and Mazoyer, 1994, Roland, 1993, Forman et al., 1995). Simulation necessitates making some assumptions about the distribution of the voxel statistic and the spatial autocorrelation or smoothness of the voxel statistic image under the null hypothesis. The null distribution of cluster area is conditional on both the smoothness W of the statistic image and the size of threshold u applied at voxel level. Therefore, cluster size distributions reported on the basis of Monte Carlo simulations are only appropriate for testing images which happen to have W and u identical to the arbitrary smoothness and voxel threshold of the simulated data (Frackowiak et al., 1996, Poline et al., 1997).

Friston et al. (1994) derived an exponential form for the null distribution of cluster size, which is more generally applicable to estimate the probability of a cluster as a function of the smoothness, voxel threshold and dimensionality of any statistic image. However, as these and other authors have commented (Roland and Gulyas, 1996, Rabe-Hesketh et al., 1997), this distribution is derived from results in Gaussian field theory that are exact only in the limit of very high thresholds, i.e., $u \approx 6$ (assuming the voxel statistic has standard Normal distribution). This is unfortunate because a test of cluster size is likely to be most useful in detecting relatively large sets of spatially connected suprathreshold voxels that can only arise if the threshold applied at voxel level is fairly low, i.e., $u \approx 2$ (Poline et al., 1997).

Here we develop and validate permutation tests for two spatial statistics estimated in brain images: i) 2-dimensional area of a suprathreshold cluster; and ii) 2-dimensional "mass"of a suprathreshold cluster. The methods are illustrated by application to MRI data acquired from 16 normal adolescents as part of a study of hyperkinetic disorder; see Bullmore et al. (1999) for further details. These 16 datasets were randomly divided into two groups of 8 images each.

3.1 MRI Data Acquisition and Preprocessing

Dual echo, fast spin echo MRI data were acquired at 1.5 Tesla in the sagittal plane parallel to the interhemispheric fissure using a GE Signa system (General Electric, Milwaukee WI, USA) at the Maudsley Hospital, London: TR = 4 s, TE_1 = 20 ms, TE_2 = 100 ms, in-plane resolution = 0.86 mm, slice thickness = 3 mm.

Extracerebral tissues were automatically identified and removed from the images by an algorithm using a linear scale-space set of features obtained from derivatives of the Gaussian kernel (Suckling et al., 1999). The proportional volume of grey matter was estimated at each intracerebral voxel by a logistic discriminant function (Bullmore et al., 1995). Maps of grey matter volume estimated for each image were registered in the standard space of Talairach & Tournoux by an affine transformation, and smoothed by a 2D Gaussian filter with full width at half maximum (FWHM) = 5.61 voxels or 6.53 mm (Brammer et al., 1997, Press et al., 1992).

3.2 Estimation of Voxel Statistics

Following registration of each individual's grey matter volume maps in standard space, we fitted the following linear model at each of approximately 300,000 voxels:

$$g_{i,j,k} = \mu_i + a_{i,j} + e_{i,j,k}. \tag{1}$$

Here $g_{i,j,k}$ is the grey matter volume estimated at the ith voxel for the kth individual in the jth group; μ_i is the overall mean volume at the ith voxel; $\mu_i + a_{i,j}$ is the mean volume for the jth group at the ith voxel; and $e_{i,j,k}$ is a residual quantity unique to the kth individual in the jth group. A possible test statistic is the standardised coefficient $A = a/\text{SE}(a)$. This statistic can, of course, be tested against a t distribution with $N_1 + N_2 - p$ df, where $N_1 = N_2 = 8$ is the number of subjects in each group, and $p = 2$ is the number of parameters to be estimated. Alternatively, we can test A by the following permutation test algorithm:

1. Calculate $A = a/\text{SE}(a)$ at each voxel of the observed data

2. Randomly reassign each individual to one of two groups and re-estimate A after each permutation

3. Repeat step two 10 times, and pool the resulting estimates over all voxels to sample a permutation distribution comprising approximately 3,000,000 estimates of A (including the observed statistics)

4. Compare the observed statistics against percentiles of the sampled permutation distribution for a two-tailed test of arbitrary size

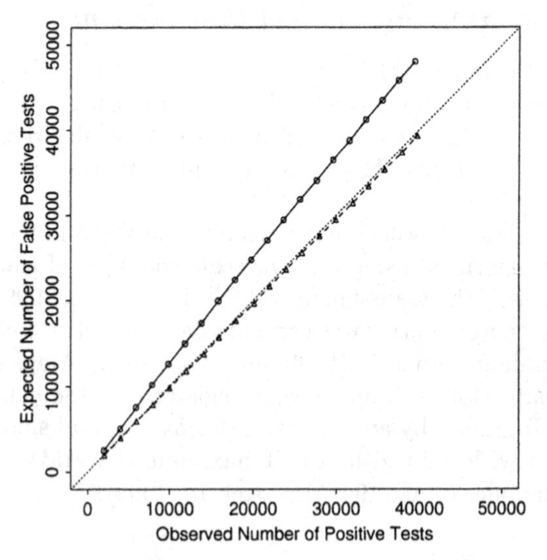

Figure 1. Type I error calibration curves for voxel level tests by theory and permutation. The *expected* number of false positive voxels over a range of sizes of test $0.005 < P \leq 0.1$ can be compared directly to the *observed* number of "significant" voxels following tests by theory (solid line; circles) and permutation (dashed line; triangles). The two groups tested were randomly decided subsets of 16 healthy volunteers, therefore the expected and observed numbers of positive tests should be equal (dotted line).

The practice of conducting a small number of permutations at each voxel, then pooling the results over all voxels, makes the process more tractable computationally. Nominal Type I error control for this procedure can be calibrated by comparing the observed number of false positive tests to the expected number of false positive tests; see Figure 1. However, note that if one wished to test voxel (or cluster) statistics in the observed data against the null distribution of the *maximum* such statistic over the whole image then clearly each permutation would only contribute one element to the null distribution, so the number of permutations would have to be much larger than 10.

Since the two groups in this case are randomly decided subsets of a presumably homogeneous sample of healthy volunteers, the observed and expected number of tests should be equal. It can be seen that the permutation test is exact, whereas the theoretical test tends to be over-conservative.

3.3 Cluster Level Tests

An arbitrary threshold of size $u = 2$ was then applied to the voxel statistic $|A|$ image to generate a set of spatially contiguous suprathreshold voxel clusters. Two test statistics were measured for each such cluster: i) the number of spatially connected voxels comprising each cluster, i.e., cluster area; and ii) the absolute sum of suprathreshold voxel statistics over all voxels comprising each cluster, i.e., cluster mass.

Two-dimensional cluster area ν can be tested against critical values of a theoretically derived exponential distribution Friston et al. (1994):

$$P(\nu = x) = \psi e^{-\psi x} \tag{2}$$

$$\psi = \frac{(2\pi)^{-3/2} W^{-2} u e^{-u^2/2}}{\Phi(-u)} \tag{3}$$

$$W = \frac{\text{FWHM}}{\sqrt{4 \log 2}} \tag{4}$$

where Φ denotes the standard Normal cumulative distribution function.

There is no well-established theoretical distribution for cluster mass.

Alternatively, both statistics can be tested by the following permutation algorithm:

1. Calculate $|A| = |a/\text{SE}(a)|$ at each voxel of the observed data; apply threshold u to the resulting statistic image
2. Measure area and mass of each observed suprathreshold cluster
3. Randomly reassign each individual to one of two groups, applying an identical permutation at each voxel to preserve the spatial correlational structure of the observed images

4. Re-estimate $|A|$ after each permutation and apply threshold u to the permuted statistic image
5. Measure area and mass of each permuted suprathreshold cluster
6. Repeat the previous three steps 10 times and pool the resulting estimates of mass and area over clusters to sample the two permutation distributions
7. Compare observed cluster statistics against percentiles of the appropriate permutation distribution for one-tailed tests of arbitrary size.

Null distributions of cluster area ascertained by permutation and theory can be compared in Figure 2; the permutation distribution of cluster mass is shown in Figure 3. Nominal Type I error control by permutation tests of both cluster area and mass was calibrated by analysis of two randomly decided subsets of the sample (as above); see Figures 4 and 5. Type I error control by permutation testing can also be compared to nominal Type I error control by a theoretical test of cluster area; see Figure 4. It can be seen that permutation tests of both area and mass are exact, whereas the theoretical test of cluster area is over-conservative.

These results show that permutation tests for spatial statistics estimated in human brain mapping can be designed and evaluated quite simply. Permutation algorithms allow a wider range of spatial statistics to be tested, and are empirically exact.

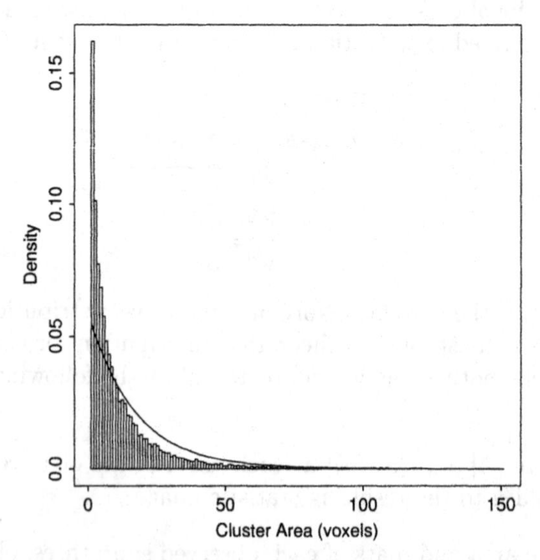

Figure 2. Theoretical and permutation distributions for the 2D area of suprathreshold voxel clusters under the null hypothesis. The histogram shows the permutation distribution; the solid line shows the theoretical distribution, Eq. (2) with parameters FWHM = 5.67 voxels and $u = 1.85$.

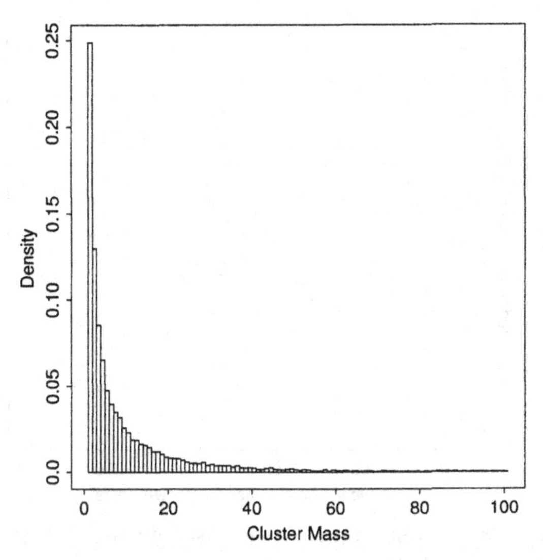

Figure 3. Permutation distribution for the mass of suprathreshold voxel clusters. There is no well-established theoretical distribution for this statistic.

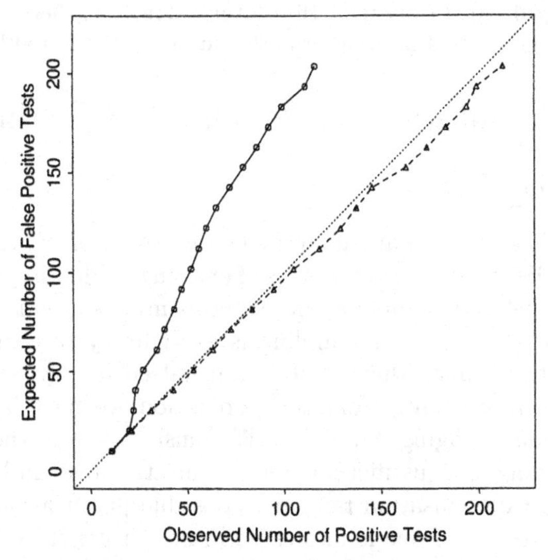

Figure 4. Type I error calibration curves for tests of cluster area by theory and permutation. The *expected* number of false positive clusters over a range of sizes of test $0.005 < P \leq 0.1$ can be compared directly to the *observed* number of "significant" clusters following tests by theory (solid line; circles) and permutation (dashed line; triangles). The two groups tested were randomly decided subsets of 16 healthy volunteers, therefore the expected and observed numbers of positive tests should be equal (dotted line).

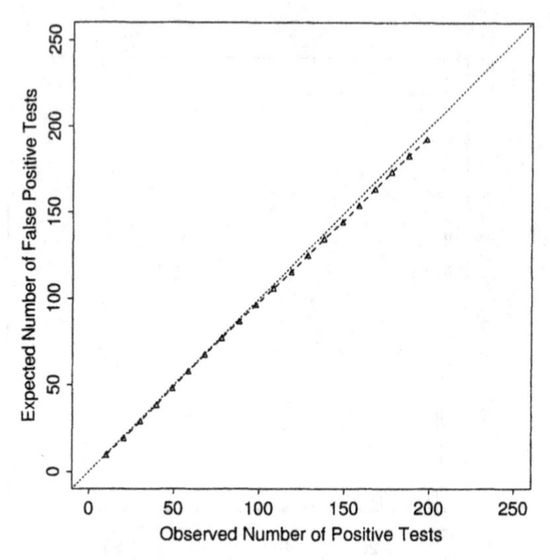

Figure 5. Type I error calibration curve for tests of cluster mass by permutation. The *expected* number of false positive clusters over a range of sizes of test $0.005 < P \leq 0.1$ can be compared directly to the *observed* number of "significant" clusters following tests by permutation (dashed line; triangles). The two groups tested were randomly decided subsets of 16 healthy volunteers, therefore the expected and observed numbers of positive tests should be equal (dotted line).

4 Some Limitations of Permutation Testing

4.1 *Exchangeability*

Exchangeability of the data guarantees exactness of a permutation test (Lehmann, 1986); and two guarantees of exchangeability implicit in Fisher's examples are random sampling and experimental randomisation. We have already noted that random sampling is exceedingly rare in human brain mapping (Vitouch and Gluck, 1997), and indeed it may be impossible to achieve random sampling from the population we most aspire to make inferences about (Edgington, 1980). We must probably therefore eschew random sampling as a justification for permutation tests in brain mapping. Experimental randomisation remains a possible justification in some situations. For example, if one is interested in measuring the effects of a drug on cerebral physiology using fMRI, one could randomly assign a group of subjects to one of two possible treatments (active drug or placebo), and test the observed difference between treatments by permutation. Similarly, if one is interested in measuring functional MR signal change in response to two or more stimulus conditions then, providing the stimuli were originally presented in random order, one can conduct an exact permutation test by randomly reassigning stimulus types to blocks of time in the experimental input function (Raz et al., 1998). But what about situations where the "treatment" effect of interest is not under experimental control? For example, one might be interested in whether patients with schizophrenia (or any other disease) have a different physiological response to some experimental stimulus compared to a group of normal volunteers. Clearly we cannot randomly assign a group of initially normal individuals to be schizophrenic or not for the duration of the experiment. In such a case we may have to justify exchangeability on the grounds that "the null hypothesis implies that the mechanism generating the data makes all of the possible orderings equally likely to arise" (Manly, 1991).

It remains questionable how much it matters if we have to justify exchangeability by recourse to a model of the "mechanism generating the data" if that mechanism is not as well known to us as, for example, the experimental design generating Darwin's pea data. If there are any doubts concerning arguments for exchangeability, it seems advisable to assess the exactness of a proposed permutation test empirically.

4.2 Null hypothesis testing only

Permutation tests do not allow assessment of the probability of a statistic under the alternative hypothesis. It may be advantageous to consider the distribution of a test statistic observed over all voxels as a mixture of a null distribution and an alternative distribution (Everitt and Bullmore, 1999). In the context of brain mapping, the estimated parameters of such a theoretically specified mixture model include an overall measure of effect size and an estimate of the proportion of voxels distributed under the alternative hypothesis, both of which may be useful as omnibus statistics. Mixture modelling also allows significant voxels to be identified by applying a posterior probability threshold which reflects some arbitrary balance between the risks of both Type I and Type II error.

5 Conclusion

On balance, the case for permutation testing in human brain mapping seems strong. Permutation tests are conceptually simple, versatile, valid on the basis of far fewer assumptions than are required for theoretical hypothesis testing, and likely to be at least as powerful as the best parametric test. They are also no longer too tedious to be of widespread use or worthy of serious statistical attention.

Acknowledgments: EB was supported by the Wellcome Trust.

6 References

Arndt, S., T. Cizadlo, N.C. Andreasen, D. Heckel, S. Gold, and D.S.O Leary (1996). Tests for comparing images based on randomisation and permutation methods. *Journal of Cerebral Blood Flow and Metabolism 16*, 1271–1279.

Blair, R.C. and W. Karniski (1994). Distribution-free statistical analyses of surface and volumetric maps. In R. Thatcher, M. Hallett, T. Zeffiro, E. Roy, and M. Huerta (Eds.), *Functional Neuroimaging: Technical Foundations*, pp. 19–28. New York: Academic Press.

Brammer, M.J., E.T. Bullmore, A. Simmons, S.C.R. Williams, P.M. Grasby, R.J. Howard, P.W.R. Woodruff, and S. Rabe-Hesketh (1997). Generic brain activation mapping in fMRI: a nonparametric approach. *Magnetic Resonance Imaging 15*, 763–770.

Bullmore, E.T., M.J. Brammer, G. Rouleau, B.S. Everitt, A. Simmons, T. Sharma, S. Frangou, R.M. Murray, and G. Dunn (1995). Computerized brain tissue classification of magnetic resonance images: A new approach to the problem of partial volume artefact. *NeuroImage 2*, 133–147.

Bullmore, E.T., M. Brammer, S. Williams, S. Rabe-Hesketh, C. Janot, A. David, J. Mellers, R. Howard, and P. Sham (1996). Statistical methods of estimation and inference for functional MR images. *Magnetic Resonance in Medicine 35*, 261–277.

Bullmore, E.T., J. Suckling, S. Overmeyer, S. Rabe-Hesketh, E. Taylor, and M.J. Brammer (1999). Global, voxel and cluster tests, by theory and permutation, for a difference between two groups of structural MR images of the brain. *IEEE Transactions on Medical Imaging 18*, 32–42.

Edgington, E.S. (1980). *Randomization Tests.* New York: Marcel Dekker.

Everitt, B.S. and E.T. Bullmore (1999). Mixture model mapping of brain activation in functional magnetic resonance images. *Human Brain Mapping 7*, 1–14.

Fisher, R.A. (1935). *The Design of Experiments.* Edinburgh: Oliver & Boyd.

Fisher, R.A. (1936). The coefficient of racial likeness and the future of craniometry. *Journal of the Royal Anthropological Society 66*, 57–63.

Forman, S.D., J.D. Cohen, M. Fitzgerald, W.F. Eddy, M.A. Mintun, and D.C. Noll (1995). Improved assessment of significant activation in functional magnetic resonance imaging fMRI: use of a cluster size threshold. *Magnetic Resonance in Medicine 33*, 636–647.

Frackowiak, R.S.J., S. Zeki, J.-B. Poline, and K.J. Friston (1996). A critique of a new analysis proposed for functional neuroimaging. *European Journal of Neuroscience 8*, 2229–2231.

Friston, K.J., K.J. Worsley, R.S.J. Frackowiak, J.C. Mazziotta, and A.C. Evans (1994). Assessing the significance of focal activations using their spatial extent. *Human Brain Mapping 1*, 214–220.

Good, P. (1994). *Permutation Tests.* Springer Series in Statistics. New York: Springer.

Holmes, A.P., R. Blair, J. Watson, and I. Ford (1996). Nonparametric analysis of statistical images from functional mapping experiments. *Journal of Cerebral Blood Flow and Metabolism 16*, 7–22.

Lange, N., J.N. Giedd, F.X. Castellanos, A.C. Vaituzis, and J.L. Rapoport (1997). Variability of human brain structure size: ages 4–20 years. *Psychiatry Research: Neuroimaging 74*, 1–12.

Lehmann, E.L. (1986). *Testing Statistical Hypotheses.* Wiley Series in Probability and Mathematical Statistics: Probability and Mathematical Statistics. New York: Wiley.

Manly, B. J. F. (1991). *Randomisation and Monte Carlo Methods in Biology.* London: Chapman and Hall.

McIntosh, A.R., F.L. Bookstein, J. Haxby, and C. Grady (1996). Multivariate analysis of functional brain images using partial least squares. *NeuroImage 3*, 143–157.

Poline, J.-B. and B.M. Mazoyer (1993). Analysis of individual tomography activation maps by clusters. *Journal of Cerebral Blood Flow and Metabolism 13*, 425–437.

Poline, J.-B. and B.M. Mazoyer (1994). Analysis of individual brain activation maps using hierarchical description and multiscale detection. *IEEE Transactions on Medical Imaging 13*, 702–710.

Poline, J.-B., K.J. Worsley, A.C. Evans, and K.J. Friston (1997). Combining spatial extent and peak intensity to test for activations in functional imaging. *NeuroImage 5*, 83–96.

Press, W.H., S.A. Teukolsky, W.T. Vetterling, and B. Flannery (1992). *Numerical Recipes: The Art of Scientific Computing in C* (2nd ed.). Cambridge: Cambridge University Press.

Rabe-Hesketh, A.S., E.T. Bullmore, and M.J. Brammer (1997). Analysis of functional magnetic resonance images. *Statistical Methods in Medical Research 6*, 215–237.

Raz, J., B. Turetsk, and C. Liu (1998). Statistical analysis of functional magnetic resonance images acquired using the auditory P300 paradigm. *NeuroImage 7*, S603.

Roland, P.E. (1993). *Brain Activation.* New York: Wiley-Liss.

Roland, P.E. and B. Gulyas (1996). Assumptions and validations of statistical tests for functional neuroimaging. *European Journal of Neuroscience 8*, 2232–2235.

Suckling, J., M.J. Brammer, A. Lingford-Hughes, and E.T. Bullmore (1999). Removal of extracerebral tissues in dual-echo magnetic resonance images via linear scale-space features. *Magnetic Resonance Imaging 17*, 247–256.

Vitouch, O. and J. Gluck (1997). Small group PETting: sample sizes in brain mapping research. *Human Brain Mapping 5*, 74–77.

10

Group Analysis of Individual Activation Maps Using 3D Scale-Space Primal Sketches and a Markovian Random Field

Olivier Coulon, Jean-François Mangin, Jean-Baptiste Poline, Vincent Frouin, and Isabelle Bloch

ABSTRACT We present here a new method of cerebral activation detection. This method is applied on individual activation maps of any sort. It aims at processing a group analysis while preserving individual information and at overcoming as far as possible problems induced by spatial normalization used to compare different subject. The analysis is made through a multi-scale object-based description of the individual maps and these descriptions are compared, rather than comparing directly the images in a stereotactic space. The comparison is made using a graph, on which a labeling process is performed. The label field on the graph is modeled by a Markovian random field, which allows us to introduce high-level rules of interrogation of the data.

1 Introduction

Understanding the neural correlates of human brain function is a growing field of research. Brain activation detection has essentially been approached so far in terms of statistical analysis (Friston et al., 1995, Worsley et al., 1996) using a common anatomical reference. Although they have been validated in a wide range of applications, these analyses lead to some problems in terms of localization and/or detection with regard to anatomy, mainly because the spatial normalization performed to compare images from different subjects only matches gross features. Moreover, anatomical information is poorly considered, and after a statistical analysis, it is generally difficult to estimate from the group result the individual activated areas in terms of shape, extent, or position with respect to anatomical landmarks. This knowledge should help to study inter-subject functional and anatomical variability and would improve localization with regard to anatomy. We propose here a new method based on a description of individual activation maps in terms of structures. This is followed by the comparison of these descriptions across subjects, rather than comparing directly the images in a stereotactic space at a pixel level. The method is designed to overcome, as far as possible, the problems induced by spatial normalization. After detection over a group of subjects, the method allows an easy way back to the individual structures, and more generally permits high level interrogation (at a structure level) and in the future more informed interrogation of functional data sets.

2 Method

The method presented here is applicable to any kind of "activation maps" PET/fMRI difference images, statistical t-maps,.... It is divided into the three following steps: first, each individual map involved in the study is described by a hierarchical object-based multiscale representation, namely the scale-space primal sketch. Second, a graph is built, which contains all of the primal sketches and represents the potential repetitions of a structure from one subject to another. Finally, a labeling process is performed on the graph, which aims at identifying the objects representing functional activations.

2.1 The Scale-Space Primal Sketch

The scale-space primal sketch is a multiscale description of an image, proposed by Lindeberg and Eklundh (1992). It is based on well-known properties of linear scale-space to describe the structure of an image (Koenderink, 1984, Koenderink and van Doorn, 1986). We present here briefly the way it is built. For more precise details, we invite the reader to refer to Lindeberg and Eklundh (1992), or to Coulon et al. (1997) for the particular three-dimensional case applied on activation maps.

From a three-dimensional activation map, we compute a linear scale-space (either by iteration of the Heat equation (Koenderink, 1984), or by convolution of the original image by Gaussian functions of increasing variance). At each level of scales, i.e., at each level of smoothing, objects—called *grey-level blobs*—are extracted in a fully automatic way, based on singular points in the image (Figure 1). One and only one grey-level blob is associated with each local maximum of the intensity.The aim is to make explicit the behavior of these objects through the scales, such that we can describe the whole structure of the image. We then link these grey-level blobs from one scale to another, and try to detect events in the scale-space. These events can be theoretically itemized in four types (annihilation, apparition, split, and merge) and are called *bifurcations*. Between two bifurcations, a single grey-level blob is evolving through the scales, and is identified as a

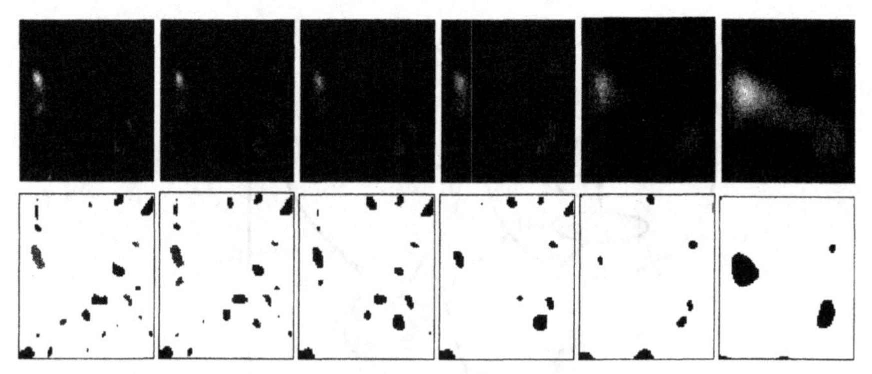

Figure 1. A slice of a 3D activation map, at scales $t = 1, 2, 4, 8, 16$, and 32, with the corresponding grey-level blobs

multiscale object called *scale-space blob*. The primal sketch is then composed of scale-space blobs linked by bifurcations, as described in Figure 2. Measurements are associated with each scale-space blob. These (multiscale) measurements are related to intensity, contrast, geometrical volume, and lifetime along the scales, of the associated scale-space blob.

It has been shown that scale-space blobs are a reliable and exhaustive way to describe objects of interest in activation maps (Coulon et al., 1997). However, measurements are not efficient enough to characterize which of these blobs are representing activations. We then need to look at several subjects to take the detection decision, and this is the aim of the process presented in the following sections.

2.2 The Comparison Graph

We want to create a *comparison graph* such that it contains the primal sketches of all the subject involved in the analysis, and such that this graph makes explicit the potential repetitions of a structure (i.e., a scale-space blob) over the subjects. At this point, we do not yet want to decide if these repetitions are real (i.e., associated to an activation repeated over subjects) or random. The decision will be made at the next step (Section 2.3). Our constraint is to be exhaustive and make the graph such that no *real* repetition is forgotten in it. Therefore, the graph nodes are scale-space blobs of the primal sketches involved in the process, and the graph links are built between blobs belonging to different primal sketches and, which may represent the same structure (activation).

To compare different primal sketches we normalize them in a common referential frame, with usual normalization procedures (Ashburner and Friston, 1997). This is the only way we have at the moment to compare different subjects, but a longer term aim is to build the spatial referential using the subjects individual anatomy (Mangin et al., 1995). The comparison graph

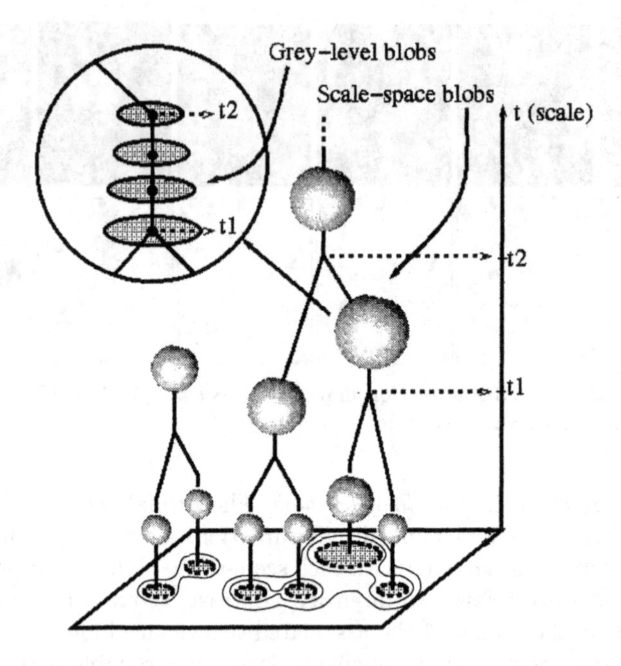

Figure 2. A simplified representation of the scale-space primal sketch (here in a two-dimensional case).

should be a convenient framework for this purpose.

The first obvious criterion to link two blobs belonging to two primal sketches is the overlap of their spatial support. If two blobs have this overlap, they may represent the same activation. We combine this spatial overlap criterion with a scale overlap (since scale-space blobs have a scale dimension), as far as the considered individual maps have the same inner scale. If these two criteria are fulfilled, we create a *direct* link between the two scale-space blobs b_1 and b_2, noted $[b_1 - b_2]_{\text{out1}}$.

Since we want to introduce some flexibility in the position of activation, to overcome potential normalization problems, we have to allow blobs without any spatial overlap but close enough to be linked in the graph. Note that very small focal activations may also have no overlap. We therefore use the fact that information becomes more stable (but grosser) as scale increases: two blobs being linked by a direct *out1* link may not represent an activation but may roughly indicate an area in which there is a more focal activation. This situation suggests that at a finer scale there might be a blob at the same location that represents an activation, this blob being in the part of the primal sketch which is "under" the first blob. This experimental observation led us to define a second type of link, that we called *induced* links (or *out2* links for mathematical notations): if b_1 and b_2 are two blobs having no direct link between them, they have an induced link

Figure 3. Direct links and the corresponding induced links

$[b_1 - b_2]_{\text{out2}}$ if they are "under" (in their primal sketches) two blobs c_1 and c_2, and there is a direct link $[c_1 - c_2]_{\text{out1}}$. Such links are created only when b_1 and b_2 overlap in the scale dimension, which reduces the combinational possibilities.

The building of the graph is illustrated in a simplified way on Figure 3. We point out the fact that allowing blobs without spatial overlap to be linked is a key point of the process, since it is the solution for a greater flexibility overcoming spatial normalization limitations.

2.3 The Detection Model: Use of a Markovian Random Field

Activation detection is performed using a labeling process that uses the inter-subject comparison graph previously described. Our aim is to associate a positive label to each activation in the graph, and a null label to structures of noninterest. An activation (i.e., a positive label) represents a localization, and can therefore have only one occurrence in each of the individual primal sketches.

We present here the model we use to perform the detection. This model is defined using the following set of rules:

(i) a blob representing an activation is likely to have high measurements.

(ii) two blobs representing the same activation must be linked in the graph and have the same positive label.

(iii) two blobs representing the same activation (same positive label) are likely to have spatial supports close to each other.

(iv) an activation is represented only once in each subject, i.e., one positive label should have only one occurrence per primal sketch.

(v) a blob representing an activation is likely to appear at a fine scale in its primal sketch.

One can see that these criteria define *local dependencies* only: the knowledge of a blob label depends only on the characteristics of the blob and on the knowledge of the labels of its neighbors in the graph. In other words, the model specifies that a blob represents an activation if it is endowed with high enough measurements in the map and if the corresponding structure is repeated over the other subjects. These dependencies allow us to model the label field in the graph as a Markovian random field. The labeling process can thus be performed by the maximization of the posterior probability $P(X \mid Y)$, X being the *label field* on the graph, and Y being the *data*, i.e., the characteristics of the blobs (measurements, scale). We want to find the optimal labeling given the data, following a *maximum a posteriori* process. The Bayes rule:

$$P(X \mid Y) = \frac{P(Y \mid X)P(X)}{P(Y)}, \tag{1}$$

tells that this maximization is equivalent to maximize the product $P(X) \times P(Y \mid X)$, $P(Y \mid X)$ expressing the knowledge on the measurements given the labels, and the prior probability $P(X)$ expressing the a priori knowledge (the model itself).

With the usual independency hypothesis, one has $P(Y \mid X) = \prod_{s=1}^{N} P(y_s \mid x_s)$, $s = 1, \ldots, N$ being the N sites (blobs) in the graph, y_s and x_s being the realisations of X and Y at site s. Finally one can write:

$$P(Y \mid X) \propto \exp(-\sum_{s=1}^{N} V(y_s \mid x_s)) = \exp(-U(Y \mid X)), \tag{2}$$

where $V(y_s \mid x_s)$ is a *potential* function.

The Markov hypothesis tells that $P(x_s \mid X_{S \setminus s}) = P(x_s \mid x_{\nu_s})$: the probability of realization of X on a site only depends on the realization in a neighborhood ν_s of this site. If the *positivity condition* is respected, i.e., no realization has a null probability ($\forall X,\ P(X) > 0$), the Hammersley–Clifford theorem (Besag, 1974, Geman and Geman, 1984) tells that the Markov field is equivalent to a Gibbs field, i.e., one can write:

$$P(X) = \frac{\exp(-U(X))}{Z}, \tag{3}$$

where Z is a normalization constant (the *partition function*); $U(X)$ is an energy function defined as a sum of potential function on the *cliques* in the graph. Cliques are sets of sites that are all neighbors in the graph. The *order* of the clique is the number of elements in this set. Thus, one has: $U(X) = \sum_{i=1}^{N_c} V_i(X)$, where $i = 1, \ldots, N_c$ are the cliques in the graph. With equations (1), (2), and (3), the maximization of the posterior probability $P(X \mid Y)$ can be written as the minimization of the following energy:

$$U(X \mid Y) = \sum_{s=1}^{N} V(y_s \mid x_s) + \sum_{i=1}^{N_c} V_i(X). \tag{4}$$

Such a representation is of great interest because we do not need to know the exact form of the label field distribution: the model is directly defined through the potential functions.

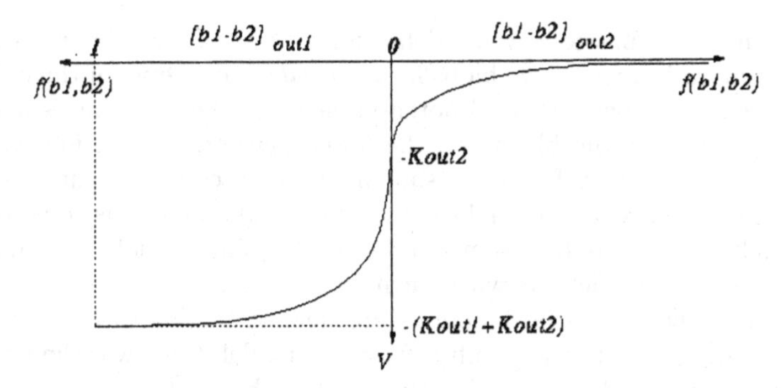

Figure 4. The potential function on second order cliques

More precisely, criterion (i) is modeled using the data-driven potential $V(y_s|x_s)$, criteria (ii) and (iii) are modeled using potential functions on 2nd order cliques, criterion (iv) is expressed on 1st order cliques (blobs), and criterion (v) appears in potential functions on cliques defined over individual primal sketches.

As an example of this modeling through potentials, we present here the definition of second order clique potentials. This potential function is particularly important since it tends to give the same label to blobs being neighbors in the graph. The idea here is to measure, for two neighboring blobs in the graph, how similar these two blobs are. The more similar they are, the lower the potential function is, and then the more it encourages them to have the same positive label. We therefore define the potential V_2 between two blobs b_1 and b_2 belonging to different primal sketches and being linked in the graph by an *out1* or *out2* link:

$$\text{if } l_{b_1} > 0,\ l_{b_2} > 0,\ l_{b_1} = l_{b_2},$$

$$\begin{cases} [b_1 - b_2]_{\text{out}_1} \implies V_2 = -K_{\text{out1}}.\frac{e^{-f(b_1,b_2)}-1}{e^{-1}-1} - K_{\text{out2}}, \\ [b_1 - b_2]_{\text{out}_2} \implies V_2 = -K_{\text{out2}}.e^{-f(b_1,b_2)}, \end{cases} \tag{5}$$

$$\text{else} \quad V_2 = 0,$$

where l_b stands for the label of blob b, K_{out1} and K_{out2} for the weights of the potential, and $f(.,.)$ for the *similarity function.* The potential function V_2 is shown in Figure 4 for the case $l_{b_1} > 0$, $l_{b_2} > 0$, $l_{b_1} = l_{b_2}$.

One can see that this definition relies on the similarity function. This similarity function is the second way, in the whole process, to overcome problems induced by spatial normalization. At the moment, we use an overlapping function for blobs being linked by an *out1* link, and an Euclidean distance function for those being linked by a *out2* link. The aim, in a longer term, is to have an anatomy-related similarity function. If one can identify any location in the brain according to individual anatomy, we will be able to define an anatomical similarity function that tells how far two blobs are to each other with respect to individual anatomical landmarks. Somehow, these similarity functions would define a new kind of spatial normalization.

To summarize, when two blobs have an *out1* links between them, if they have the same label, the local potential associated to the link is negative. The more similar the blobs are, the lower the potential is, following the function presented in Figure 4 (similarity increases from right to left). If the two blobs have different labels, the associated potential is equal to zero. This definition tends to assign same labels, to similar blobs over different primal sketches, which fits with our model.

All the other potential functions are defined in this way: the closer a local configuration in the graph is from the model, the lower the potential function associated to this situation is. On the contrary, some unlikely situations are associated to high positive potential functions. For instance, a positive potential controls the presence of several occurrences of the same positive label in one primal sketch, to satisfy rule (iv). The minimization process is then a competition between negative and positive potential values to reach the global configuration that fits the best with the model.

Once the potential functions are defined, the whole energy function is defined and is parameterized by weighting each potential function.

To minimize the energy function, we use a stochastic algorithm, the *Gibbs sampler* with annealing (Geman and Geman, 1984), shown to provide a good convergence.

3 Results

The process presented here has been tested on a PET motor protocol, including 10 subjects and 12 images per subject. For each subject, an individual statistical t-map was first computed using the *SPM* software (Functional Imaging Laboratory, London, Friston et al., 1995). A primal sketch is then built from each of the maps, and the 10 primal sketches are compared using the labeling process.

Three different conditions (4 images per conditions) were performed by the subjects: a rest condition (R), a periodic auditory-cued right hand movement (M_1), and a self-paced right hand movement (M_2). Individual maps were made using the 12 images of each subject, and looking at the contrast ($M_1 - R$). A group analysis was performed using the *SPM* software, and used as a reference to validate our results. Numerous activations were found at a very significant level in the group map in expected brain regions.

After the labeling process, 13 positive labels were found. Expected activations were detected together with two false positive, both outside the brain and caused by border effects during the blob extraction (they are therefore easy to eliminate and do not prevent the correct interpretation of the maps). Some of the activations (positive labels) had no occurrence in one or two of the individual maps, and we could check that these had very

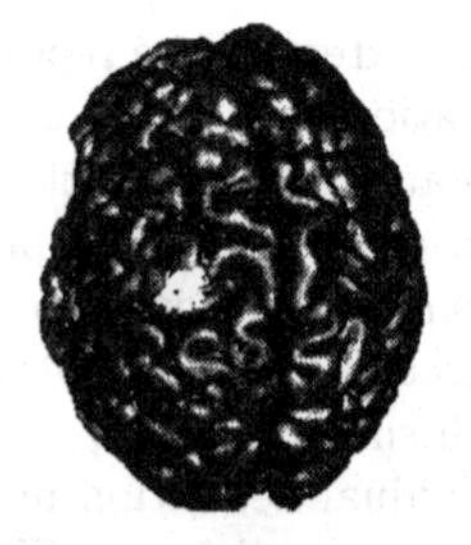
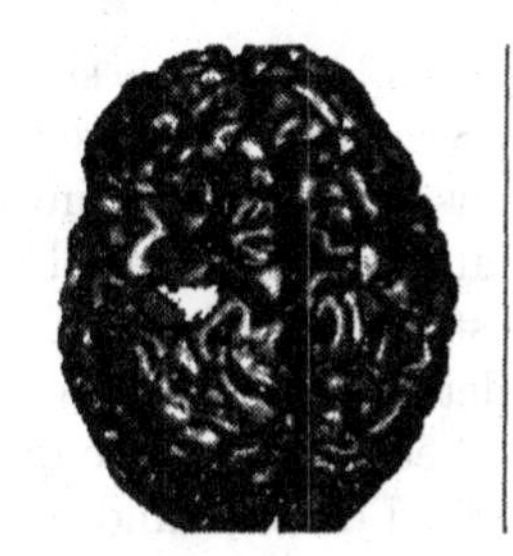
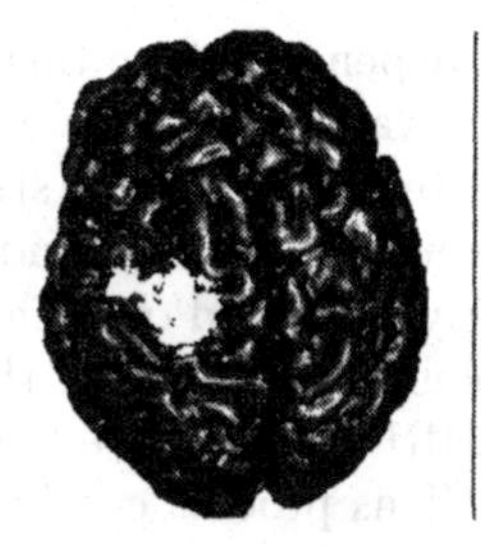

Figure 5. The sensory-motor right-hand activation of 3 subjects (in white) on their individual anatomy

low intensity in the corresponding individual maps. A classical threshold on the individual maps yielded poor results, either too selective, or too noisy, and this shows a crucial advantage of our process: the detection is processed for each subject taking into account not only the intensity in the map but also the knowledge of the other subjects maps.

This study also showed that the local energy function associated with each label is a good measure of the adequacy with the model. This was further confirmed by simulations. Particularly, when a positive label corresponds to no real activation, its local energy is high enough to discriminate it from labels associated with real activations.

Figure 5 shows the occurrence of the right-hand motor activation on 3 subjects, on a rendering of their individual anatomy (Mangin et al., 1998). One can see that the shape and extent are quite variable, although the 3 corresponding blobs have been detected as representing the same activation (same positive label in the graph). The location is very sensible since they are all in the right-hand sensori-motor area (along the left central sulcus (Yousry et al., 1997)).

4 Conclusion

We have presented here a new method to analyze brain functional images that considers the detection at a structural level, and permits a way back to individual results after the detection over a group of subjects. It uses the power and comprehensiveness of multiscale methods to describe images structure, by looking at their whole scale-space without any a priori about scales of interest, and without any "coarse-to-fine" strategy. The process has been proved to be able to efficiently detect expected activations on a PET dataset. It is promising for functional MRI studies, since fMRI provides better individual maps than PET. Further research still has to be undertaken to solve outstanding questions, particularly concerning the choice of the optimal scale used to *represent* (as opposed to detect) a scale-space blob. Detection and interpretation of the results are in fact not completely independent from this representation scale since the extent of the reported activations depends on it. Secondly, a precise evaluation of the data-driven potential functions still has to be further investigated. For the latter, it is important to provide data-driven potential functions directly and properly evaluated from the distribution of the measurements associated to the blobs. In that way, the detection will be more accurate, especially with fMRI for which we have a better confidence in each subject map.

This process can be applied on any kind of individual activation maps: PET difference images, individual statistical maps computed from PET or fMRI sequences. It is particularly interesting in the latter case, because good individual analyses are possible with fMRI, while group analysis are theoretically more difficult. Furthermore, we showed that spatial normalization spurious effects could be reduced by means of the comparison graph and of an appropriate definition of similarity between blobs from different subjects.

Although it is difficult to relate the proposed analysis to standard statistical analyses, it is worth noting that there is some kind of analogy with analyses that use random effects linear models, in the sense that activation detection is performed using a subject by subject variability rather than a scan by scan variability. We also note that if an experimentwise error was to be defined on the proposed procedure, it would be analogous to an error rate based on the size of a region rather than on its peak height. The choice of the representation scale also relates in some (loose) sense to the choice of the threshold that defines the regions to be tested in a "spatial extent" test, that itself speaks to the issue of regional specificity (how sure are we that all the brain structure(s) represented are involved in the brain responses to the imaging experimental paradigm?).

Note that such an approach is not opposed to classical statistical methods, since it uses individual maps made with these methods. We advocate that both approaches are complementary. Statistical methods present the great advantage to summarize the data, a model of the experimental design, and the information related to a particular question in a single image. At an individual level, with no inter-subject variability, these maps are very relevant. Our method provides a efficient group analysis using these individual maps.

Finally, we would like to point out the fact that using a Markovian model for the detection allows the user of such a system to interrogate the data in ways that can be designed according to what he is looking for. It is very easy to define new potential functions in which one can introduce, for instance, a priori information about a precise expected location, or about the search for a network of activations instead of isolated activations. That makes the system able to investigate functional data sets in a much higher level way than what was done so far.

Acknowledgments: The authors would like to thanks D. Romanov, M. Zilbovicius, and Y. Samson, for providing data, for their help, and for the interest they payed to this work.

5 References

Ashburner, J. and K. Friston (1997). Spatial transformation of images. In R. Frackowiak, K. Friston, C. Frith, R. Dolan, and J. Mazziotta (Eds.), *Human Brain Function*, pp. 43–58. New York: Academic Press.

Besag, J. (1974). Spatial interaction and the statistical analysis of lattice systems (with discussion). *Journal of the Royal Statistical Society. Series B. Methodological 36*, 192–236.

Coulon, O., I. Bloch, V. Frouin, and J.-F. Mangin (1997). Multiscale measures in linear scale-space for characterizing cerebral functional activations in 3D PET difference images. In B. Haar Romeny, L. Florack, J. Koenderink, and M. Viergever (Eds.), *Scale-Space Theory in Computer Vision*, Volume 1252 of *Lecture Notes in Computer Science*, Utrecht, 1997, pp. 188–199. Springer.

Friston, K.J., A.P. Holmes, J.-B. Poline, K.J. Worsley, C.D. Frith, and R.S.J. Frackowiak (1995). Statistical parametric maps in functional imaging: a general linear approach. *Human Brain Mapping 2*, 189–210.

Geman, S. and D. Geman (1984). Stochastic relaxation, Gibbs distributions, and the Bayesian restoration of images. *IEEE Transactions on Pattern Analysis and Machine Intelligence 6*, 721–741.

Koenderink, J.J. (1984). The structure of images. *Biological Cybernetics 50*, 363–370.

Koenderink, J.J. and A.J. van Doorn (1986). Dynamic shape. *Biological Cybernetics 53*, 383–396.

Lindeberg, T. and J.-O. Eklundh (1992). Scale-space primal sketch: construction and experiments. *Image and Vision Computing 10*, 3–18.

Mangin, J.-F., O. Coulon, and V. Frouin (1998). Robust brain segmentation using histogram scale-space analysis and mathematical morphology. In S. Delp, W. Wells, and A. Colchester (Eds.), *Medical Image Computing and Computer-Assisted Intervention—MICCAI '98*, Volume 1496 of *Lecture Notes in Computer Science*, Cambridge, MA, 1998, pp. 1230–1241. Springer.

Mangin, J.-F., Regis J, I. Bloch, V. Frouin, Y. Samson, and J. Lopez-Krahe (1995). A MRF based random graph modelling the human cortical topography. In *Computer Vision, Virtual Reality and Robotics in Medicine*, Volume 905 of *Lecture Notes in Computer Science*, Nice, 1995, pp. 177–183. Springer.

Worsley, K.J., S. Marrett, P. Neelin, A.C. Vandal, K.J. Friston, and A.C. Evans (1996). A unified statistical approach for determining significant signals in images of cerebral activation. *Human Brain Mapping 4*, 58–73.

Yousry, T.A., U.D. Schimd, H. Alkadhi, D. Schmidt, A. Peraud, A. Buettner, and P. Winkler (1997). Localisation of the motor hand area to knob on the precentral gyrus, a new landmark. *Brain 120*, 141–157.

11

Some Statistical Aspects of Magnetoencephalography

Bernd Lütkenhöner

ABSTRACT Owing to the fact that any magnetic field distribution measured outside the head can be explained by an infinite number of current distributions inside the head, a great diversity of analysis techniques for magnetoencephalographic (MEG) data evolved, each having its specific advantages and disadvantages. This article reviews selected statistical aspects of such estimation techniques. For the sake of simplicity, only the analysis of a single time slice of MEG data will be considered. After a brief outline of MEG source analysis, characteristic features of the background noise are explained by considering the model of random dipoles homogeneously distributed in the brain. Thereafter some basic issues of linear estimation theory are reviewed, focusing on the fact that unique averages of model parameters can be estimated even though a unique solution of the underlying inverse problem does not exist. Finally some aspects of nonlinear parameter estimation problems are considered, as for example dipole localization accuracy and dipole separability.

1 Introduction

Magnetoencephalography (MEG) is a non-invasive functional imaging technique (Committee on the Mathematics and Physics of Emerging Dynamic Biomedical Imaging, National Research Council, 1996) which allows to study brain functions with almost unlimited temporal resolution. This is a remarkable advantage compared to other functional imaging techniques, like positron emission tomography (PET) or functional magnetic resonance imaging (fMRI), the temporal resolution of which cannot surpass the limits set by the time constants of the metabolic processes involved. Since any magnetic field distribution measured outside the head can be explained by an infinite number of current distributions inside the head, a successful source analysis of MEG data is dependent on constraints (either based on prior information about the sources or on suitable assumptions). There is still an ongoing debate as to what constraints are most advantageous under what circumstances. Thus, it is not surprising that the spectrum of available analysis procedures continuously broadens.

In view of this situation, only a few exemplary analysis techniques can be treated in the text that follows. Since the main emphasis is placed on statistical aspects of MEG source analysis, the techniques themselves will be described here only in broad outline (more methodological details can be found in the cited literature). For the sake of simplicity, only the analysis of a single time slice of MEG data will be considered. But this confinement is not so restrictive as it may seem at first glance, because in many situations the parameter optimization problem disintegrates, at least partially, into optimization problems for the individual time slices. A full disintegration evidently occurs if the parameter sets for the individual time slices are independent. If this holds for only part of the parameters, meaning that some parameters have to be optimized simultaneously for the whole analysis window, the latter are often calculated iteratively, and the optimization problem for the former parameters disintegrates, in each iteration step, as described above. To avoid misunderstandings it shall be emphasized, however, that the transition from a single time instant to a full time-series model is, in general, not that simple, because the time series estimation and the spatial estimation may become very interweaved.

2 Outlines of MEG Source Analysis

2.1 Forward Problem

Sensitive magnetic measurement devices placed close to the surface of the head are capable of recording the magnetic field arising from intracerebral currents. It is commonly supposed that the dominant generators of the magnetic field are postsynaptic currents, though some early responses to external stimuli may well represent coherent volleys of action potentials (Williamson and Kaufman, 1987, 1990). For modeling the relationship between intracerebral currents and magnetic field it is convenient to distinguish *primary currents* from *volume currents.* The latter are of a passive nature and obey (on a macroscopic scale) Ohm's law. Currents not being volume currents are considered as primary currents (Hämäläinen et al., 1993). From this definition it follows that primary currents are flowing mainly inside and in the immediate vicinity of a cell. Consequently, the sites of the primary currents correspond to the sites of neural activity.

Since intracerebral currents and magnetic field are linearly related, as are primary and volume currents (so that it is possible to express the volume currents in terms of the primary currents), the data value recorded by a specific sensor, say the ith sensor ($1 \leq i \leq N$), can be modeled as

$$d_i = \int_V \vec{G}_i(\vec{r}\,') \cdot \vec{J}^p(\vec{r}\,')\, d\vec{r}\,' \qquad (1)$$

where $\vec{J}^p(\vec{r}\,')$ is the density of the primary currents, $\vec{G}_i(\vec{r}\,')$ is the leadfield of the ith sensor, and V is the volume containing the sources (i.e., the volume corresponding to the brain). The leadfield is a vector field describing the sensitivity distribution of a sensor (Hämäläinen et al., 1993, Malmivuo and Plonsey, 1995). Its mathematical form depends on both the volume conductor model (i.e., the model describing the relationship between primary currents and volume currents) and the sensor configuration. As an example, the leadfield of a magnetometer coil at position $\vec{r}$, with its axis oriented radially to the surface of a homogeneous spherical volume conductor, shall be given by (Lütkenhöner and Grave de Peralta Menendez, 1997):

$$\vec{G}(\vec{r}') = \frac{\mu_0/4\pi}{|\vec{r}-\vec{r}'|^2}\left(\frac{\vec{r}-\vec{r}'}{|\vec{r}-\vec{r}'|} \times \frac{\vec{r}}{|\vec{r}|}\right) \tag{2}$$

If the simple model of a homogeneous sphere is replaced by a more realistic volume conductor model, the calculation of the leadfield becomes more laborious, of course, and numerical techniques like the boundary element method (Fuchs et al., 1998, Hämäläinen and Sarvas, 1989, Menninghaus et al., 1994) or the finite element method (Bertrand et al., 1991, Buchner et al., 1997, Thevenet, 1992, Thevenet et al., 1991) may be required.

Most numerical techniques make use of a discrete current distribution, which means that $\vec{J}^p$ is assumed to have the structure

$$\vec{J}^p(\vec{r}') = \sum_{l=1}^{L} \vec{m}_l \delta(\vec{r}_l' - \vec{r}'), \tag{3}$$

where δ is the delta distribution. The L centers of localized activity assumed in (3) represent current dipoles: $\vec{r}_l'$ is the location of the lth current dipole, and $\vec{m}_l$ is its moment. Inserting assumption (3) into (1) yields

$$d_i = \sum_{l=1}^{L} \vec{G}_i(\vec{r}_l') \cdot \vec{m}_l. \tag{4}$$

The system of equations resulting for all N channels obviously has the structure

$$\mathbf{d} = \mathbf{G}(\mathbf{x})\mathbf{m}, \tag{5}$$

where $\mathbf{d}$ is a column vector composed of the N simultaneously measured data values, $\mathbf{m}$ is a column vector consisting of the $M = 3L$ components of the dipole moments, and $\mathbf{G}$ is a $N \times M$ matrix. The elements of matrix $\mathbf{G}$, called the data kernel, are nonlinear functions of a parameter vector $\mathbf{x}$ which represents all the dipole locations.

2.2 Inverse Problem

Using (1) or its discrete counterparts (4) or (5), the magnetic field can be predicted for arbitrary current distributions in the brain. However, in practice the inverse problem has to be solved: The current distribution in the brain has to be estimated from the measured data.

Only discrete current distributions shall be considered here. Whether the inverse problem (5) is overdetermined or underdetermined is evidently dependent on the number of dipoles, L. However, the total number of parameters, and thus the determinacy of the problem, is affected by other aspects as well. In the most general case, each dipole is associated with six parameters: the three coordinates specifying the dipole location and the three components of the dipole moment. But in practice there are often less than six parameters per dipole. In the case of a spherical volume conductor, for instance, the radial component of the dipole moment does not contribute to the magnetic field (Hämäläinen et al., 1993, Sarvas, 1987) so that the effective number of parameters per dipole is reduced to five. Only three parameters per dipole are sufficient if the dipoles are assumed to be located on a given surface (namely the cortical surface) with a dipole moment oriented perpendicular to that surface Fuchs et al. (1994), Lütkenhöner et al. (1995a,b). An even higher reduction of the number of parameters is achieved if certain parameters are assumed to be time-invariant Scherg (1990), Scherg and von Cramon (1985). While the meaning of the parameter vectors $\mathbf{x}$ and $\mathbf{m}$ is crucially dependent on such aspects, the general structure of the inverse problem always corresponds to (5).

In most practical applications the source model is consisting of either a few dipoles (so that the inverse problem to be solved is, in principle, overdetermined) or a large number of dipoles (so that the inverse problem is clearly underdetermined). In the overdetermined case, it is convenient to use an iterative optimization procedure for the parameter vector $\mathbf{x}$ and to solve, in each iteration step, the resulting linear inverse problem for the parameter vector $\mathbf{m}$. The least-squares solution of this problem is

$$\mathbf{m}^{\text{est}} = \mathbf{G}^{\dagger}(\mathbf{x})\mathbf{d}, \tag{6}$$

with

$$\mathbf{G}^{\dagger} = \left[\mathbf{G}^T\mathbf{G}\right]^{-1}\mathbf{G}^T \tag{7}$$

representing the generalized inverse Menke (1984). In the underdetermined case, the parameter vector $\mathbf{x}$ is usually specified in advance, i.e., no attempt is made to estimate it from the data.[1] Since there are more parameters than data, equation (5) does not provide enough information to determine the model parameters $\mathbf{m}$ in a unique way. Thus, some extra information has to be added or certain expectations about the solution have to be taken into account. It is often requested, for instance, that the solution has to explain the data while minimizing the L_2 norm of $\mathbf{m}$. The resulting solution,

[1]It is assumed, for instance, that the dipoles are located on a rectangular grid with a specified spacing (e.g., 1 cm) between adjacent grid points or on a grid restricted to the cortical regions where the sources are assumed to originate.

called the *minimum-norm solution*, is again given by equation (6), but the generalized inverse now has the form (Menke, 1984)

$$\mathbf{G}^{\dagger} = \mathbf{G}^T \left[\mathbf{G}\mathbf{G}^T\right]^{-1}. \tag{8}$$

Since measurements are usually contaminated by noise, it must be considered as a disadvantage of the minimum-norm solution that it explains the data exactly. Furthermore, the L_2 norm of $\mathbf{m}$ is not necessarily the most appropriate measure of solution simplicity. These problems are avoided if the solution $\mathbf{m}^{\text{est}}$ is derived by minimizing the expression

$$\Phi(\mathbf{m}) = (\mathbf{d} - \mathbf{G}\mathbf{m})^T \mathbf{W}_e (\mathbf{d} - \mathbf{G}\mathbf{m}) + \epsilon^2 \mathbf{m}^T \mathbf{W}_m \mathbf{m}. \tag{9}$$

The first term corresponds to a generalized prediction error, whereas the second term can be interpreted as a generalized measure of solution simplicity. The parameter ϵ^2 determines the relative importance given to the prediction error and the model simplicity, and the matrices $\mathbf{W}_e$ and $\mathbf{W}_m$ represent some kind of metric for the data and the model parameter space, respectively. Minimization of (9) results in the *weighted damped least-squares solution*, which has the generalized inverse

$$\mathbf{G}^{\dagger} = \left(\mathbf{G}^T \mathbf{W}_e \mathbf{G} + \epsilon^2 \mathbf{W}_m\right)^{-1} \mathbf{G}^T \mathbf{W}_e \tag{10}$$

or, equivalently (supposed that all inverses exist),

$$\mathbf{G}^{\dagger} = \mathbf{W}_m^{-1} \mathbf{G}^T \left(\mathbf{G}\mathbf{W}_m^{-1}\mathbf{G}^T + \epsilon^2 \mathbf{W}_e^{-1}\right)^{-1} \tag{11}$$

(see Menke, 1984, p. 54f, p. 97f). A great variety of discrete inverse solutions developed for the analysis of MEG data have either the structure (10) or the structure (11), as pointed out in Grave de Peralta Menendez and Gonzalez Andino (1998, 1999). If $\mathbf{W}_e$ and $\mathbf{W}_m$ are identity matrices and $\epsilon = 0$, the equations (10) and (11) evidently reduce to (7) and (8). If $\mathbf{W}_e$ is not the identity matrix, but corresponds to the inverse of the data covariance matrix $[\operatorname{cov}\mathbf{d}]$, i.e.

$$\mathbf{W}_e = [\operatorname{cov}\mathbf{d}]^{-1}, \tag{12}$$

the maximum likelihood solution is obtained (see, e.g., Menke, 1984, p. 82). Supposed that the covariance matrix of the model parameters, $[\operatorname{cov}\mathbf{m}]$, is known as well, an appropriate choice for $\epsilon^2 \mathbf{W}_m$ would be

$$\epsilon^2 \mathbf{W}_m = [\operatorname{cov}\mathbf{m}]^{-1} \tag{13}$$

(see,e.g., Dale and Sereno, 1993).

2.3 Bayesian Viewpoint of Discrete Inverse Theory

The above formulas can be derived also from a Bayesian viewpoint of discrete inverse theory. The consideration that follows is based on Menke (1984, p. 79ff), in spite of certain differences in the terminology. According to the Bayesian viewpoint, the conditional probability of the model parameters $\mathbf{m}$ given the data $\mathbf{d}$ is

$$P(\mathbf{m} \mid \mathbf{d}) = P(\mathbf{m}) \cdot P(\mathbf{d} \mid m), \tag{14}$$

where $P(\mathbf{m})$ is the a priori distribution of the the model parameters and $P(\mathbf{d} \mid m)$ is the likelihood of the data given the model parameters. A solution of the underlying inverse problem is obtained by maximizing $P(\mathbf{m} \mid \mathbf{d})$.

It shall be assumed now that both $P(\mathbf{m})$ and $P(\mathbf{d} \mid \mathbf{m})$ are multivariate Gaussian distributions,

$$P(\mathbf{m}) \propto \exp\left(-\tfrac{1}{2}(\mathbf{m} - \langle\mathbf{m}\rangle)^T [\operatorname{cov}\mathbf{m}]^{-1}(\mathbf{m} - \langle\mathbf{m}\rangle)\right), \tag{15}$$

$$P(\mathbf{d} \mid m) \propto \exp\left(-\tfrac{1}{2}(\mathbf{d} - \mathbf{G}\mathbf{m})^T [\operatorname{cov}\mathbf{d}]^{-1}(\mathbf{d} - \mathbf{G}\mathbf{m})\right), \tag{16}$$

where $\langle\mathbf{m}\rangle$ denotes the vector of most probable model parameters, and $[\operatorname{cov}\mathbf{d}]$ and $[\operatorname{cov}\mathbf{m}]$ are again the covariance matrices of the data and the model parameters, respectively. Being the product of two multivariate Gaussian distributions, $P(\mathbf{m}|\mathbf{d})$ is multivariate Gaussian as well. The model parameter vector maximizing $P(\mathbf{d} \mid \mathbf{m})$ is (see Menke, 1984, p. 93)

$$\begin{aligned} \mathbf{m}^{\text{est}} &= \mathbf{G}^\dagger(\mathbf{d} - \mathbf{G}\langle\mathbf{m}\rangle) + \langle\mathbf{m}\rangle \\ &= \mathbf{G}^\dagger\mathbf{d} + (\mathbf{I} - \mathbf{R})\langle\mathbf{m}\rangle, \end{aligned} \tag{17}$$

where $\mathbf{I}$ is the identity matrix, $\mathbf{R}$ is the model resolution matrix defined as

$$\mathbf{R} = \mathbf{G}^\dagger\mathbf{G} \tag{18}$$

(considered in more detail below), and $\mathbf{G}^\dagger$ represents the generalized inverse. The latter can be written in two equivalent forms:

$$\mathbf{G}^\dagger = (\mathbf{G}^T[\operatorname{cov}\mathbf{d}]^{-1}\mathbf{G} + [\operatorname{cov}\mathbf{m}]^{-1})^{-1}\mathbf{G}^T[\operatorname{cov}\mathbf{d}]^{-1} \tag{19}$$

$$\mathbf{G}^\dagger = [\operatorname{cov}\mathbf{m}]\mathbf{G}^T(\mathbf{G}[\operatorname{cov}\mathbf{m}]\mathbf{G}^T + [\operatorname{cov}\mathbf{d}])^{-1}. \tag{20}$$

The difference between model and theory, $\mathbf{m} - \langle\mathbf{m}\rangle$, represents the observational error, supposed that an exact theory is available. Otherwise it is the sum of observational and modelization error (see also Tarantola, 1987, p. 58, where it is shown that observational and modelization errors combine to a single error even if the forward problem is nonlinear). The equations (19) and (20) evidently reduce to (10) and (11), with $\mathbf{W}_e$ and $\epsilon^2\mathbf{W}_m$ as specified in (12) and (13).

2.4 Parameter Estimation in a Transformed World

Provided that the matrices $\mathbf{W}_e$ and $\mathbf{W}_m$ are symmetric and positive definite, lower triangular matrices $\mathbf{T}_e$ and $\mathbf{T}_m$ exist such that

$$\mathbf{W}_m = \mathbf{T}_m \mathbf{T}_m^T, \tag{21}$$

$$\mathbf{W}_e = \mathbf{T}_e \mathbf{T}_e^T. \tag{22}$$

One such factorization, which can be referred to as "taking the square root" of a matrix, is the Cholesky decomposition (Golub and Loan, 1989, Press et al., 1992).

Equation (9) can be rewritten now as

$$\Phi(\mathbf{m}') = (\mathbf{d}' - \mathbf{G}'\mathbf{m}')^T(\mathbf{d}' - \mathbf{G}'\mathbf{m}') + \epsilon^2 \mathbf{m}'^T \mathbf{m}' \tag{23}$$

with

$$\mathbf{m}' = \mathbf{T}_m^T \mathbf{m}, \tag{24}$$

$$\mathbf{d}' = \mathbf{T}_e^T \mathbf{d}, \tag{25}$$

$$\mathbf{G}' = \mathbf{T}_e^T \mathbf{G} (\mathbf{T}_m^T)^{-1} \tag{26}$$

(see also Menke, 1984, p. 117f). By means of the transformations (24)–(26) a "transformed world" is formed so that a consideration of the unweighted minimization problem (23) is completely equivalent to a consideration of the weighted problem (9) in the real world. The special case $\mathbf{W}_m = \mathbf{I}$ was considered already by Sarvas (1987), and the resulting transformation was applied in Yamazaki et al. (1992) to remove the spatial correlation of the background electroencephalogram.

Conceptually it is important that Cholesky decomposition does not provide a unique "square root" of a positive definite matrix. To make this evident, (21) is rewritten as

$$\mathbf{W}_m = \mathbf{T}_m \mathbf{U}_m^T \cdot \mathbf{U}_m \mathbf{T}_m^T, \tag{27}$$

where $\mathbf{U}_m$ denotes an arbitrary orthonormal matrix (which has the property $\mathbf{U}_m^T \mathbf{U}_m = \mathbf{U}_m^{-1} \mathbf{U}_m = \mathbf{1}$). Equation (22) can be rewritten correspondingly. Thus, an infinite number of "transformed worlds" exist.

2.5 Basic Statistical Issues

First the general case shall be considered that $\mathbf{d}$ is a vector of random variables with mean $\langle \mathbf{d} \rangle$ and covariance $[\operatorname{cov} \mathbf{d}]$. Furthermore, $\mathbf{m}$ is assumed to be a linear function of $\mathbf{d}$, i.e., $\mathbf{m} = \mathbf{M}\mathbf{d} + \mathbf{v}$, where $\mathbf{M}$ and $\mathbf{v}$ are an arbitrary matrix and an arbitrary vector, respectively, of appropriate dimensions. Then the mean and the covariance of $\mathbf{m}$ are (see, e.g. Menke, 1984, p. 28)

$$\langle \mathbf{m} \rangle = \mathbf{M} \langle \mathbf{d} \rangle + \mathbf{v} \tag{28}$$

$$[\mathrm{cov}\,\mathbf{m}] = \mathbf{M}[\mathrm{cov}\,\mathbf{d}]\mathbf{M}^T \tag{29}$$

Regarding the model parameters calculated according to (6) this means that, given the covariance matrix for the noise in the measured data, $[\mathrm{cov}\,\mathbf{d}]$, the covariance matrix of $\mathbf{m}^{\mathrm{est}}$ can be calculated as

$$[\mathrm{cov}\,\mathbf{m}^{\mathrm{est}}] = \mathbf{G}^{\dagger}[\mathrm{cov}\,\mathbf{d}](\mathbf{G}^{\dagger})^T. \tag{30}$$

Equation (30) describes how noise in the measured data propagates to the estimated parameters.

Besides that, $\mathbf{G}^{\dagger}$ itself may depend on statistical aspects of the data. In the equations (10) and (11) only the data kernel $\mathbf{G}$ is purely deterministic by nature, whereas the choice of the weighting factor matrices $\mathbf{W}_e$ and $\mathbf{W}_m$ may depend on prior statistical information, as already exemplified in the equations (12) and (13).

3 Spatial Statistics of the Measurement Noise

3.1 Random Dipole Model

A source analysis of MEG data is generally founded on the assumption that the recorded data are composed of a signal (often related to a certain event, for instance an external stimulus) and independent background noise. The latter is usually dominated by the spontaneous activity of intracerebral sources so that it exhibits a high spatial correlation (Kuriki et al., 1994). A simple, but useful model for this kind of "noise" activity was devised by Cuffin and Cohen (1977): the random-dipole model. The underlying assumption is that dipoles having random orientations are distributed on a line, on a surface, or in a volume. An example is shown in Figure 1(a). The shaded area represents the brain, whereas the arrows represent the random dipoles. The figure also illustrates a stereotyped measurement configuration. In Figure 1(b), a 37-channel magnetometer system (resembling the MAGNES system of Biomagnetic Technologies Benzel et al., 1993) is shown as a projection into the x–y plane. The 7 coils with the coordinate $y = 0$ appear also in Figure 1(a), where the coil axes are indicated in addition. The latter are oriented to the center of sphere, which means that only the radial component of the magnetic field is measured.

The random dipole model has been used in several simulation studies (Kaufman et al., 1991, Lütkenhöner, 1991, 1992a, Lütkenhöner et al., 1991). A detailed statistical analysis considering electrical potentials as well as magnetic fields was presented by de Munck et al. (1992). Their study was taken up in Lütkenhöner (1994), though confined to the magnetic field. It was shown that, for the conditions illustrated in Figure 1 (random dipoles homogeneously distributed in a sphere), the magnetic field values recorded

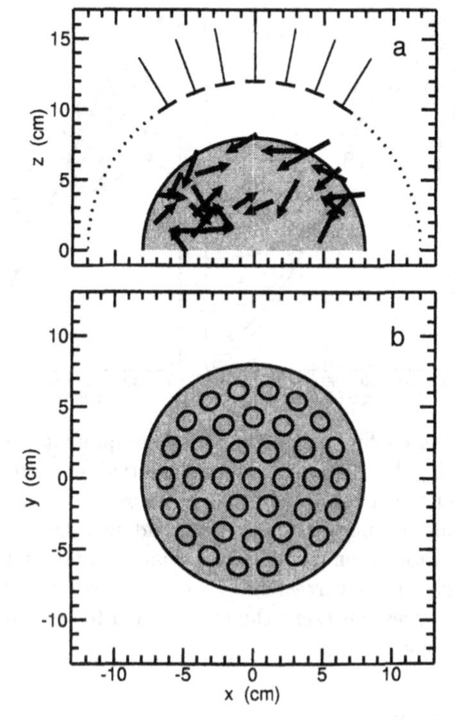

Figure 1. Geometry of the 37-channel magnetometer system used for model simulations. The shaded area represents the brain. In the upper panel, the brain is "filled" with random dipoles.

by two magnetometer coils with an angular distance θ (angle between the two coil centers as seen from the center of sphere) have the covariance

$$\mathrm{cov}^{\mathrm{sphere}}(R, r, \theta) = \left(\frac{\mu_0}{4\pi}\right)^2 \sigma_{\mathrm{RD}}^2 \cdot \sum_{l=1}^{\infty} \frac{l(l+1)}{(2l+1)(2l+3)} \frac{R^{2l}}{r^{2(l+2)}} P_l(\cos\theta), \quad (31)$$

where r is the distance between coil centers and center of sphere, R is the radius of the sphere containing the random dipoles, $\mu_0/4\pi$ is a constant,[2] and σ_{RD} is the "total strength" of the random dipoles. The latter quantity is defined as follows. Supposed that there are random dipoles at N different locations, then the three Cartesian components of each dipole moment are assumed to have independent Gaussian amplitude distributions with standard deviation $\sigma_{\mathrm{RD}}/(3N)$. A similar formula was derived for the case that the random dipoles are distributed on a spherical surface with radius

[2]Supposed that the covariance of the magnetic field is measured in $(\mathrm{pT})^2$, σ_{RD} in nAm, and the distances r and R in cm, then the constant $\mu_0/(4\pi)$ can be replaced by the value 1.

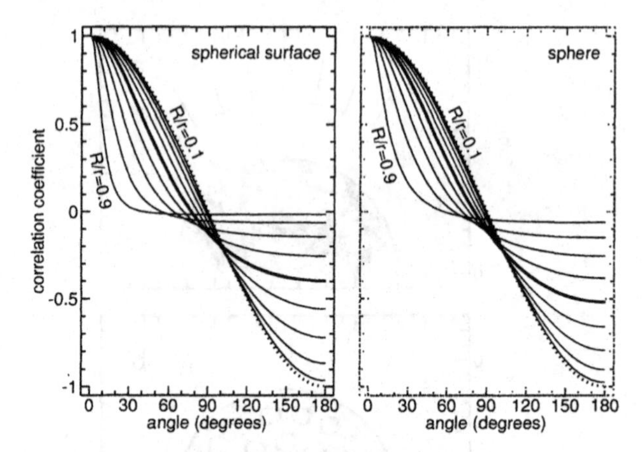

Figure 2. Correlation coefficient for the radial component of the magnetic field as a function of the angular distance between the two measurement locations. The ratio R/r is varied between 0.1 and 0.9, in steps of 0.1 (R is the radius of the sphere with the random dipoles, and r is the radius of the measurement surface). Left half: random dipoles on a spherical surface. Right half: random dipoles in a sphere. The thick curves correspond to $\rho = 0.5$, whereas the cosine functions plotted as dotted curves represent the expectation for $\rho \to 0$ (from Lütkenhöner (1994) with permission).

R rather than in a sphere:

$$\operatorname{cov}^{\text{surface}}(R,r,\theta) = \frac{1}{3}\left(\frac{\mu_0}{4\pi}\right)^2 \sigma_{\text{RD}}^2 \cdot \sum_{l=1}^{\infty} \frac{l(l+1)}{2l+1} \frac{R^{2l}}{r^{2(l+2)}} P_l(\cos\theta). \tag{32}$$

Numerically more efficient formulas, making use of elliptic integrals, are given in Lütkenhöner (1994). For $\theta = 0$ the equations (31) and (32) yield the variance of the magnetic field generated by the random dipoles. For this special case, the right-hand sides of the equations can be replaced by simple analytical expressions ($\rho = R/r$):

$$\operatorname{var}^{\text{sphere}}(R,r) = \left(\frac{\mu_0}{4\pi}\right)^2 \frac{\sigma_{\text{RD}}^2}{8(rR)^2} \cdot \left(\frac{3-\rho^2}{1-\rho^2} - \frac{3+\rho^2}{\rho} \operatorname{arc\,tanh}\rho\right) \tag{33}$$

$$\operatorname{var}^{\text{surface}}(R,r) = \left(\frac{\mu_0}{4\pi}\right)^2 \frac{\sigma_{\text{RD}}^2}{12r^4} \left(\frac{1+\rho^2}{(1-\rho^2)^2} - \frac{1}{\rho} \operatorname{arc\,tanh}\rho\right). \tag{34}$$

These two formulas are quite useful in Monte Carlo simulations as they allow to adjust the "total strength" of the random dipoles, σ_{RD}, so that the variance of the resulting magnetic field has a specified value.

Figure 2 shows the correlation coefficient $\operatorname{cov}(R,r,\theta)/\operatorname{var}(R,r)$ as a function of the angular distance θ, for different values of R/r. The curves on the left were obtained for random dipoles homogeneously distributed on a spherical surface with radius R, those on the right were obtained for random dipoles homogeneously distributed in a sphere with radius R. Qualitatively, the two sets of curves are similar. If the coils are located very close to the sphere or the spherical surface with the random dipoles ($R/r \to 1$), the correlation coefficient rapidly drops within the first 30°. In the other extreme when all random dipoles are located very close to the center of sphere ($R/r \to 0$), the graph of the correlation coefficient assumes a cosine shape (dotted curve). While these two extreme cases are quite unrealistic, the case $R/r = 0.5$ (thick curve) can be considered as a good reference. The correlation coefficient is relatively high for $\theta < 30°$, it becomes zero around 80°, and significant negative correlations can be observed for $\theta > 120°$.

3.2 Covariance Matrix and Weighting Factor Matrix

Owing to the inherent symmetry of the model shown in Figure 1, the data covariance matrix $[\text{cov}\,\mathbf{d}]$ derived from equation (31) or equation (32) has a quite symmetrical structure. This is illustrated in Figure 3(a), which shows a plot doubly organized in a sensor layout form. This means that both the positions of the clusters of circles as well as the positions of the circles within a cluster reflect the positions of the corresponding coils in the magnetometer array. Thus, to get an impression of the covariance of two channels, first the cluster of circles at the position of the one channel has to be selected. Within that cluster, then the circle at the position of the second channel has to be found. The area of that circle finally reflects the magnitude of the covariance. In the case considered here, the random dipoles were assumed to be located on a spherical surface with a radius of 6 cm. In addition to the noise generated by the random dipoles, spatially uncorrelated Gaussian noise was assumed, with a standard deviation being 1/10 of the standard deviation of the noise generated by the random dipoles.

Figure 3(b) is organized in exactly the same way as Figure 3(a), but visualizes the inverse of the covariance matrix. As mentioned above (see equation (12)), this inverse corresponds to the weighting factor matrix $\mathbf{W}_e$ of the maximum likelihood solution. Filled circles now indicate negative values. The weighting factor pattern is more complex than the covariance pattern, but there is again a well-developed symmetry. The circles with the largest diameters correspond to the diagonal elements of the weighting factor matrix. Except for borderline measurement locations, each of these circles is surrounded by six filled circles representing negative weighting factors. With regard to (9) this means that the prediction error is not obtained by adding up the prediction errors for the individual channels, but the differences between these individual prediction errors and the prediction errors of the neighboring channels.

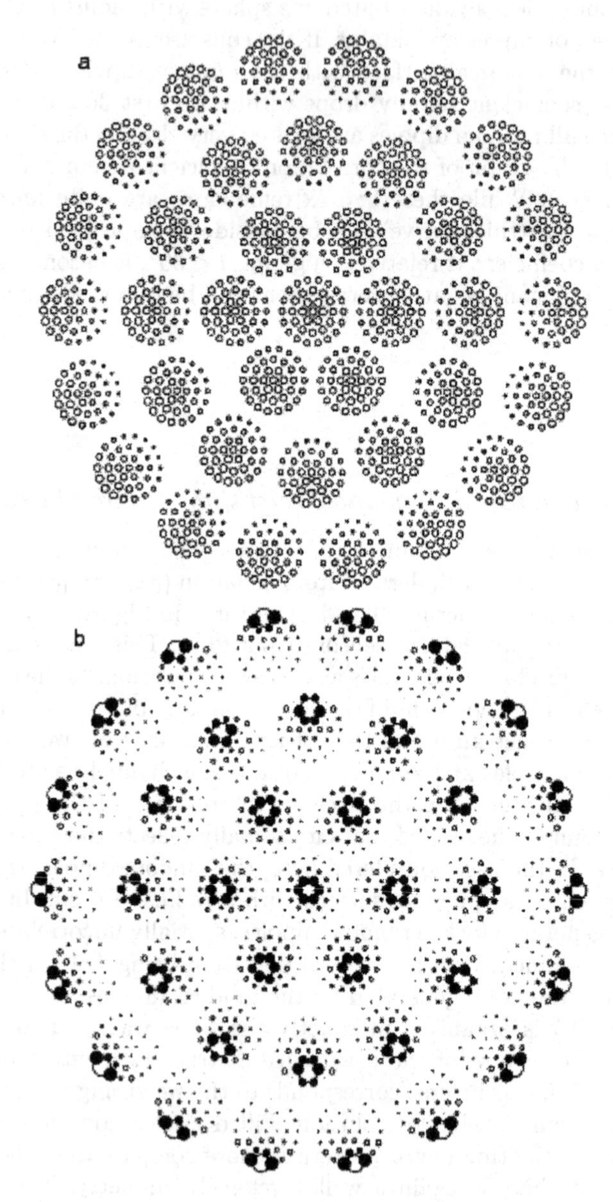

Figure 3. (a) Theoretical covariances. (b) Theoretical weighting factors. Both figures are doubly organized in a sensor-layout form. Each row corresponds to one cluster of symbols. Elements belonging to the same column are found at identical positions within the clusters. The area of a circle corresponds to the magnitude of a matrix element. Open circles: positive values. Filled circles: negative values (from Lütkenhöner (1998c) with permission).

3.3 An Example of a Transformed World

Provided that $\mathbf{W}_e = [\text{cov}\,\mathbf{d}]^{-1}$, it is advantageous to apply the Cholesky decomposition (22) to the covariance matrix $[\text{cov}\,\mathbf{d}]$ rather than the weighting factor matrix $\mathbf{W}_e$. This means, a lower triangular matrix $\mathbf{D}_e$ is calculated such that

$$[\text{cov}\,\mathbf{d}] = \mathbf{D}_e\mathbf{D}_e^T \tag{35}$$

thus,

$$\mathbf{W}_e = (\mathbf{D}_e^{-1})^T\mathbf{D}_e^{-1}. \tag{36}$$

Comparing the latter equation with (22) finally yields

$$\mathbf{T}_e^T = \mathbf{D}_e^{-1}. \tag{37}$$

Figure 4 visualizes the transformation matrix $\mathbf{T}_e$ representing the Cholesky decomposition of the weighting factor matrix displayed in Figure 3(b). As in the previous figure, the areas of the open and filled circles visualize the magnitudes of the matrix elements, with open circles corresponding to positive and filled circles corresponding to negative values (the plot of matrix elements with very small magnitudes was suppressed). Each column of the matrix $\mathbf{T}_e$ corresponds to one cluster of circles. Matrix elements belonging to the same row are found at identical positions within the clusters. As $\mathbf{T}_e$ is a lower triangular matrix, the number of circles belonging to a cluster

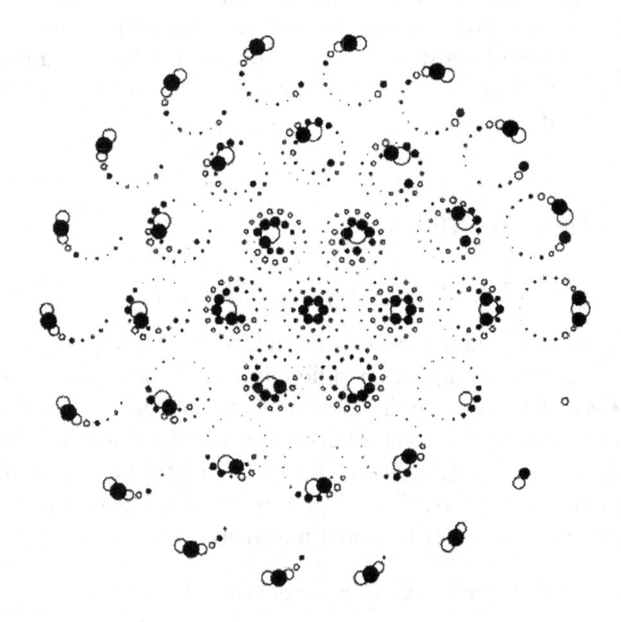

Figure 4. Visualization of the transformation matrix $\mathbf{T}_e$ derived from the covariance matrix shown in Figure 3(a).

varies between 37 (1st coil, located in the center) and 1 (37th coil, located in the outermost ring of coils, at an angle corresponding to about 4 o'clock).

Making use of (25), (29), (35), and (37) it follows that the covariance of the transformed data is

$$[\operatorname{cov} \mathbf{d}'] \mathbf{T}_e^T \cdot [\operatorname{cov} \mathbf{d}] \cdot \mathbf{T}_e = \mathbf{T}_e^T \mathbf{D}_e \mathbf{D}_e^T \mathbf{T}_e = \mathbf{1}. \tag{38}$$

Thus, the transformation (25) can be interpreted as follows: Synthetic channels are calculated so that any correlation between channels is removed. Figure 4 explains how these synthetic channels are built up. As $\mathbf{T}_e$ is a 37×37 lower triangular matrix, the cluster associated with the 37th coil consists of only one circle, representing the 37th diagonal element of $\mathbf{T}_e$. This means that, except for a constant weighting factor, the 37th synthetic channel is identical with the original 37th channel. By moving in a clockwise direction on the outermost circle, the clusters corresponding to the coil numbers $36, 35, \ldots, 20$ are found. As $\mathbf{T}_e$ is a lower triangular matrix, the ith synthetic channel is formed from the original channels with the numbers $i, i+1, \ldots, 37$. The figure shows that, for the coils on the outer ring, the weighting factor associated with the ith channel is always positive, whereas that associated with the $(i+1)$th channel is negative, and that associated with the $(i+2)$th channel is positive again (for $i < 36$). The complexity of the pattern of weighting factors increases considerably as soon as coils from the inner two rings are involved. The pattern obtained for the 1st coil, finally, has a quite symmetrical structure (see center of figure). It is remarkable that the diagonal elements of the transformation matrix do not necessarily have the largest magnitudes. In general, the magnitudes of the weighting factors decrease with increasing Euclidean distance from the coil under consideration.

4 Localized Averages

4.1 *Estimates Versus Averages of Model Parameters*

Though the generalized inverse $\mathbf{G}^\dagger$ resembles in some way the ordinary matrix inverse, there are important differences: It is not square, and neither $\mathbf{G}^\dagger \mathbf{G}$ nor $\mathbf{G}\mathbf{G}^\dagger$ need equal an identity matrix (Menke, 1984, p. 62). Regarding the interpretation of a generalized inverse, there is a type of dualism (Menke, 1984, p. 105)): $\mathbf{G}^\dagger \mathbf{d}$ can be interpreted either as an estimate of the model parameter vector $\mathbf{m}^{\text{est}}$ (cf. equation (6)) or as a vector consisting of weighted averages of the true model parameters,

$$\langle \mathbf{m} \rangle = \mathbf{G}^\dagger \mathbf{d} = \mathbf{G}^\dagger \mathbf{G} \cdot \mathbf{m} = \mathbf{R} \cdot \mathbf{m}, \tag{39}$$

where $\mathbf{R}$ is the model resolution matrix already defined in (18). Though the numerical values of $\mathbf{m}^{\text{est}}$ and $\langle \mathbf{m} \rangle$ are identical, the underlying philosophies are quite different. In the latter case, the solution can be viewed as a unique quantity that exists independently of any a priori information applied to the inverse problem (Menke, 1984, p. 105).

The ith element of the vector $\langle \mathbf{m} \rangle$ $(1 \leq i \leq M)$ is obtained by multiplying the ith row of the model resolution matrix with the actual (but unknown) model parameters $\mathbf{m}$. Thus, for each element of $\langle \mathbf{m} \rangle$ an average of the type

$$\langle m \rangle = \mathbf{a}^T \mathbf{m} \tag{40}$$

is calculated, where $\mathbf{a}$ is some $M \times 1$ vector, called the averaging vector. An average of the type (40) may be a unique estimate even if unique estimates of the parameters themselves do not exist. More precisely, an average is unique if and only if the averaging vector $\mathbf{a}$ can be represented as a linear combination of the rows of the data kernel $\mathbf{G}$ (Menke, 1984, p. 105), this means

$$\mathbf{a} = \mathbf{G}^T \mathbf{w}, \tag{41}$$

where $\mathbf{w}$ is a $N \times 1$ vector with weighting factors. If $\mathbf{a}$ represents one row of the model resolution matrix, $\mathbf{w}$ corresponds to one row of the generalized inverse. However, in principle any vector $\mathbf{w}$ having at least one non-zero element can be used to form a unique averaging vector. A different issue is, of course, the interpretation of the resulting average, which may be problematic if the average is not "localized": An average is said to be localized, if the elements of the averaging vector have a small magnitude except for some "closely related" parameters. With regard to MEG source analysis this means, for instance, that an element of an averaging vector should have a significant magnitude if and only if it is associated with a dipole located in the vicinity of some particular brain structure.

In general, the weighting factors will be chosen so that the average $\langle m \rangle$ represents an estimate of some activity of interest (AOI). In this case it it is useful to normalize the weights so that

$$\mathbf{w}^T \mathbf{d}^{AOI} = 1, \tag{42}$$

where $\mathbf{d}^{AOI}$ is the data vector arising from an AOI with amplitude one. The activity of interest may correspond, for example, to the amplitude of a dipole at a specific location with well-defined and known direction or to the amplitude of a certain quadrupole component or to the mean current density in an extended patch of the cortical surface.

While (40) is quite suggestive regarding the interpretation of $\langle m \rangle$, it obscures its calculation from given data (assumed to be noiseless here). To focus on the latter point, (41) is inserted into (40), yielding

$$\langle m \rangle = \mathbf{w}^T \mathbf{G} \mathbf{m} = \mathbf{w}^T \mathbf{d}. \tag{43}$$

Thus, the average $\langle m \rangle$ is simply a weighted[3] average of the data, i.e., the estimation of the AOI can be imagined as a projection of the measured data into the region of interest. Since there are concurrent activities at other locations, the task is to focus the data such that the AOI is reproduced as well as possible, while interferences from concurrent sources are minimized. This idea corresponds to the concept of a software lens developed by Freeman (1980a,b). A consequence of the normalization (42) is that the AOI is retrieved without error if there are no other sources in the brain and the sensors (magnetometer coils) pick up no noise at all.

4.2 Resolution Field

If $\mathbf{w}$ corresponds to one row of the generalized inverse, an inspection of the corresponding row of the model resolution matrix can show whether the average (40) is localized or not. However, if the weights are not derived from a generalized inverse, a model resolution matrix is not available. To overcome this problem the forward model (1) is used to rewrite (43) as

$$\langle m \rangle = \int_V \vec{\jmath}^{\,p}(\vec{r}\,') \cdot \vec{R}(\vec{r}\,' \mid \mathbf{w})\, d\vec{r}\,', \tag{44}$$

where

$$\vec{R}(\vec{r}\,' \mid \mathbf{w}) = \sum_{i=1}^{N} w_i \vec{G}_i(\vec{r}\,') \tag{45}$$

is the resolution field (Lütkenhöner and Grave de Peralta Menendez, 1997). The resolution field is a vector field similar to the lead fields $\vec{G}_i$ $(1 \leq i \leq N)$. Lead field and resolution field have in common that, by calculating the scalar product with the dipole moment m, a measure of the impact of a hypothetical dipole at $\vec{r}\,'$ is obtained. However, the meaning of this measure of impact is different for the two fields: in the case of the lead field it quantifies the contribution of a hypothetical dipole to the data recorded in a specific measurement channel, whereas in the case of the resolution field it quantifies the contribution of a hypothetical dipole to the average $\langle m \rangle$. Thus, the resolution field can be considered as the lead field of a synthetic virtual sensor (Robinson, 1989).

4.3 Trade-Off Between Model Resolution and Model Variance

In the spirit of Backus and Gilbert (1970, 1967, 1968) the spread of an averaging vector $\mathbf{a}$ can be quantified as

---

[3]The weights have a physical dimension (like Am/T), because a magnetic field has to be converted into a dipole moment or a current density or whatsoever, depending on the definition of the AOI.

$$\text{spread}(\mathbf{a}) = \sum_{i=1}^{M} \Delta_i a_i^2, \tag{46}$$

where a_i is the ith element of $\mathbf{a}$ and Δ_i measures the distance between the dipole location associated with the ith model parameter and some reference location. Supposed, for example, that the averaging vector corresponds to the kth row of the model resolution matrix $\mathbf{R}$, the reference location corresponds to the location of the dipole which is associated with the kth model parameter. After inserting (41), equation (46) can be rewritten as

$$\text{spread}(\mathbf{a}) = \sum_{i=1}^{N}\sum_{j=1}^{N} w_i w_j S_{ij} \tag{47}$$

with

$$S_{ij} = \sum_{l=1}^{M} \Delta_l G_{il} G_{jl} \tag{48}$$

(see, e.g., Menke, 1984, p. 73).

As a rule, there is a trade-off between the model resolution (quantified, e.g., in terms of the spread of the averaging vector) and the variance of the average $\langle m \rangle$. The latter can be calculated analogously to (30), though that equation itself can be applied only if the weights $\mathbf{w}$ represent a row of the generalized inverse. In the more general case of arbitrary weights, the variance of the average $\langle m \rangle$ is evidently

$$\text{var}\{\langle m \rangle\} = \mathbf{w}^T[\text{cov}\,\mathbf{d}]\mathbf{w}. \tag{49}$$

Backus and Gilbert (1970, 1967, 1968) developed a formal procedure for finding a reasonable compromise between the two opposing goals of having on the one hand a "localized" average (associated with a small spread of the averaging vector $\mathbf{a}$) and on the other hand a small variance. But their procedure is rarely used in the context of MEG source analysis.

Though enhancing the model resolution usually increases the variance of $\langle m \rangle$, it can happen nevertheless that the model with the lower resolution results in a higher variance than the model with the higher resolution. Examples will be given in the next subsection. The reason for this seeming paradox is that $\text{var}\{\langle m \rangle\}$ depends on the covariance of the data, in contrast to spread($\mathbf{a}$), which depends only indirectly, if at all, on the covariance of the data (via a possible dependence of $\mathbf{w}$).

4.4 Examples of Weighting Factor Patterns

Equation (43) shows that each element of $\mathbf{w}$ is associated with a specific location in space: the location of the respective measurement coil. Thus, a set of weighting factors can be be interpreted as a spatial pattern. Four examples are shown in Figure 5. These examples will be explained now, and the variances of the resulting averages $\langle m \rangle$ will be compared for different noise conditions.

The measurement device considered here is exactly the same as in Lütkenhöner (1998a), Lütkenhöner et al. (2000). As in Figure 1, the magnetometer coils are arranged on a spherical surface with a radius of 12 cm, and an inner coil on the z axis ("north pole") is surrounded by equidistantly spaced rings of coils. However, now an instrument consisting of 148 rather than 37 coils is considered. It represents a geometrical simplification of a whole-head magnetometer system. Each symbol in Figure 5(a)–(d) represented one coil. The values of the associated weighting factors are indicated by type and size of the symbol plotted. A weighting factor is represented by a dot if its value is small compared to the largest weighting factors, otherwise it is represented by a circle with an area corresponding to its absolute value. A filled circle indicates a weighting factor with negative sign.

The four examples have in common that they are somehow related to an imaginary current dipole located on the z axis, having a distance of 6 cm from the center of sphere and pointing into the x direction. The amplitude of the moment of this imaginary dipole represents the AOI to be estimated. In the case of Figure 5(a) the estimate is simply the field value measured by the coil with the best signal-to-noise ratio, except for a scaling factor derived from equation (42). Thus, all weighting factors except one are zero. In the case of Figure 5(b) the least-squares estimate of the amplitude of the dipole moment is considered (assuming exact knowledge of the dipole location and the direction of the dipole moment). A similar estimate is considered in Figure 5(c), but the weighting factors correspond to a maximum likelihood estimation accounting for a certain data covariance matrix (specified below). The weighting factors in Figure 5(d) represent one particular row of a generalized inverse corresponding to a regularized[4] minimum-norm estimation. The weights were normalized so that they comply with (42). For the sake of simplicity, the four estimates shall be denoted as the best-channel, the least-squares, the maximum-likelihood and the minimum-norm estimate, respectively.

A comparison of Figure 5(b) and Figure 5(c) illustrates that assumptions about the statistical properties of the noise can substantially change the spatial pattern of the weighting factors. In the latter figure the noise was assumed to arise from current dipoles randomly distributed in a sphere with a radius of 8 cm (random dipole model). To account for the fact that measured data typically contain also spatially uncorrelated instrumental noise, the diagonal elements of the noise covariance matrix (calculated by applying a numerically more efficient version of equation (31)) were multiplied

[4] A pseudo-inverse of the data kernel $\mathbf{G}$ was calculated using a singular value decomposition. Only the first 100 singular values were taken into account.

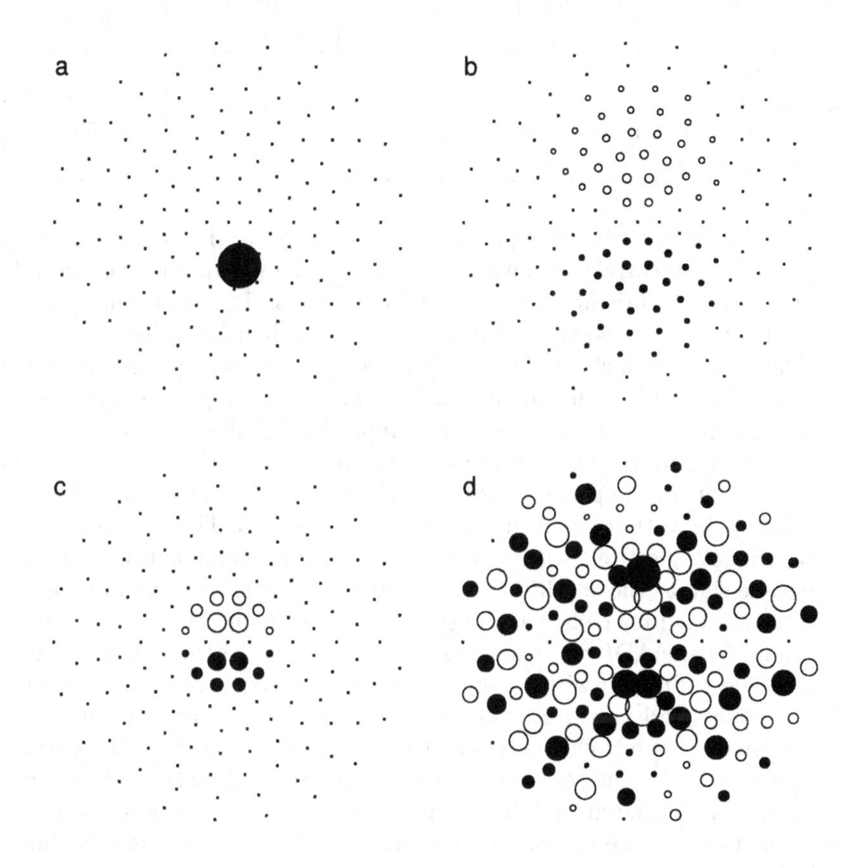

Figure 5. Visualization of the weights for the individual sensors of a 148-channel system. (a) Best-channel estimate. (b) Least-squares estimate. (c) Maximum-likelihood estimate. (d) Minimum-norm estimate. The coil locations were projected into a plane so that x and y axis of the three-dimensional space were mapped to abscissa and ordinate, respectively. The area of a symbol is proportional to the absolute value of the weighting factor for the respective coil, except that weighting factors with an absolute value smaller than a certain limit are represented by dots. Positive weighting factors indicated by open symbols, negative weighting factors indicated by filled symbols. (From Lütkenhöner and Grave de Peralta Menendez (1997) with permission)

Table 1. Variance of the average $\langle m \rangle$ for four different averages $\langle m \rangle$ and three different noise models (arbitrary units).

Estimate \ type of noise		var{⟨m⟩}	
	spatially uncorrelated	random dipoles	mixed
best-channel	201	201	201
least-squares	8	160	122
maximum-likelihood	36	49	55
minimum-norm	936	62	281

by the factor $\frac{4}{3}$. This corresponds to the assumption that the variance of the spatially uncorrelated noise is one third of the variance of the spatially correlated noise arising from the random dipoles. The covariance matrix was finally rescaled so that its diagonal elements had the value one.

The weighting factor patterns obtained for the least-squares estimate (Figure 5(b)) and the maximum-likelihood estimate (Figure 5(c)) are similar: positive weighting factors in the upper half of the plot ($y > 0$), and negative weighting factors in the bottom half ($y < 0$). This result reflects the fact that, in the present case, the AOI corresponds to a dipole located on the z axis, with its moment pointing in x direction. This means that for measurement locations with a positive y coordinate the magnetic field has the same sign as the amplitude of the dipole moment, whereas an opposite sign is obtained for coils with a negative y coordinate. Thus, to obtain an estimate for the AOI, contributions from coils with a negative y coordinate have to be multiplied with a negative weight. A striking difference between least-squares and maximum-likelihood estimate is that the latter is essentially based on a few channels close to the central coil (z axis). This result suggests that, for the source considered here, magnetometer systems covering only a limited area of the scalp have about the same performance as a whole-head magnetometer system, provided that a maximum-likelihood estimate is used.

A completely different pattern is obtained for the minimum-norm estimate. Figure 5(d) shows that positive and negative weighting factors are intermixed. This means that the minimum-norm estimate performs complex difference operations with the data from the different channels.

The variances of the four estimates are compared in Table 1 for three different noise models: spatially uncorrelated noise ($[\operatorname{cov} \mathbf{d}] = \sigma^2 \mathbf{I}$), noise arising from random dipoles in the brain, and the mixed case. The latter case corresponds to the noise model used for the derivation of the weighting factors of the maximum-likelihood model (see above). With spatially uncorrelated noise it is, of course, the least-squares estimate which has the lowest variance, because uncorrelated noise with identical standard deviations in all channels is the assumption underlying the least-squares estimation procedure. The value obtained for the maximum-likelihood estimate is more than four times greater than that obtained for the least-squares estimate. Nevertheless, the maximum-likelihood estimate performs quite reasonably compared to the other two estimates. The noise sensitivity of the minimum-norm estimate is extremely bad in the example considered here, but this feature depends, of course, on the choice of the regularization parameters.

If the noise arises exclusively from the random dipoles, it is the maximum-likelihood estimate which has the best performance, whereas the variance obtained for the least-squares estimate is almost as high as that of the best-channel estimate. A remarkable finding is that the variance of the minimum-norm estimate is not much higher than that of the maximum-likelihood estimate. However, the performance of the minimum-norm estimate deteriorates if spatially uncorrelated noise is added to the noise arising from the random dipoles (rightmost column in Table 1).

Three-dimensional visualizations of the resolution fields associated with the weighting factor patterns shown in Figure 5 are provided in Lütkenhöner and Grave de Peralta Menendez (1997). It is certainly not surprising that the best-channel estimate is not at all specific for the AOI. Though the resolution field of the least-squares estimate has a completely different appearance, it is "broadly tuned" as well. A remarkable improvement of the spatial specificity is obtained by switching from a least-squares to a maximum-likelihood estimate. Despite the great differences in the weighting factor patterns of the maximum-likelihood estimate and the minimum-norm estimate, their resolution fields turned out to be very similar. This similarity is in striking contrast to the finding (Table 1) that the variances of the associated estimates $\langle m \rangle$ may differ considerably. The results obtained for the maximum-likelihood estimate exemplify, furthermore, that a superior spatial resolution (compared to the best-channel and the least-squares estimate) is not necessarily associated with a higher variance. An explanation for this finding was already given in the previous subsection.

5 Selected Aspects of Nonlinear Parameter Estimation Problems

Equation (5) indicated that the data kernel **G** may depend in a non-linear fashion on parameters **x**. However, the subsequent considerations were basically confined to linear inverse problems. In this section, selected aspects of non-linear inverse problems will be treated.

5.1 Linear Approximation

Equation (5) can be considered as a special case of an equation

$$\mathbf{d} = \left(g_1(\boldsymbol{\Theta}), g_2(\boldsymbol{\Theta}), \ldots, g_N(\boldsymbol{\Theta})\right)^T = \mathbf{g}(\boldsymbol{\Theta}), \tag{50}$$

where $\mathbf{\Theta}$ denotes the vector $[\mathbf{x}, \mathbf{m}]^T$, comprising all N_Θ model parameters. A maximum likelihood estimation of the parameters $\mathbf{\Theta}$ requires the minimization of the prediction error

$$\Phi(\mathbf{\Theta}) = \left(\mathbf{d} - \mathbf{g}(\mathbf{\Theta})\right)^T \mathbf{W}_e \left(\mathbf{d} - \mathbf{g}(\mathbf{\Theta})\right). \tag{51}$$

In the vicinity of the true parameter values, $\langle\mathbf{\Theta}\rangle$, the linear approximation

$$\mathbf{g}(\mathbf{\Theta}) \simeq \mathbf{g}(\langle\mathbf{\Theta}\rangle) + \mathbf{F}(\mathbf{\Theta} - \langle\mathbf{\Theta}\rangle) \tag{52}$$

can be applied, where $\mathbf{F}$ is the Jacobian matrix of $\mathbf{g}$ at $\langle\mathbf{\Theta}\rangle$, i.e. the matrix elements are

$$F_{ij} = \frac{\partial g_i}{\partial \Theta_j}(\langle\mathbf{\Theta}\rangle). \tag{53}$$

Inserting (52) into (51) yields

$$\Phi(\mathbf{\Theta}) \simeq \mathbf{\Delta}^T \mathbf{W}_e \mathbf{\Delta} \tag{54}$$

with

$$\mathbf{\Delta} = \{\mathbf{d} - \mathbf{g}(\langle\mathbf{\Theta}\rangle)\} - \mathbf{F}\{\mathbf{\Theta} - \langle\mathbf{\Theta}\rangle\}. \tag{55}$$

Braces were used to emphasize that the structure of (54) corresponds to that of the special case $\epsilon = 0$ of (9). As a consequence of this analogy, the parameter set minimizing (54) can be found by rewriting (6) and (10):

$$\mathbf{\Theta}^{\text{est}} = \langle\mathbf{\Theta}\rangle + \left((\mathbf{F}^T\mathbf{W}_e\mathbf{F})^{-1}\mathbf{F}^T\mathbf{W}_e\right) \cdot \left(\mathbf{d} - \mathbf{g}(\langle\mathbf{\Theta}\rangle)\right). \tag{56}$$

Supposed that the theory is exact, $\mathbf{d} - \mathbf{g}(\langle\mathbf{\Theta}\rangle)$ represents the measurement noise.

In the case of zero-mean noise with covariance $[\text{cov}\,\mathbf{d}]$, the mean and the covariance of $\mathbf{\Theta}^{\text{est}}$ can be calculated according to (28) and (29). The mean of $\mathbf{\Theta}^{\text{est}}$ is evidently $\langle\mathbf{\Theta}\rangle$, and the covariance is

$$[\text{cov}\,\mathbf{\Theta}] = \left((\mathbf{F}^T\mathbf{W}_e\mathbf{F})^{-1}\mathbf{F}^T\mathbf{W}_e\right) \cdot [\text{cov}\,\mathbf{d}] \cdot \left((\mathbf{F}^T\mathbf{W}_e\mathbf{F})^{-1}\mathbf{F}^T\mathbf{W}_e\right)^T. \tag{57}$$

This equation reduces to

$$[\text{cov}\,\mathbf{\Theta}] = (\mathbf{F}^T[\text{cov}\,\mathbf{d}]^{-1}\mathbf{F})^{-1} \tag{58}$$

if $\mathbf{W}_e = [\text{cov}\,\mathbf{d}]^{-1}$.

As shown above, inverse problems with a symmetric, positive definite weighting factor matrix $\mathbf{W}_e$ can be transformed so that, in the "transformed world", both the weighting factor matrix and the data covariance matrix correspond to the identity matrix. In this case, the confidence ellipsoid for $\mathbf{\Theta}$, given the solution $\mathbf{\Theta}^{\text{est}}$, is

$$\mathcal{E}_\Theta = \{\mathbf{\Theta} : (\mathbf{\Theta} - \mathbf{\Theta}^{\text{est}})^T(\mathbf{F}^T\mathbf{F})^{-1}(\mathbf{\Theta} - \mathbf{\Theta}^{\text{est}}) \leq \mu^2\}, \tag{59}$$

where μ^2 is the p-percentage point of the $\chi^2_{N_\Theta}$ distribution (Sarvas, 1987).

5.2 Dipole Source Localization Errors

In general, the covariance matrix $[\mathrm{cov}\,\mathbf{\Theta}]$ has to be investigated numerically. But under especially simple conditions an analytical treatment is feasible. Such an analysis was presented in Lütkenhöner (1996), where the most fundamental aspects of a MEG based dipole source localization were explored. Some results of this study will be outlined here briefly. A related analysis was given by Mosher et al. (1993), who derived Cramer–Rao lower error bounds for EEG and MEG dipole source localization.

In the case of a single current dipole in a spherical volume conductor the parameter vector $\mathbf{\Theta}$ can be written as $(X, Y, Z, m_\xi, m_\upsilon)^T$, where $(X, Y, Z)^T$ is the dipole location, and m_ξ and m_υ are the two tangential components of the dipole moment. The latter refer to directions $\vec{u}_\xi$ and $\vec{u}_\upsilon$, which are approximately identical with the x and the y direction in Figure 1, provided that the dipole is located in the vicinity of the z axis (see Lütkenhöner, 1996 for details). To calculate the elements of the matrix $\mathbf{F}^T\mathbf{F}$ it is obviously required to add up N terms, one for each channel. However, for a sufficiently dense grid of sensors this summation can be approximated by an integration over the measurement surface.

In Lütkenhöner (1996) the measurement surface corresponded to a spherical cap with radius r and span 2Θ (angle as seen from the center of sphere), and all sensors had a radial orientation. The magnetic field was assumed to result from a current dipole located on the z axis. Its moment had the amplitude J^p and pointed into the x direction. Thus, according to the nomenclature suggested in Hari et al. (1988), X represents the longitudinal coordinate (same direction as the dipole moment), Z the radial coordinate (measuring the distance between the center of sphere and the dipole location), and Y the transverse coordinate (perpendicular to radial and longitudinal coordinate). The noise was assumed to have mean zero and covariance $[\mathrm{cov}\,\mathbf{d}] = \sigma^2\mathbf{I}$. After having derived an approximation for the matrix $\mathbf{F}^T\mathbf{F}$, the covariances of the model parameters were calculated using (58). The resulting covariance matrix turned out to have the structure

$$[\mathrm{cov}\,\mathbf{\Theta}^{\mathrm{est}}] = \frac{2\pi(1-\cos\Theta)}{N} r^6 \left(\frac{\mu_0}{4\pi}\frac{J^p}{\sigma}\right)^{-2} \times \begin{pmatrix} \Gamma_{X,X} & 0 & 0 & 0 & 0 \\ 0 & \Gamma_{Y,Y} & 0 & 0 & 0 \\ 0 & 0 & \Gamma_{Z,Z} & \frac{J^p}{r}\Gamma_{Z,m_\xi} & 0 \\ 0 & 0 & \frac{J^p}{r}\Gamma_{Z,m_\xi} & \left(\frac{J^p}{r}\right)^2\Gamma_{m_\xi,m_\xi} & 0 \\ 0 & 0 & 0 & 0 & \left(\frac{J^p}{r}\right)^2\Gamma_{m_\upsilon,m_\upsilon} \end{pmatrix} \tag{60}$$

where the quantities $\Gamma_{X,X}$, $\Gamma_{Y,Y}$, $\Gamma_{Z,Z}$, Γ_{Z,m_ξ}, Γ_{m_ξ,m_ξ}, and $\Gamma_{m_\upsilon,m_\upsilon}$ represent standardized (dimensionless) covariances. Explicit formulas are given in Lütkenhöner (1996). The structure of the matrix shows that the estimation errors for the dipole parameters X, Y, and m_υ are independent of other estimation errors, whereas the estimation error for the Z coordinate is correlated with the estimation error for the component m_ξ. The effect of this correlation is that an overestimation of the dipole depth (i.e., an underestimation of Z in the present case) is accompanied by an overestimation of the dipole amplitude. Similar conclusions were obtained in Hari et al. (1988). It is noteworthy, furthermore, that the estimation errors for the two components of the dipole moment (m_ξ and m_υ) are independent of the actual amplitude of the dipole moment, J^p.

The square-roots of the diagonal terms of the covariance matrix (60) can be interpreted as the expected estimation errors for the five dipole parameters. These errors exhibit fundamental differences with respect to their asymptotic behavior for deep dipoles ($R = \sqrt{X^2 + Y^2 + Z^2} \to 0$). If the RMS value of the magnetic field is kept constant (requiring a depth-dependent definition of the dipole moment J^p), the error for Y is proportional to the distance R between dipole and center of sphere, the errors for the other dipole coordinates and the relative error for the transverse component of the dipole moment are constant, and the relative error for the longitudinal component of the dipole moment follows a $1/R$ law.

It should be emphasized that the above results were derived under the assumption that the measurement noise exhibits no spatial correlation. A simulation study with spatially correlated noise (Lütkenhöner, 1998b) showed that the statistical properties of the estimation errors are crucially dependent on the assumptions underlying the estimation procedure. In contrast to the theoretical analysis for spatially uncorrelated noise (Lütkenhöner, 1996), where extremely small estimation errors were found for the transverse coordinate Y, the two tangential coordinates (X and Y) obtained by least-squares fit exhibited about the same scattering. However, conditions similar to those predicted in the theoretical analysis were obtained when using a maximum likelihood estimation accounting for the covariance of the noise. This finding suggests that the very special position of the transverse coordinate can be observed only if the fit algorithm operates with an appropriate noise model.

5.3 Separability of Two Dipoles

In the presence of a second dipole, the parameter estimation errors for the first dipole are crucially dependent on the spatial relationship between the two dipoles. Mosher et al. (1993) derived Cramer–Rao lower error bounds for this situation. They showed that the localization accuracy for a dipole can be affected by a second dipole as far as 4 cm, depending on the angle between the two dipole moments. The most favorable condition was an orthogonal orientation of the two dipole moments, whereas a parallel orientation turned out to be the worst condition. This result was confirmed in Lütkenhöner (1998a), where minimum requirements for a successful separation of two dipoles were explored by investigating the ability of a one-dipole model to explain the magnetic field actually resulting from two dipoles. In the latter study it was shown, furthermore, that the separability of two parallel dipoles is moderately enhanced if both moments assume an orientation perpendicular to the line connecting the dipole locations. Two dipoles in different depths generally require a much higher signal-to-noise ratio than two dipoles in the same depth. The separability of two anti-parallel dipoles is not limited by concurrence with a one-dipole model, but by the low signal amplitudes resulting from a mutual cancellation of the fields arising from the two dipoles, and by concurrence with a quadrupole model.

A typical problem in iterative two-dipole source estimation procedures is that the dipole moments "explode", while the dipole locations approach each other (Lehnertz et al., 1989, Lütkenhöner, 1992b, Lütkenhöner et al., 1992). In the end, the two dipoles have almost identical locations and moments, except that they point into opposite directions.[5] Huge dipole moments pointing into opposite directions are to be expected even if exact knowledge of the dipole locations is available, supposed that the distance between the two dipoles is small (Lütkenhöner, 1992a). The following analysis provides an explanation for this phenomenon. For a single channel, (5) reduces to $d_i = \mathbf{g}_i^T(\mathbf{x})\mathbf{m} = \mathbf{m}^T\mathbf{g}_i(\mathbf{x})$, where $\mathbf{g}_i$ is a column vector of dimension M which corresponds to the ith row of the data kernel $\mathbf{G}(\mathbf{x})$. Supposed that the measured magnetic field arises from a *single* current dipole, $\mathbf{m}$ represents the dipole moment, whereas $\mathbf{x}$ represents the dipole location. Thus, in the absence of noise, the forward model for *two* current dipoles is

$$\mathbf{d}_i = \mathbf{m}_1^T\mathbf{g}_i(\mathbf{x}_1) + \mathbf{m}_2^T\mathbf{g}_i(\mathbf{x}_2). \tag{61}$$

Provided that the distance between the two dipoles is sufficiently small, it is convenient to substitute $\mathbf{x}_1 = \mathbf{x} + \boldsymbol{\delta}\mathbf{x}/2$ and $\mathbf{x}_2 = \mathbf{x} - \boldsymbol{\delta}\mathbf{x}/2$, where $\mathbf{x} = (\mathbf{x}_1 + \mathbf{x}_2)/2$ is the mean location and $\boldsymbol{\delta}\mathbf{x} = \mathbf{x}_1 - \mathbf{x}_2$ is the location difference. In the vicinity of $\mathbf{x}$, the linear approximation

$$\mathbf{g}_i(\mathbf{x} \pm \boldsymbol{\delta}\mathbf{x}/2) \simeq \mathbf{g}_i(\mathbf{x}) \pm \mathbf{D}_i(\mathbf{x})\boldsymbol{\delta}\mathbf{x}/2 \tag{62}$$

can be applied, where $\mathbf{D}_i$ is the Jacobian matrix derived from $\mathbf{g}_i$ (compare (52)). Inserting (62) into (61) yields

$$\mathbf{d}_i \simeq \mathbf{m}^T\mathbf{g}_i(\mathbf{x}) + \boldsymbol{\delta}\mathbf{m}^T\mathbf{D}_i(\mathbf{x})\boldsymbol{\delta}\mathbf{x}, \tag{63}$$

where $\mathbf{m} = \mathbf{m}_1 + \mathbf{m}_2$ and $\boldsymbol{\delta}\mathbf{m} = \mathbf{m}_1 - \mathbf{m}_2$. With $\alpha = ||\boldsymbol{\delta}\mathbf{m}|| \cdot ||\boldsymbol{\delta}\mathbf{x}||$, $\mathbf{u}_{\boldsymbol{\delta}\mathbf{m}} = \boldsymbol{\delta}\mathbf{m}/||\boldsymbol{\delta}\mathbf{m}||$, and $\mathbf{u}_{\boldsymbol{\delta}\mathbf{x}} = \boldsymbol{\delta}\mathbf{x}/||\boldsymbol{\delta}\mathbf{x}||$ this equation can be rewritten as

$$\mathbf{d}_i \simeq \mathbf{m}^T\mathbf{g}_i(\mathbf{x}) + \alpha\mathbf{u}_{\boldsymbol{\delta}\mathbf{m}}^T\mathbf{D}_i(\mathbf{x})\mathbf{u}_{\boldsymbol{\delta}\mathbf{x}}, \tag{64}$$

[5]Analogously, the matrix inversion in (7) becomes numerically ill-conditioned, particularly in the presence of noise, such that the generalized inverse cannot be formed reliably without regularization.

Thus, under conditions where the approximation (63) is applicable, it is impossible to derive separate estimates for the dipole distance and the magnitude of the difference of the moments. Only their product, α, can be estimated. A small dipole distance is consequently associated with huge dipole moments with opposite signs.

6 Concluding Remarks

Model selection is the decisive first step of any MEG source analysis, because only by adding supplementary information or by making certain assumptions about the solution it is possible to overcome the problem that any magnetic field pattern registered outside the head has an infinite number of possible interpretations. This kind of non-uniqueness is primarily of a deterministic and not of a statistical nature, because basically the same problem would arise if the data are measured without noise and the analysis is based on a theoretical model establishing a precise relationship between intracerebral current density and measured data (exact forward model). In that hypothetical case, two classes of models could be distinguished: those which are capable of explaining the data and those which are not. Models explaining the data are not necessarily plausible from a physiological point of view. In a quadrupolar source configuration, for example, large dipole moments with nearly opposite directions (as considered in the context of equation (63)) are generally not plausible, because the required intracerebral current densities are beyond the usual physiological limits. In principle, such plausibility criteria can be implemented using statistical techniques (a priori probability distributions for the model parameters). As briefly outlined in Section 2.3, the Bayesian approach may be used to view the covariance matrices introduced in Section 2.2 as Gaussian priors on source and observational noise probability distributions. Other more focal priors are considered in Baillet and Garnero (1997), Phillips et al. (1997), Schmidt et al. (1999), generally leading to more complicated solution procedures. As in any Bayesian approach, the construction of prior probabilities is a controversial matter (Gouveia and Scales, 1997) as long as suitable and reliable experimental data are lacking. Besides that, there are many alternative ways to control the properties of the solution. Large dipole moments in quadrupolar source configurations, for instance, are avoided also by accounting for the "solution simplicity" in the expression to be minimized (second term on the right of equation (9)). It is still an open debate as to which approach is most advantageous under what circumstances.

In practice, measurement noise is unavoidable, and the theoretical model underlying the data analysis procedure is usually imperfect. Such circumstances evidently aggravate the principle problem of non-uniqueness, since conditions which would be distinct under ideal circumstances become indistinguishable if the differences between the associated field patterns are buried in noise or if they are concealed by inadequacies of the forward model. For example, two dipoles can be distinguished from a single dipole (i.e., they can be "resolved") only if the field pattern arising from the two dipoles differs sufficiently from that of a single dipole.

How to proceed if statistical information about the covariance of the noise, $[\mathrm{cov}\,\mathbf{d}]$, and the data, $[\mathrm{cov}\,\mathbf{m}]$, is available has been outlined above. But it is, first of all, required to *provide* these covariance matrices. Whereas it is straightforward to estimate $[\mathrm{cov}\,\mathbf{d}]$ from the data themselves (supposed that some simplifying assumptions about the properties of the noise are justified), the estimation of $[\mathrm{cov}\,\mathbf{m}]$ is problematic. Dale and Sereno (1993) suggested to express the correlation between cortical patches as a function of their distance and to derive this function from invasive measurements. However, this procedure is probably too simplistic, because neighboring patches may be involved in different tasks, while distant patches may closely cooperate. The alternative, estimation of $[\mathrm{cov}\,\mathbf{m}]$ from the data (Sekihara and Scholz, 1995, 1996), represents an inverse problem quite similar to the one considered here so that the involved problems are presumably also similar.

It is not surprising that, regarding the *variance* of the estimated parameters, estimation techniques accounting for the statistical properties of the recorded noise generally outperform methods using oversimplified noise models (Lütkenhöner, 1998b,c, Sekihara et al., 1992). However, the effect of the noise model on the solution can be subtle, with the consequence that solutions obtained with different noise models are not necessarily consistent. An example was given in Lütkenhöner and Steinsträter (1998), where the tonotopic organization of the human auditory cortex was investigated with utmost precision. Data recorded with a 37-channel magnetometer were analyzed using the model of a single current dipole. The locations obtained by least-squares fit (assumption of spatially uncorrelated noise) and maximum likelihood estimation (applying the covariance matrix estimated from the measured data) systematically differed by several millimeters. These discrepancies can be put down to the fact that the assumption of a current dipole in a homogeneous spherical volume conductor does not provide a completely satisfactory model for the underlying generators in the brain. Since such inconsistencies do not affect all channels to the same extent, the effect on the solution is evidently dependent on the weighting factors associated with the channels, and the latter are dependent on the noise model, as described above (Figure 5).

Acknowledgments: The author thanks Pedro Valdes and Rolando Biscay (Havana), John C. Mosher (Los Alamos, NM), and Kensuke Sekihara (Tokyo) for their constructive comments on a prior version of this article.

7 References

Backus, G.E. and F. Gilbert (1967). Numerical applications of a formalism for geophysical inverse problems. *Geophysical Journal of the Royal Astronomical Society 13*, 247–276.

Backus, G.E. and F. Gilbert (1968). The resolving power of gross earth data. *Geophysical Journal of the Royal Astronomical Society 16*, 169–205.

Backus, G. and F. Gilbert (1970). Uniqueness in the inversion of inaccuarate gross earth data. *Philosophical Transactions of the Royal Society of London. Series A. Mathematical, Physical Sciences and Engineering 266*, 123–192.

Baillet, S. and L. Garnero (1997). A Bayesian approach to introducing anatomo-functional priors in the EEG/MEG inverse problem. *IEEE Transactions on Biomedical Engineering 44*, 374–385.

Benzel, E.C., J.D. Lewine, R.D. Bucholz, and W.W. Orrison (1993). Magnetic source imaging: a review of the Magnes system of Biomagnetic Technologies Incorporated. *Neurosurgery 33*, 252–259.

Bertrand, O., M. Thevenet, and F. Perrin (1991). 3D finite element method in brain electrical activity studies. In J. Nenonen, H. Rajala, and T. Katila (Eds.), *Biomagnetic Localization and 3D Modelling*, pp. 154–171. Helsinki: Helsinki University of Technology. Report TKK-F-A689.

Buchner, H., G. Knoll, M. Fuchs, A. Rienacker, R. Beckmann, M. Wagner, J.Silny, and J. Pesch (1997). Inverse localization of electric dipole current sources in finite element models of the human head. *Electroencephalography and Clinical Neurophysiology 102*, 267–278.

Committee on the Mathematics and Physics of Emerging Dynamic Biomedical Imaging, National Research Council (1996). *Mathematics and Physics of Emerging Biomedical Imaging.* Washington, D.C: National Academy Press. Available on the Internet at the URL `http://www.nas.edu/`.

Cuffin, B. and D. Cohen (1977). Magnetic fields produced by models of biological current sources. *Journal of Applied Physics 48*, 3971–3980.

Dale, A. and M. Sereno (1993). Improved localization of cortical activity by combining EEG and MEG with MRI cortical surface reconstruction: a linear approach. *Journal of Cognitive Neuroscience 5*, 162–176.

de Munck, J., P.C. Vijn, and F. Lopes da Silva (1992). A random dipole model for spontaneous brain activity. *IEEE Transactions on Biomedical Engineering 39*, 791–804.

Freeman, W. (1980a). A software lens for image reconstruction of the EEG. In H. Kornhuber and L. Deecke (Eds.), *Motivation, Motor and Sensory Processes of the Brain: Electrical Potentials, Behaviour and Clinical Use*, pp. 123–127. Amsterdam: Elsevier.

Freeman, W. (1980b). Use of spatial deconvolution to compensate for distortion of EEG by volume conduction. *IEEE Transactions on Biomedical Engineering 27*, 421–429.

Fuchs, M., R. Drenckhahn, H. Wischmann, and M. Wagner (1998). An improved boundary element method for realistic volume-conductor modeling. *IEEE Transactions on Biomedical Engineering 45*, 980–997.

Fuchs, M., M. Wagner, H. Wischmann, K. Ottenberg, and O. Dössel (1994). Possibilities of functional brain imaging using a combination of MEG and MRT. In *Oscillatory Event Related Brain Dynamics*, pp. 435–457. New York: Plenum.

Golub, G. and C. Loan (1989). *Matrix computations* (2nd ed.). Baltimore, MD: Johns Hopkins Univ.

Gouveia, W. and J. Scales (1997). Resolution of seismic waveform inversion: Bayes versus Occam. *Inverse Problems 13*, 323–349.

Grave de Peralta Menendez, R. and S. Gonzalez Andino (1998). A critical analysis of linear inverse solutions to the neuroelectromagnetic inverse problem. *IEEE Transactions on Biomedical Engineering 45*, 440–448.

Grave de Peralta Menendez, R. and S. Gonzalez Andino (1999). Distributed source models: standard solutions and new developments. In C. Uhl (Ed.), *Analysis of Neurophysiological Brain Functioning.* Berlin: Springer.

Hämäläinen, M., R. Hari, R. Ilmoniemi, J. Knuutila, and O. Lounasmaa (1993). Magnetoencephalography—theory, instrumentation, and applications to noninvasive studies of the working human brain. *Reviews of Modern Physics 65*, 413–497.

Hämäläinen, M. and J. Sarvas (1989). Realistic conductivity geometry model of the human head for interpretation of neuromagnetic data. *IEEE Transactions on Biomedical Engineering 36*, 165–171.

Hari, R., S. Joutsiniemi, and J. Sarvas (1988). Spatial resolution of neuromagnetic records: theoretical calulations in a spherical model. *Electroencephalography and Clinical Neurophysiology 71*, 64–72.

Kaufman, L., J.H. Kaufman, and J.Z. Wang (1991). On cortical folds and neuromagnetic fields. *Electroencephalography and Clinical Neurophysiology 79*, 211–226.

Kuriki, S., F. Takeuchi, and T. Kobayashi (1994). Characteristics of the background fields in multichannel-recorded magnetic field responses. *Electroencephalography and Clinical Neurophysiology 92*, 56–63.

Lehnertz, K., B. Lütkenhöner, M. Hoke, and C. Pantev (1989). Considerations on a spatio-temporal two-dipole model. In S. Williamson, M. Hoke, G. Stroink, and M. Kotani (Eds.), *Advances in Biomagnetism*, pp. 563–566. New York: Plenum.

Lütkenhöner, B. (1991). A simulation study of the resolving power of the biomagnetic inverse procedure. *Clinical Physics and Physiological Measurements 12, Suppl. A*, 73–78.

Lütkenhöner, B. (1992a). On the biomagnetic inverse procedure's capability of separating two current dipoles with a priori known locations. In M. Hoke, S. Erné, Y. Okada, and G. Romani (Eds.), *Biomagnetism: Clinical Aspects*, pp. 687–692. Amsterdam: Excerpta Medica.

Lütkenhöner, B. (1992b). *Möglichkeiten und Grenzen der neuromagnetischen Quellenanalyse.* Münster: Lit.

Lütkenhöner, B. (1994). Magnetic field arising from current dipoles randomly distributed in a homogeneous spherical volume conductor. *Journal of Applied Physics 75*, 7204–7210.

Lütkenhöner, B. (1996). Current dipole localization with an ideal magnetometer system. *IEEE Transactions on Biomedical Engineering 43*, 1049–1061.

Lütkenhöner, B. (1998a). Dipole separability in a neuromagnetic source analysis. *IEEE Transactions on Biomedical Engineering 45*, 572–581.

Lütkenhöner, B. (1998b). Dipole source localization by means of maximum likelihood estimation. I. Theory and simulations. *Electroencephalography and Clinical Neurophysiology 106*, 314–321.

Lütkenhöner, B. (1998c). Dipole source localization by means of maximum likelihood estimation. II. Experimental evaluation. *Electroencephalography and Clinical Neurophysiology 106*, 322–329.

Lütkenhöner, B., T. Elbert, E. Menninghaus, O. Steinsträter, and C. Wienbruch (1995a). Electro- and magnetoencephalographic source analysis: current status and future requirements. In H. Herrmann, D. Wolf, and W. Pöppel (Eds.), *Workshop on Supercomputing in Brain Research: From Tomography to Neural Networks*, pp. 175–192. Singapore: World Scientific.

Lütkenhöner, B. and R. Grave de Peralta Menendez (1997). The resolution-field concept. *Electroencephalography and Clinical Neurophysiology 102*, 326–334.

Lütkenhöner, B., R. Greenblatt, M. Hämäläinen, J. Mosher, M. Scherg, C. Tesche, and P. Valdes Sosa (2000). Comparison between different approaches to the biomagnetic inverse problem: Workshop report. In C. Aine, Y. Okada, G. Stroink, S. Swithenby, and C. Wood (Eds.), *Biomag96*, Volume 1. New York: Springer.

Lütkenhöner, B., K. Lehnertz, M. Hoke, and C. Pantev (1991). On the biomagnetic inverse problem in the case of multiple dipoles. *Acta Oto-Laryngologica. Supplement (Stockholm) 491*, 94–105.

Lütkenhöner, B., E. Menninghaus, O. Steinsträter, C. Wienbruch, H. Gißer, and T. Elbert (1995b). Neuromagnetic source analysis using magnetic resonance images for the construction of source and volume conductor model. *Brain Topography 7*, 291–299.

Lütkenhöner, B., C. Pantev, and M. Hoke (1992). Interpretation of auditory evoked magnetic fields in terms of two simultaneously active dipoles. *Biomed. Res. 13*, 17–22.

Lütkenhöner, B. and O. Steinsträter (1998). High-precision neuromagnetic study of the functional organization of the human auditory cortex. *Audiology and Neuro-Otology 3*, 191–213.

Malmivuo, J. and R. Plonsey (1995). *Bioelectromagnetism: principles and applications of bioelectric and biomagnetic fields.* New York: Oxford University Press.

Menke, W. (1984). *Geophysical Data Analysis: Discrete Inverse Theory* (revised ed.). San Diego, CA: Academic Press.

Menninghaus, E., B. Lütkenhöner, and S. Gonzalez (1994). Localization of a dipolar source in a skull phantom: Realistic versus spherical model. *IEEE Transactions on Biomedical Engineering 41*, 986–989.

Mosher, J.C., M.E. Spencer, R.M. Leahy, and P.S. Lewis (1993). Error bounds for EEG and MEG dipole source localization. *Electroencephalography and Clinical Neurophysiology 86*, 303–321.

Phillips, J., R. Leahy, and J. Mosher (1997). MEG-based imaging of focal neuronal current sources. *IEEE Transactions on Medical Imaging 16*, 338–348.

Press, W.H., S.A. Teukolsky, W.T. Vetterling, and B. Flannery (1992). *Numerical Recipes: The Art of Scientific Computing in C* (2nd ed.). Cambridge: Cambridge University Press.

Robinson, S. (1989). Theory and properties of lead field synthesis analysis. In S. Williamson, M. Hoke, G. Stroink, and M. Kotani (Eds.), *Advances in Biomagnetism*, pp. 599–602. New York: Plenum.

Sarvas, J. (1987). Basic mathematical and electromagnetic concepts of the biomagnetic inverse problem. *Physics in Medicine and Biology 32*, 11–22.

Scherg, M. (1990). Fundamentals of dipole source potential analysis. In F. Grandori, M. Hoke, and G. Romani (Eds.), *Auditory Evoked Magnetic Fields and Electric Potentials*, Volume 6 of *Advances in Audiology*, pp. 40–69. Basel: Karger.

Scherg, M. and D. von Cramon (1985). Two bilateral sources of the late AEP as identified by a spatio-temporal dipole model. *Electroencephalography and Clinical Neurophysiology 62*, 32–44.

Schmidt, D., J. George, and C. Wood (1999). Bayesian inference applied to the electromagnetic inverse problem. *Human Brain Mapping 7*, 195–212.

Sekihara, K., Y. Ogura, and M. Hotta (1992). Maximum-likelihood estimation of current-dipole parameters for data obtained using multichannel magnetometer. *IEEE Transactions on Biomedical Engineering 39*, 558–562.

Sekihara, K. and B. Scholz (1995). Average-intensity reconstruction and Wiener reconstruction of bioelectric current distribution based on its estimated covariance matrix. *IEEE Transactions on Biomedical Engineering 42*, 149–157.

Sekihara, K. and B. Scholz (1996). Generalized Wiener estimation of three-dimensional current distribution from biomagnetic measurements. *IEEE Transactions on Biomedical Engineering 43*, 281–291.

Tarantola, A. (1987). *Inverse Problem Theory.* New York: Elsevier.

Thevenet, M. (1992). *Modelisation de l'activité électrique cérébrale par la méthode des éléments finis.* Ph.D. thesis, Institut National des Sciences Appliquées, Lyon.

Thevenet, M., O. Bertrand, F. Perrin, T. Dumont, and J. Pernier (1991). The finite element method for a realistic head model of electrical brain activities: preliminary results. *Clinical Physics and Physiological Measurements 12, Suppl. A*, 89–94.

Williamson, S. and L. Kaufman (1987). Analysis of neuromagnetic signals. In A. Gevins and A. Remond (Eds.), *Methods of Analysis of Brain Electrical and Magnetic Signals*, Volume 1 of *Handbook of Electroencephalography and Clinical Neurophysiology—Revised Series*, pp. 405–448. Amsterdam: Elsevier.

Williamson, S. and L. Kaufman (1990). Theory of neuroelectric and neuromagnetic fields. In F. Grandori, M. Hoke, and G. Romani (Eds.), *Auditory Evoked Magnetic Fields and Electric Potentials*, Volume 6 of *Advances in Audiology*, pp. 1–39. Basel: Karger.

Yamazaki, T., B. van Dijk, and H. Spekreijse (1992). Confidence limits for the parameter estimation in the dipole localization method on the basis of spatial correlation of background EEG. *Brain Topography 5*, 195–198.

12

Statistics of Nonlinear Spatial Distortions in Histological Images

Thorsten Schormann
Andreas Dabringhaus

ABSTRACT Local spatial distortions in histological images hinder an exact global registration with corresponding magnetic resonance images (MRI) of the same object. In order to estimate appropriate reference points for an optimized least-square affine transformation matrix, the statistics of deformations is investigated. It is shown, that in the case of correlated and anisotropic histological procedures, local spatial distortions are Rayleigh–Bessel distributed according to the eigenvalues of matrix $\mathbf{M}$ describing variance and covariance of the distortions. For uncorrelated, anisotropic procedures, the probablity density function is given by a Rayleigh–Bessel function with corresponding variances and in the case of an uncorrelated and isotropic treatment the density can be described by a Rayleigh function. An advantage of this generalized theory is, that the information about the eigensystem of spatial deformations introduced by angle ϑ can be included in the Rayleigh–Bessel distribution which fits the experimental data more accurately over the entire histogram range and it is not necessary to rotate the images into the eigensystem of $\mathbf{M}$. The application of the theory to histological and corresponding MR images demonstrates an improved registration quality.

1 Introduction

The integration of micro- and macrostructural information derived from histological sections and MR images (MR: Magnetic Resonance) of the same object requires theories for linear and nonlinear transformations in order to achieve an one-to-one correspondence between images resulting from both imaging modalities. Therefore, the aim of an accurate registration is to establish a corrrespondence between voxels, such that for each voxel of the MR volume it will be possible to identify the corresponding voxel in the histological volume of the same object, e.g., the human brain. Unfortunately, this accurate alignment is hindered by various spatial linear and nonlinear distortions resulting from the histological preparation procedures which also destroy the 3-D integrity of the object during the histological cutting procedure. The aim of the present approach is to investigate the statistics of nonlinear deformations $\vec{\Delta}^{(i)}$ in order to achieve an accurate global alignment, thereby preserving the morpholgy. The latter may be changed by subsequent application of nonlinear deformation techniques (see below). In addition, local nonlinear deformations prevent an exact determination of global linear least-square transformation parameters derived from an automated procedure for the definition and correlation of landmarks Schormann et al. (1995).

The problem of finding appropriate global affine transformation parameters is solved by estimating a threshold on the basis of the probability density function $N(\vec{0}, \mathbf{M})$ (with covariance matrix $\mathbf{M}$) which determine the landmarks that should be rejected because they are expected to be misregistrations or pronounced nonlinear local deformations leading to a significant global linear geometrical misalignment.

1.1 Statistics of Nonlinear Distortions and the Motivation for Image Processing

Computerized atlases enable to superimpose information derived from different imaging modalities, e.g., such as MRI (Magnetic Resonance Imaging) and histological volumes which are 3-D reconstructed from serial sections. With this superimposition, it is then possible to improve the MR resolution by a factor of up to 10^3 using the microscopical information of the histological sections. Thus, it is possible to identify specific brain structures such as, e.g., fibre tracts on a microscopical level in order to analyze their variability which is then a reliable basis for the interpretation of functional data (fMRI). However, for this purpose it is required to transform the histological sections with highest accuracy onto the reference. This is accomplished by the application of linear and nonlinear transformations in a hierarchical manner. In a first step, a global affine transformation is applied, whereby the accuracy of this linear alignment is an important prerequisite for the second step, the application of nonlinear transformations (see below, Figure 2). Unfortunately, an accurate determination of affine parameters for the alignment of histological sections with their corresponding MR- sections is hindered by local deformations resulting from the histological preparation procedures. As a consequence, both sections reveal global misalignment (e.g., rotation, Figure 1) whereby local deformations remain. This is due to the fact, that the global least-square affine transformation parameters are estimated from landmarks which can be devided into either exact landmarks or those which are located in nonlinear spatial deformations. The total least-square cost function then consists of one part resulting from the exact landmarks (including those with a small degree of distortion) and from severely distorted landmarks leading to global

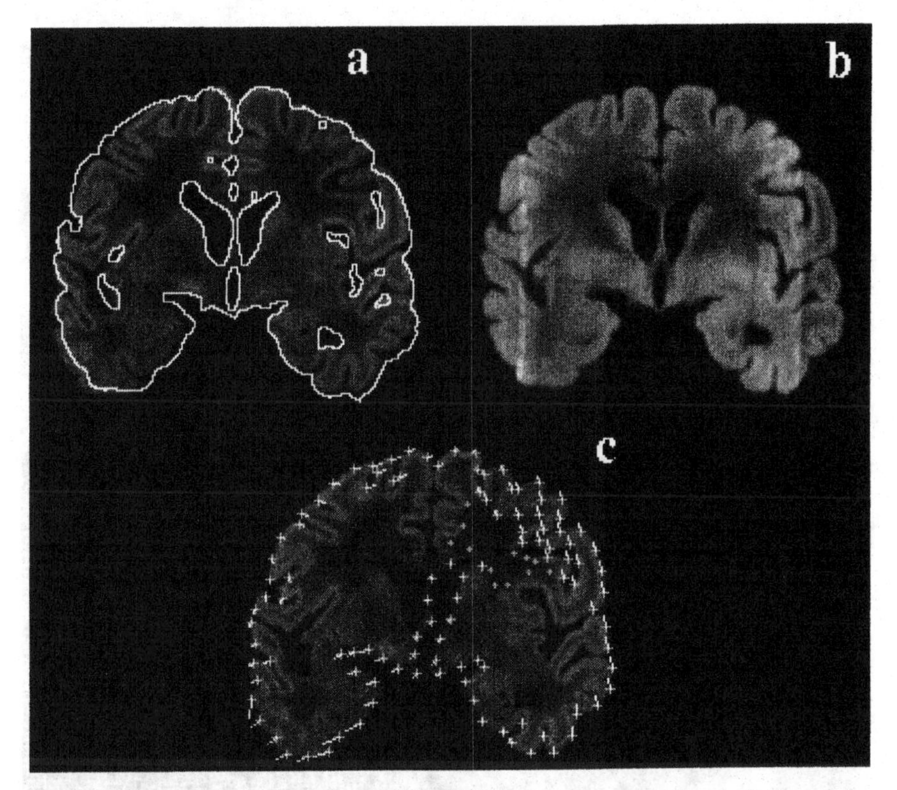

Figure 1. Example of an improved global alignment using the statistics of deformations. The crosses indicate the location of landmarks and the small lines originating from some crosses reveals the extent of nonlinear deformations after linear transformation. These are the deformation vectors $\vec{\Delta}_n^{(i)}$. The definition and correlation of corresponding reference points is accomplished automatically (for details see Schormann et al. (1995)). The histological (a) and the corresponding MR-section (b) are rotated with respect to each other after least-square transformation without using a statistical cost function ($\sim 6^\circ$ in this example, the rotational error may achieve values up to $\sim 10^\circ$). For a comparison, the outer contour of the MR-section is superimposed onto the histological section. (c) Superimposed reference points and deformation vectors $\vec{\Delta}_n^{(i)}$ for all possible locations i of the result shown in (a). The small crosses indicate reference points for which the crosscorrelation is less than a threshold.

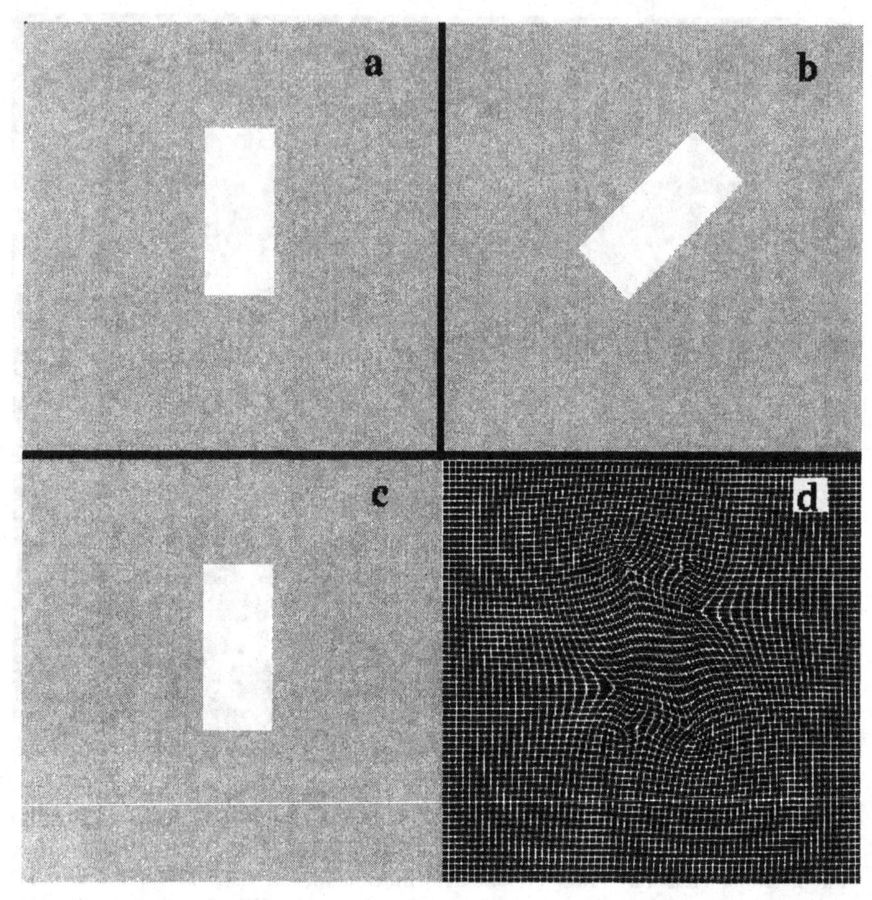

Figure 2. Difference between a (desired) global rotation and a "pseudo-global" rotation which is assembled by local nonlinear transformations. Source image (b) is nonlinearly transformed to the reference image (a), whereby the deformation is visualized by application to a regular grid (d). (c) Result after nonlinear transformation. A global transformation, e.g. rotation, is generally not achieved by nonlinear deformation techniques, which can be seen by close inspection of the deformation field (d) and thus, the morphology of the object is destroyed. This effect can be minimized by analyzing the statistics of deformations.

misalignment (Figure 1).

In principle, this global linear misalignment and the nonlinear distortions can be corrected by nonlinear deformation techniques (Bajcsy and Kovacic., 1989, Schormann et al., 1996). However, the resulting movement of the source image is generally not a global transformation, which is required in order to maintain the morphology of the object. The example in Figure 2 reveals by inspection of the deformation field, that the apparently global rotation is assembled by local compressions and expansions of the object. This is an undesirable effect, since rigid motions are achieved by local transformations, whereby the morphology of the object is destroyed. In addition, the computational effort for the determination of the deformation field which is derived from a system of coupled partial differential equations with up to 1.2×10^5 unknowns is tremendously high compared to an optimized linear alignment using the statistics of deformations.

A further requirement for the application of a nonlinear deformation technique based on the minimization of the gray value difference between source and reference is an—at least minimal—overlap of corresponding regions, in order to correlate homologous strucures. The theory presented here restricts the correction of local deformations to small parts of the image, thereby avoiding a global misalignment, which then does not need to be corrected locally.

2 Theory

In order to estimate optimal affine transformation parameters from a set of corresponding reference points $\{\vec{x}^{\,(\nu)}\}$ in the histological section (nonlinearly distorted) and $\{\vec{y}^{(\nu)}\}$ in the corresponding undistorted MR reference section with $\nu = 1, \ldots, N$ where $N \geq 3$ is the total number of reference points, it is required to analyze the probability of nonlinear deformations such that small degrees of deformations are tolerated and pronounced distortions are suppressed. The geometrical difference $\vec{\Delta}$ between histological and MR reference points after global, linear transformation with least-square matrix $\mathbf{A}$ and translation vector $\vec{b}$ is given by $\vec{\Delta}^{(\nu)} = \vec{y'}^{(\nu)} - \vec{y}^{(\nu)}$ where y' is calculated by $y' = \mathbf{A}\vec{x} + \vec{b}$. This situation is also visualized in Figure 1(c), whereby the crosses indicate the location of the reference points and the small lines originating from some crosses reveal the extent of nonlinear deformations relative to the undistorted reference section. In a first step, the total mean μ and covariance $\mathbf{M}$ can be determined from all N locations.

The purpose of the theory is to derive a probability density function which uses all the information about the preparation procedures without any assumptions or approximations (Satterthwaite, 1946, Worsley et al., 1995) concerning $\mathbf{M}$, which describes variance and covariance of deformations. Although the generalization of the theory produces apparently only slightly different results compared to Schormann et al. (1995), the theoretical and practical consequences are far-reaching, with improved correspondence between experimental and theoretical data, that is, the information about the eigensystem of **M** can be included in the Rayleigh–Bessel distribution, such that it is not required to rotate the images into the eigensystem of **M** and simultaneously inflicting no rigid restrictions on the preparation procedures.

However, the total deformation $\vec{\Delta}_n^{(i)}$ at location i is given by the sum of n single independent geometrical deformations (for the physical formation of deformations and "independent" see Section 3), denoted by random vectors $\vec{\Delta}^{(i)}(j)$:

$$\vec{\Delta}_n^{(i)} = \sum_{j=1}^{n} \vec{\Delta}^{(i)}(j) \tag{1}$$

with mean $\vec{\mu} = \langle \vec{\Delta}_n \rangle = \frac{1}{N} \sum_{i=1}^{N} \vec{\Delta}^{(i)} = \vec{0}$ after least-square transformation and covariance matrix **M**

$$\mathbf{M} = \begin{pmatrix} \sigma_{11}^2 & \sigma_{12}^2 \\ \sigma_{12}^2 & \sigma_{22}^2 \end{pmatrix} \tag{2}$$

with $\sigma_{ij}^2 = \langle (\Delta_i - \mu_i) \cdot (\Delta_j - \mu_j) \rangle$ where $\langle * \rangle$ denotes averaging of quantity $*$ with respect to N, the total number of reference points and $\sigma_{12} = \sigma_{21}$.

In order to apply the central limit theorem (CLT; (Bronstein and Semendjajew, 1981, Fabian and Hannan, 1984, Papoulis, 1965)) to the random vectors $\vec{\Delta}^{(i)}(j)$, it is required to consider the multidimensional CLT: Let $\vec{z}_1, \vec{z}_2, \ldots, \vec{z}_n$ be a sequence of independent random vectors, e.g., $\vec{z}_n = (z_{n1} \ldots z_{nd})^t$ with corresponding densities $p_{jk}(\vec{z}), j = 1, \ldots, n; k = 1, \ldots, d$, whereby n denotes the number of independent preparation procedures, d the spatial dimension. Then, p_k is given by

$$p_k(\vec{z}) = p_{1k}(\vec{z}) * \cdots * p_{nk}(\vec{z}) \quad k = 1, \ldots, d \tag{3}$$

where $*$ denotes the convolution and with

$$\vec{z} = \sum_{j=1}^{n} \vec{z}_j, \quad \vec{\mu} = \sum_{j=1}^{n} \vec{\mu}_j, \quad \vec{\sigma^2} = \sum_{j=1}^{n} \vec{\sigma^2}_j \tag{4}$$

resulting approximately in a normal distribution

$$p(\vec{z})\, d\vec{z} \sim \frac{1}{2\pi\sqrt{\det \mathbf{M}}} \exp(-\tfrac{1}{2}\vec{z}^t \mathbf{M}^{-1} \vec{z})\, d\vec{z} \tag{5}$$

where t denotes the transposed of $\vec{z}$ if the Lyapunov condition (Bauer, 1972) is satisfied

$$\lim_{n\to\infty} \sum_{j=1}^{n} \sigma_{jk}^2 \to \infty \quad \forall k = 1, \ldots, d$$

and

$$\int_{-\infty}^{\infty} \cdots \int_{-\infty}^{\infty} |\vec{z}|^{\alpha} p_j(\vec{z})\, dz_1 \cdots dz_d < C \quad \forall j; \alpha > 2 \tag{6}$$

where C is a constant and $|\vec{z}|^{\alpha} = (z_1^2 + \cdots + z_d^2)^{\alpha/2}$, $\alpha > 2$.

Assuming, that the random vectors $\vec{\Delta}_n^{(i)}$ are independent (see below) with mean $\vec{\mu} = \vec{0}$ after least-square tranformation, the assumptions for applying the multidimensional CLT are fullfilled:

(i) The deformations are approximatively the same, such that the mean $\langle \vec{\Delta}_n^{(i)} \vec{\Delta}_n^{t(j)} \rangle$ converges to $\mathbf{M}$ for $n \to \infty$. The number of preparation procedures is $n \geq 17$ as described below.

(ii) The largest deformation is limited to the finite size of the image, such that the Lyapunov condition is satisfied. An orthonormal basis $\mathcal{B}$ with elements $\vec{b}$ can be calculated from $\mathbf{M}$ so that $\vec{b}^t \cdot \vec{\Delta}^{(i)}(j) = \vec{y}^{(i)}(j)$ fullfills the Lyapunov condition for every $\vec{b}$ in $\mathcal{B}$ (Fabian and Hannan, 1984).

From these considerations, it follows that the probability $p(\vec{\Delta})$ of deformation $\vec{\Delta}$ in the interval $d\vec{\Delta}$ can be approximatively written as a multidimensional normal distribution:

$$p(\vec{\Delta})\, d\vec{\Delta} = \frac{1}{2\pi\sqrt{\det \mathbf{M}}} \exp(-\tfrac{1}{2}\vec{\Delta}^t \mathbf{M}^{-1} \vec{\Delta})\, d\vec{\Delta} \tag{7}$$

2.1 The General Case: Correlated and Anisotropic M

In the most general case for correlated and anisotropic preparation procedures ($\sigma_{12} \neq 0$, $\sigma_{11} \neq \sigma_{22}$), it is required to consider the coordinate system for the determination and the eigensystem for the formation of nonlinear distortions:

In the general bivariate normal case for deformation vectors $\vec{\Delta}$, a rotational transformation is necessary to diagonalize the distribution covariance matrix $\mathbf{M}$. However, the determination of deformations $|\vec{\Delta}|$ should be invariant against rotations. Therefore, one may expect that re-expression of the probability densities relative to the new set of rotated axes cannot affect the Rayleigh–Bessel distribution of $|\vec{\Delta}|$. The derivation below shows, that the information about the eigensystem of distortions during their formation can be included by new constants δ and $\langle \sigma^2 \rangle$ and it is therefore part of the Rayleigh–Bessel distribution. In this way, the orientation of the eigensystem (formation of distortions) is part of the Rayleigh–Bessel distribution which in turn is affected by this rotation, whereas the eigensystem is rotational invariant such that then the determination is not influenced by a rotational transformation of the images (Figure 3).

The first step is to diagonalize covariance matrix $\mathbf{M}$ and re-express the new set of transformations into the multidimensional normal distribution.

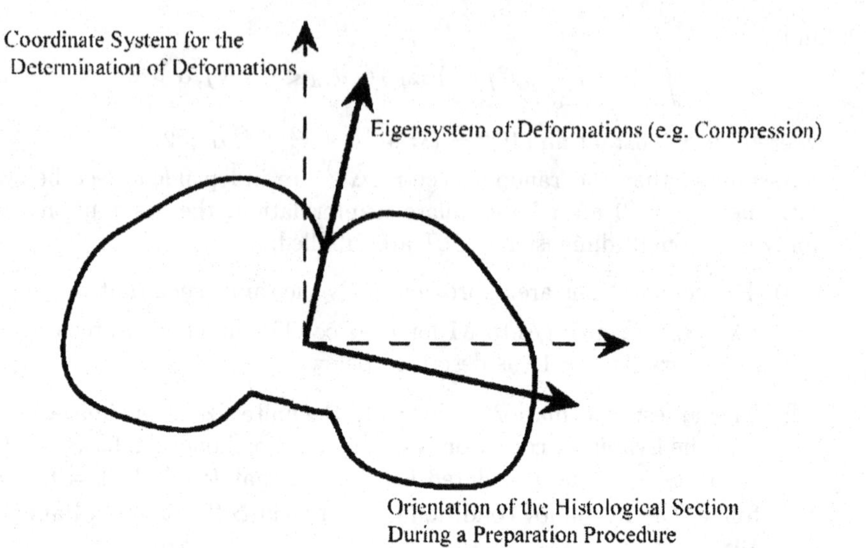

Figure 3. Eigensystem of local deformations caused by the histological preparation procedures and coordinate system for the determination of local deformations. The information about the eigensystem is part of the probability density function (see Eqs. (15)–(17), (22)–(23)), whereas the coordinate system for the determination of distortions is rotationally invariant. Note, that the latter is only true if the information about the eigensystem is included. Otherwise, the determination is valid only in the eigensystem of $\mathbf{M}$.

Assuming, that the inverse of $\mathbf{M}$ exists ($\det \mathbf{M} \neq 0$), the eigensystems of $\mathbf{M}$, $\mathbf{M}^{-1}$ are calculated by

$$\mathbf{M} = \mathbf{R}^t \mathbf{\Lambda} \mathbf{R}, \quad \mathbf{M}^{-1} = \mathbf{R}^t \mathbf{\Lambda}^{-1} \mathbf{R} \tag{8}$$

whereas $\mathbf{R}$ is a rotation and $\mathbf{\Lambda}$ an eigenvalue matrix given by

$$\mathbf{R} = \begin{pmatrix} \cos\vartheta & -\sin\vartheta \\ \sin\vartheta & \cos\vartheta \end{pmatrix}, \quad \mathbf{\Lambda} = \begin{pmatrix} \lambda_1^2 & 0 \\ 0 & \lambda_2^2 \end{pmatrix}, \quad \mathbf{\Lambda}^{-1} = \begin{pmatrix} \lambda_1^{-2} & 0 \\ 0 & \lambda_2^{-2} \end{pmatrix}. \tag{9}$$

Combining Eqs. (7)–(9), introducing polar coordinates with

$$\vec{\Delta} = r \cdot \begin{pmatrix} \cos\varphi \\ \sin\varphi \end{pmatrix}, \quad r = |\vec{\Delta}| \tag{10}$$

and using the trigonometric identities for $\sin(\varphi + \vartheta)$, $\cos(\varphi + \vartheta)$ yields

$$p(r, \varphi)\, dr\, d\varphi = \frac{1}{2\pi\sqrt{\det \mathbf{M}}} \exp(-\tfrac{r^2}{2} \vec{e}^t \mathbf{\Lambda}^{-1} \vec{e}) r\, dr\, d\varphi \tag{11}$$

with $\vec{e} = \begin{pmatrix} \cos(\varphi+\vartheta) \\ \sin(\varphi+\vartheta) \end{pmatrix}$, $\det \mathbf{M} = \det(\mathbf{R}^t \mathbf{\Lambda} \mathbf{R}) = \lambda_1^2 \lambda_2^2$.

Equation (11) can be written as

$$p(r,\varphi)\,dr\,d\varphi = \frac{1}{2\pi\sqrt{\det \mathbf{M}}} \exp\left(-\frac{r^2}{2}\frac{\lambda_2^2-\lambda_1^2}{\lambda_1^2\lambda_2^2}\cos^2(\varphi+\vartheta)\right) \cdot \exp\left(-\frac{1}{2}\frac{r^2}{\lambda_2^2}\right) r\,dr\,d\varphi. \quad (12)$$

Substituting

$$\cos^2(\varphi+\vartheta) = \tfrac{1}{2}(1+\cos[2\cdot(\varphi+\vartheta)]) \quad (13)$$

and with the definition of $u := 2(\varphi+\vartheta)$ yields

$$p(r,u)\,dr\,du = \frac{1}{4\pi\sqrt{\det \mathbf{M}}} \exp\left(-\frac{r^2}{4}\frac{\lambda_2^2+\lambda_1^2}{\lambda_1^2\lambda_2^2}\right) \cdot \exp\left(-\frac{r^2}{4}\frac{\lambda_2^2-\lambda_1^2}{\lambda_1^2\lambda_2^2}\cos u\right) r\,dr\,du. \quad (14)$$

Defining an anisotropic term, which takes into account the different variances along the principal axes

$$\delta = \frac{1}{4}\cdot\frac{\lambda_2^2-\lambda_1^2}{\lambda_2^2\lambda_1^2} \quad (15)$$

and the generalized variance

$$\langle\sigma^2\rangle = 2\frac{\lambda_2^2\lambda_1^2}{\lambda_2^2+\lambda_1^2} \quad (16)$$

yields

$$p(r,u)\,dr\,du = \frac{1}{4\pi\sqrt{\det \mathbf{M}}} \exp\left(-\frac{r^2}{2\langle\sigma^2\rangle}\right) \exp(-\delta r^2\cos u) r\,dr\,du. \quad (17)$$

Using the definition of the Bessel function (Rottmann, 1960)

$$J_0(x) = \frac{1}{\pi}\int_0^{\pi} \mathrm{e}^{ix\cos t}\,dt \quad (18)$$

and taking into account the limits of integration with respect to the polar coordinate φ and the definition of u, yields

$$\frac{1}{\pi}\int_{2\vartheta}^{4\pi+2\vartheta} \mathrm{e}^{-\delta r^2\cos u}\,du = \frac{4}{\pi}\int_0^{\pi} \mathrm{e}^{-\delta r^2\cos u}\,du = 4\cdot J_0(i\delta r^2) \quad (19)$$

This is an important result, since the integral runs over all angles and it is not expected to be affected by the rotation of the coordinate system. However, the information about the eigensystem (ϑ) is included in δ and

$\langle\sigma^2\rangle$, which in turn depend on ϑ. Therefore, the Rayleigh–Bessel distribution includes the information about the anisotropy in a compact form due to the fact that the integral can be solved analytically since the integrand of Eq. (19) is periodical. As a consequence, it is not required to transform the images accordingly.

Substituting Eq. (19) into (17) results in

$$p(r)\,dr = \frac{1}{\sqrt{\det \mathbf{M}}} R(r) J_0(i\delta r^2)\,dr \tag{20}$$

where the Rayleigh function $R(r)$ is given by

$$R(r) = r \exp\left(-\frac{1}{2}\frac{r^2}{\langle\sigma^2\rangle}\right). \tag{21}$$

Equation (20) indicates that local, geometrical deformations in histological sections are exactly Rayleigh–Bessel distributed with eigenvalues derived from covariance matrix (Eq. (2)) from which λ_1^2, λ_2^2 are determined according to

$$\lambda_1^2 = \sigma_{11}^2 \cos^2\vartheta + \sigma_{22}^2 \sin^2\vartheta - 2\sigma_{12}^2 \sin\vartheta\cos\vartheta \tag{22}$$
$$\lambda_2^2 = \sigma_{11}^2 \sin^2\vartheta + \sigma_{22}^2 \cos^2\vartheta + 2\sigma_{12}^2 \sin\vartheta\cos\vartheta \tag{23}$$

with

$$\vartheta = \frac{1}{2}\arctan\left(\frac{2\sigma_{12}^2}{\sigma_{11}^2 - \sigma_{22}^2}\right). \tag{24}$$

In this context, it has to be mentioned, that the Rayleigh-function corresponds to a chi-square distribution with $n = 2$ degrees of freedom. In the most general case for correlated and anisotropic preparation procedures, the decomposition of the exact probability density function into a product of a Rayleigh- and a Bessel-function has the further advantage that a complex estimate as proposed by Mathai and Provost (1992) is avoided.

Figure 4(a) shows the Rayleigh–Bessel density as a function of σ_{22}^2 and constant σ_{11}^2. With increasing σ_{22}^2 and anitropic term δ the theoretical Rayleigh–Bessel distribution becomes wider, indicating that more pronounced deformations are in the image (without loss of generality, σ_{22}^2 is chosen since the function is symmetrical with respect to the variances). Figure 4(b) exhibits, that the maximum of the Rayleigh–Bessel function is shifted to small deformations, which also demonstrates the effects of the eigensystem with respect to the form of the density.

In the special case of uncorrelated, anisotropic preparation procedures ($\sigma_{12} = 0$, $\sigma_{11} \neq \sigma_{22}$), it follows, that the eigenvalues λ_1^2 and λ_2^2 tend toward the variances and thus the deformations are Rayleigh–Bessel distributed according to σ_{11}^2 and σ_{22}^2, respectively. In the case of isotropic and uncorrelated procedures ($\sigma_{11} = \sigma_{22} = \sigma$ and $\sigma_{12} = 0$), it can be shown, that local

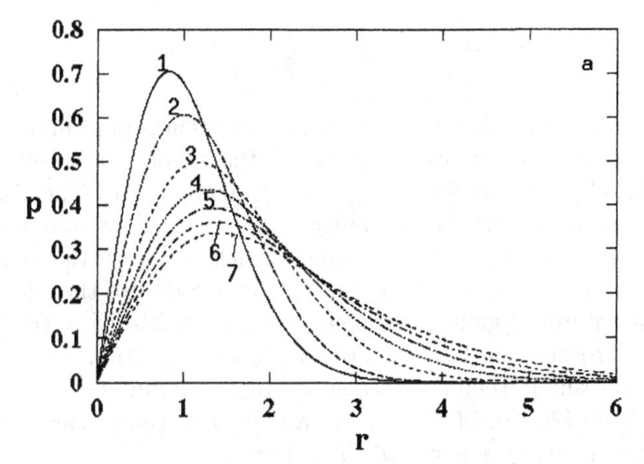

(a) Rayleigh–Bessel distribution for $\sigma_{12} = 0$, $\sigma_{11}^2 = 1 =$ const and different $\sigma_{22}^2 = 0.5$ (curve 1), $= 0.99$ (curve 2), $= 2$ (curve 3), $= 3$ (curve 4), $= 4$ (curve 5), $= 5$ (curve 6) and $= 6$ (curve 7), respectively. The anisotropic term δ increases also from -0.25 (curve 1) to 0.2 in curve 6

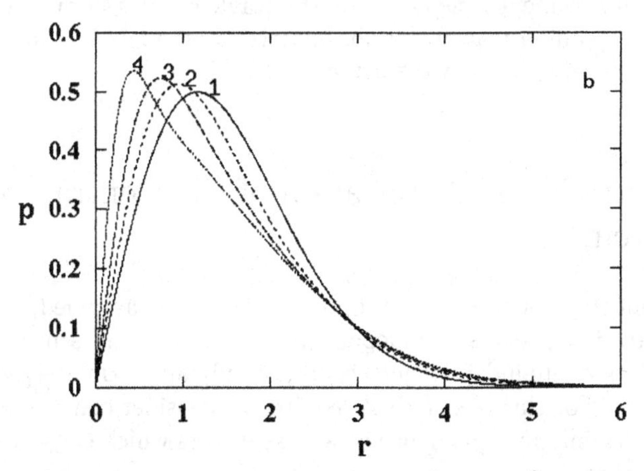

(b) Rayleigh–Bessel distribution for $\sigma_{11}^2 = 1$, $\sigma_{22}^2 = 2$ as a function of σ_{12} ($\sigma_{12} = 0, 1.3, 1.5$, and 1.65 corresponds to curves 1–4, respectively). The orientation of the eigensystem is given by $\vartheta \sim -36°$ for curves 2–4. The maximum of the probability density function is shifted to lower deformations with significantly different eigenvalues (e.g., $\lambda_1^2 = 2.9$, $\lambda_2^2 = 0.06$, $\vartheta = -37°$, curve 4) for an increasingly anisotropic term $|\delta| = 3.7$.

Figure 4.

deformations are Rayleigh distributed according to

$$p(r)\,dr = \frac{1}{\sigma^2}\exp\left(-\frac{1}{2}\frac{r^2}{\langle\sigma^2\rangle}\right) r\,dr. \tag{25}$$

Figure 5 shows the difference between experimental determined distortions (histogram) and the expectations of the theoretical Rayleigh and Rayleigh–Bessel functions for a large histological section. The Rayleigh–Bessel function demonstrates the ability to fit the experimental data more accurately over the entire histgram range which is due to the fact of less restrictive assumptions concerning **M**. The application of the theory with respect to an improved global alignment is accomplished by selection of an appropriate threshold on the basis of the density function. Let p^* be an appropriate threshold, then the corresponding maximum deformation d^* is determined from (20) for $p(d^*) = p^*$. In a next step, those reference points for which the distortion r is $r \geq d^*$ are rejected.

Figure 6 shows the registration result of a histological section generated without (a) and with (b) application of the theory. The qualitative improvement of alignment from (a) $\rightarrow$ (b) can be perceived by inspection of certain image features such as the Sylvian Fissure and the temporal lobes. The improvement is achieved by the Rayleigh–Bessel function, since the least-square affine transformation matrix is calculated from reference points which are not severely distorted.

3 Concerning the Application of the Central Limit Theorem

For the application of the central limit theorem, it is assumed, that each single preparation process is independent in a stochastic sense. The assumption of complete independence is not, or only approximately, achieved in real systems. For this reason, it is required to consider constraints which lead to a deterministic dependence as well as factors which cause a stochastic independence.

Spatial deformations due to histological processing are given by an overall brain shrinkage during 1. fixation, 2. dehydration and 3. paraffin embedding, 4. compression of the brain sections during cutting, 5. distortion resulting from the transport from the microtome to glass slides, 6. deformation of the sections during mounting on the slides and 7. stretching the section at least ten times on the slide. These procedures introduce a high stochastic degree, since the shrinkage is caused by unpredictable variations of concentrations in water during fixation and dehydration. Deformations from paraffin embedding are also influenced by stochastic variations in temperature besides deterministic changes in temperature.

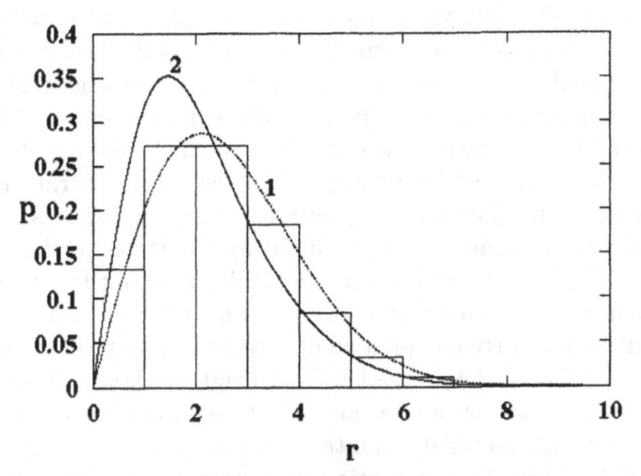

Figure 5. Comparison of the theoretical Rayleigh–Bessel (curve 1), Rayleigh (curve 2) function and experimental data (histogram). The theoretical functions are estimated from the experimental data **M** and σ^2, respectively.

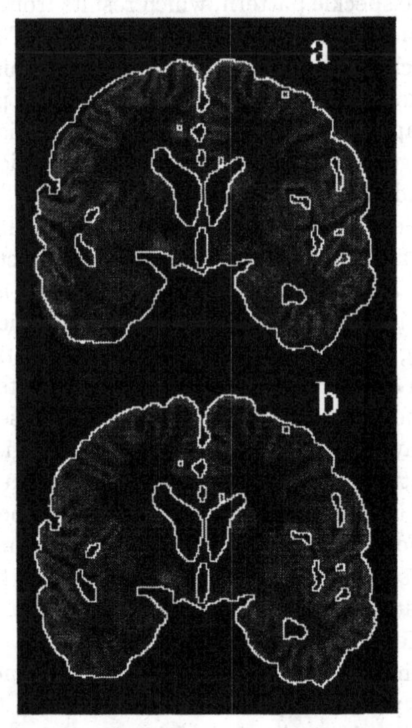

Figure 6. Comparison between linear alignment without (a) and with (b) Rayleigh–Bessel function. The histological section is transformed onto the corresponding MR reference section indicated by the outer contour superimposed onto both results.

In addition, the sections (thickness: $\sim 30\mu m$) are mounted onto slides, whereby the sections are smoothed down not only due to effects of temperature but mainly due to mechanical stretching. The degree of mechanical deformation depends on the different technical assistants and also on random factors during cutting such as the local thickness as well as the local varying requirements of stretching. In this way, an arbitrary point of the section is moved from its original position to various distances, which can be compared with a "random walk" caused by the steps cutting, embedding and especially by the repeated mechanical stretching (10 times). In this model, it has to be taken into consideration, that points in a close neighbourhood are similarly compressed or streched, such that the deformation decreases physically determined in this neighbourhood. However, due to the stochastic interference of a number of preparation processes, a significant stochastic characteristic is introduced which remains approximatively unchanged between the preparation procedures, since the sections are processed by different technical assistants in different rooms over a long time period. The formation of local deformations is comparable to the formation of a stochastic speckle pattern, which results from the interference of a coherent laser beam reflected on an uneven surface. The resulting intensity of each single speckle on a screen decreases deterministically, whereas the formation of the total speckle pattern is a stochastic process.

In the present approach, local, nonlinear deformations are estimated by a sequence of image processing techniques for definition and correlation of corresponding landmarks (Schormann et al., 1995). Taking into account the spatial coherence of nonlinear deformations, it follows, that the distance between adjacent reference points in a given section must exceed the extent of single local deformations in order to ensure an approximately independent determination (observation) of deformartions. This problem is overcome by using the distribution of local deformations which indicates that the minimum distance of adjacent reference points in a section is given by 7 voxels. Furthermore, reference points, even those with a far distance between them, may be correlated within the histological section by global compressing and stretching resulting from, e.g., the cutting procedure. This movement of the histological section can be described by the motion of an affine group. For this reason, the linear correlations between reference points in a given section are decoupled by an affine transformation, such that local deformations remain and the determination is not influenced by global spatial correlations.

For the determination of a probability density function of local deformations for an improved registration, it is assumed that (i) the random variables and (ii) the determination of deformations are approximately independent. Therefore, the central limit theorem is not applied under the assumptions of ideal, theoretical conditions but with the assumptions given above. In addition, the number of histologically induced distortions is 17, which can be considered as a good basis for the application of the central limit theorem.

4 Conclusion

It is shown, that global misalignment can be corrected by local deformations whereby generally the morphology of the object is not maintained. Since local spatial distortions in histological images hinder an exact global alignment, the probablity density function is determined analytically in order to estimate accurate global least-square affine transformation parameters. The theory shows, that the information about the eigensystem (ϑ) can be included in the anisotropic constant δ and extended variance $\langle\sigma^2\rangle$. As one important result, the Rayleigh–Bessel distribution includes the total information about $\mathbf{M}$ without any restrictive assumptions. This is due to the fact that the integral (19) can surprisingly be solved analytically since the integrand is periodical. In experiments it is demonstrated, that the application of this theory leads to an improved registration quality required for the subsequent application of nonlinear deformation techniques.

Acknowledgments: The authors wish to thank Prof. K. Janssen for his helpful advice. This study was supported by a grant from the Deutsche Forschungsgemeinschaft (SFB 194/A6).

5 References

Bajcsy, R. and S. Kovacic. (1989). Multiresolution elastic matching. *Computer Vision, Graphics, and Image Processing 46*, 1–21.

Bauer, H. (1972). *Probability Theory and Elements of Measure Theory.* International Series in Decision Processes. New York: Holt, Rinehart, and Winston.

Bronstein, I.N. and K.A. Semendjajew (1981). *Taschenbuch der Mathematik.* Leipzig: Teubner.

Fabian, V and J. Hannan (1984). *Introduction to Probability and Mathematical Statistics.* Wiley Series in Probability and Mathematical Statistics: Probability and Mathematical Statistics. New York: Wiley.

Mathai, A.M. and S.B. Provost (1992). *Quadratic Forms in Random Variables*, Volume 126 of *Statistics: Textbooks and Monographs.* New York: Marcel Dekker.

Papoulis, A. (1965). *Probability, Random Variables, and Stochastic Procersses.* New York: McGraw Hill.

Rottmann, K. (1960). *Mathematische Formelsammlung.* Mannheim: B.I. Wissenschaftsverlag.

Satterthwaite, F.E. (1946). An approximate distribution of estimates of variance components. *Biometrics 2*, 110–114.

Schormann, T., A. Dabringhaus, and K. Zilles (1995). Statistics of deformations in histology and application to improved alignment with MRI. *IEEE Transactions on Medical Imaging 14*, 25–35.

Schormann, T., S. Henn, and K. Zilles (1996). *A New Approach to Fast Elastic Alignment with Application to Human Brains*, Volume 1131 of *Lecture Notes in Computer Science.* Springer.

Worsley, K., J-B. Poline, A.C. Vandal, and K.J. Friston (1995). Tests for distributed non-focal brain activations. *NeuroImage 2*, 183–194.

13

Statistical Analysis of Brain Maps: Some Lessons from Morphometrics

Fred L. Bookstein

ABSTRACT Current methodologies for functional brain images, which inhabit data sets of ridiculously high dimension, can likely be enhanced by techniques borrowed from morphometrics, a companion methodology suiting the lower-dimensional modality of structural images. Over the last decade, modern morphometrics has converged on an important algebraic/statistical core that incorporates an explicitly theorem-driven representation of "shapes" (our principal subject of study) distinct from merely normalized images, realistic models for noise distributions in these derived spaces, permutation tests for practical issues of scientific inference, and least-squares methods for extracting predictions and patterns wherever scientific theory is weak. Variants of these tactics should apply as well in the more complicated arena of functional image analysis. This essay touches on three areas of common concern in regard to which morphometric tactics may ease comparable perplexities in the sister domain: issues of symmetry, the meanings of "localization" for image phenomena, and correlations of shape with other aspects of shape or with behavior. For some aspects of these topics, such as the description of focal phenomena, analyses of functional images already exploit stereotyped methods for which morphometrics provides interesting alternatives. For others, such as the modeling of noise after image normalization, morphometric methodology seems well ahead of functional methodology in prototyping sharp new tools.

1 Introduction

This essay extends comments made during the course of the workshop on statistics of brain mapping organized by Keith Worsley toward the end of the CRM 1997–1998 season. Keith divided the workshop into four principal tracks: "EEG," "Function," "Fields," and "Structure." Over that considerable range of scientific concerns one main methodological framework was maintained: wherever geometry entered into models or analyses, usually it was by way of arrays of Cartesian coordinates. Upon these subscripting schemes are erected fields variously of scalar "heights" or of vector "displacements."

Thereafter, the mathematical structure of those underlying label spaces themselves—the structure of descriptors of 2D or 3D Cartesian space—is rarely considered. For instance, in an analysis of displacement fields, that a voxel is shifted from (x, y, z) to $(x+3, y-2, z+1)$ is treated as a property of the voxel (x, y, z)—namely, a "shift vector" $(3, -2, 1)$—rather than any association of voxel (x, y, z) with voxel $(x+3, y-2, z+1)$, the voxel "shifted onto." Even though image normalization is, formally, the replacement of an image by an orbit under the action of some pixel relabeling group, methodologies like these make no further reference to that group structure once a representative is selected from each orbit to stand for the "normalized" image.

This essay conveys a range of techniques that overturn this unnecessary restriction by embracing the explicit geometric normalization of Cartesian shifts as an integral part of any quantitative analysis to follow. If the standard methods systematically confuse images with equivalence classes of images, the new tools may be thought of as devices for the systematic correction of this error: rules for selecting representatives of orbits that permit certain subsequent statistical manipulations without fear of confounding findings with selection rules, and a toolkit of statistical maneuvers suitably free of confounding in that sense. Most of these techniques have already been published in the literature of *morphometrics*, the multivariate statistics of biological shape variation, but that literature is not as well-known as it should be outside of a small circle of specialists. I shall sketch some salient stratagems underlying morphometric manipulations and note analogous concerns of brain mapping to which they might be pertinent. In the text that follows, purely in the interest of semantic efficiency, when I say "you" I am referring to the brain mapping/functional imaging community; when I say "we," I mean morphometricians.

2 The Core of Modern Morphometrics: An Overview

Morphometrics is the quantitative summary and statistical comparison of shapes from biological and biomedical images—their averages, their variation, their covariations. By formalizing comparisons of form in a finite-dimensional way, morphometrics makes rigorous an important channel of quantitative measurement related to organismal growth, response, and symptomatology. In addition, it offers other biomedical investigators these same powerful quantifications as covariates that increase the precision of findings (such as localized features) or help to verify empirical hypotheses (such as grouped image differences) in smaller samples than would otherwise be required. The contemporary morphometric toolkit offers a matched trio of tools for these purposes, one a statistical method for shape (the linearized Procrustes approach), one a visualization technique (the thin-plate spline), and the last a multivariate strategy for simplifying the high-dimensional covariance structures that the first two of these generate in the course of scientific investigations.

2.1 Procrustes Fits and Coordinates for Landmarks

What we mean by the *shape* of a set of labeled points (landmarks) in an image is the information in such a figure after we ignore location and orientation and set aside scale as a separate scalar for later use (the quantity called Centroid Size). Kendall (1984) shows how sets of figures that are all "the same shape" in this sense make up a Riemannian manifold, *Kendall shape space*, under a natural distance measure of its own that derives from the original Euclidean distance formula for the plane or space in which the landmark points were actually taken. The distance between any two shapes should be taken as *Procrustes distance*, root-sum-square of the Euclidean distances between the landmarks when each configuration is centered at the origin and scaled to sum of variances 1 and then one is rotated over the other to the position of least such sum of squares. Technically, then, a shape is not a normalized image but an equivalence class of images, for the special case of "images" of numbered Cartesian points on a blank background. Most medical images include subimages of exactly this nature; that is how morphometrics becomes relevant to medical image analysis.

To any sample of landmark configurations corresponds their *Procrustes average*, the derived shape having the least summed squared Procrustes distances to the original shapes of the sample. And for any sample there is an essentially unique set of *Procrustes shape coordinates*, locations of the landmarks after each configuration is fitted to their average in the translating-scaling-rotating sequence just described. Over the last ten years a small community of investigators—Bookstein, Goodall, Kent, Rohlf—have shown how *all* the conventional strategies of multivariate biometric analysis go forward using this common set of shape features, with inference by permutation test on Procrustes distance in most applications. The principal reason for the effectiveness of this representation is the astonishing degree of symmetry of the underlying shape geometry, a highly counterintuitive symmetry that obtains regardless of how lopsided the typical form happens to be.

The special features of morphometrics that make it so interesting a methodological specialty all begin right here. While the following analysis is not rigorous, it may serve to convey the essence of these *Procrustes tangent coordinates* to you. I will limit the discussion to two-dimensional data, so that the notation can exploit complex numbers when they simplify matters, with the overbar $\bar{}$ standing for complex conjugation.

Consider a sample of landmark configurations of some scientific interest that all show a substantial similarity of overall shape. It will be useful to have a model of "no signal" for analysis of this sample as it varies around a template. That model can be expressed after a Procrustes normalization as a form $z^0 = (z_1^0, \ldots, z_k^0)$ that is uninformatively centered and scaled: $\sum_i z_i^0 = 0$, $\sum_i z_i^0 \bar{z}_i^0 = 1$. The nothing-doing model can be set out as variations z around this form z^0 by errors $\epsilon = (\epsilon_1, \ldots, \epsilon_k)$ that are independently, identically, and isotropically distributed. There is no harm in modeling all the ϵ's as Gaussian of the same small variance σ^2 in each Cartesian coordinate.

To comprehend the effect of these uninformative perturbations on the shapes of the specimens z, it is enough to construct representatives of the corresponding equivalence classes. We do this by Procrustes fit of the perturbed z onto the grand mean template z^0. In this setup, finding the particular combination of translation, rotation, and rescaling that restores z to a least-squares fit to z^0 is a univariate complex regression (Bookstein, 1991). The translation component centers ϵ by subtracting off their average $\sum_i \epsilon_i / k$. The new error vector $\epsilon' = (\epsilon_1', \ldots, \epsilon_k')$ now sums to the zero vector, and so, now, do the perturbed z's. Once they have been recentered in this way, the rest of the Procrustes fit is the configuration $\hat{z} = \beta z$ for which β is the simple regression coefficient of z^0 on z, the quantity $\sum_i z_i^0 \bar{z}_i / \sum_i z_i \bar{z}_i$. Here we have

$$\sum_i z_i^0 \bar{z}_i = \sum_i z_i^0 (z_i^0 + \bar{\epsilon}_i') = 1 + \sum_i z_i^0 \bar{\epsilon}_i',$$

$$\sum_i z_i \bar{z}_i \sim \sum_i (z_i^0 \bar{z}_i^0 + z_i^0 \bar{\epsilon}_i' + \bar{z}_i^0 \epsilon_i') = 1 + \sum_i z_i^0 \bar{\epsilon}_i' + \sum_i \bar{z}_i^0 \epsilon_i'.$$

Drop the primes, since $\sum_i \bar{z}_i^0 = 0$. Then

$$\beta \sim \frac{1 + \sum_i z_i^0 \bar{\epsilon}_i}{1 + \sum_i z_i^0 \bar{\epsilon}_i + \sum_i \bar{z}_i^0 \epsilon_i} \sim 1 - \sum_i \bar{z}_i^0 \epsilon_i.$$

Regarding this "correction term" $\sum_i \bar{z}_i^0 \epsilon_i$, the imaginary part is a turn, the real part a scale change (expansion or contraction). Each of these can be further decomposed into a sum of azimuthal (resp. radial) contributions at each landmark separately. The azimuthal contribution, in particular, is familiar as a net torque.

Nearly all the geometry we will need is right here in this approximation. In terms of the multivariate distributions that will occupy us at length below, one can think of this fitting operation—the construction of a representative of an equivalence class of shapes—as a single *vector constraint* of the original i.i.d. distribution of Gaussians by a total of four equations: $\sum_i \epsilon_i = 0$ (two equations), $\sum_i \bar{z}_i^0 \epsilon_i = 0$ (two more). The distribution of the Procrustes fit coordinates βz corresponds to a *central section*, codimension 4, of the original $2k$-dimensional spherical Gaussian pertaining to the ϵ's. *By definition*, the Procrustes fit step that started from the original perturbation vector ϵ produced the element of this subspace closest to the original perturbation in the same sum-of-squares metric. Hence the Procrustes fit is (approximately) an orthogonal projection onto that central subspace. (Note, too, that the four equations are geometrically orthogonal, since $\sum_i z_i^0 = 0$: the partial projections can be carried out in any order.) But orthogonal projections of spherical distributions onto central sections continue to be spherically symmetric within the corresponding reduced-rank subspace. Hence, to this degree of approximation, the joint distribution of the shapes of the perturbed landmark configurations—the distribution of Procrustes fits to their common template—is spherically symmetric in its $(2k-4)$-dimensional subspace whenever the original data arose from those independent, identically distributed, isotropic perturbations.

The argument here is patently an approximation for "small" σ^2. For a rigorous redaction of the same point, see Dryden and Mardia (1998). For 2D data, the exact distribution of the shapes of the z's, the *Mardia–Dryden distribution*, can be proved to be spherically symmetric, though not precisely Gaussian, regardless of the value of σ^2.

2.1.1 Implications for Functional Image Analysis

The problem of modeling the variation of normalized images is not a new one (cf. the work of Strother and others), but the response needs to be a great deal more profound and formal than is customary in your field. Normalized images look like they are elements of a space of images, but they live instead in a space of equivalence classes having a subtly different geometry. The geometry of that space needs to be investigated, and exact distributions of physically, psychophysically, or biometrically based noise need to be modeled *there*, not in the original domain of photons, positrons, or what have you. For instance, if normalization is by outline position, the distribution of image contents in the vicinity of the outline can no longer be presumed to resemble a Markov random field; how should it be modeled instead? We know how this goes for Procrustes analysis; it is the previous exegesis, the explicit construction of a plausible noise model for representatives of equivalence classes, that needs to be imitated.

2.2 Thin-Plate Splines for Visualizing Shape Differences

We need not just to measure but also to *see* patterns or distinctions of shape. For this purpose there is another standard tool of the new morphometrics, the *thin-plate spline* (Bookstein, 1991). Let U be the function $U(\vec{r}) = r^2 \log r$, and consider a reference shape (in practice, a sample Procrustes average) with landmarks $P_i = (x_i, y_i)$, $i = 1, \ldots, k$. Writing

$U_{ij} = U(P_i - P_j)$, build up matrices

$$K = \begin{pmatrix} 0 & U_{12} & \dots & U_{1k} \\ U_{21} & 0 & \dots & U_{2k} \\ \vdots & \vdots & \ddots & \vdots \\ U_{k1} & U_{k2} & \dots & 0 \end{pmatrix}, \quad Q = \begin{pmatrix} 1 & x_1 & y_1 \\ 1 & x_2 & y_2 \\ \vdots & \vdots & \vdots \\ 1 & x_k & y_k \end{pmatrix}, \quad L = \begin{pmatrix} K & Q \\ Q^t & O \end{pmatrix},$$

where O is a 3×3 matrix of zeros. The thin-plate spline $f(P)$ having heights (values) h_i at points $P_i = (x_i, y_i)$, $i = 1, \dots, k$, is the function $f(P) = \sum_{i=1}^{k} w_i U(P - P_i) + a_0 + a_x x + a_y y$ where $W = \left(w_1, \dots, w_k, a_0, a_x, a_y\right)^t = L^{-1} H$ with $H = \left(h_1, h_2, \dots, h_k, 0, 0, 0\right)^t$. Then we have $f(P_i) = h_i$, all i: f interpolates the heights h_i at the landmarks P_i. Moreover, the function f has minimum *bending energy* of all functions that interpolate the heights h_i in that way: the minimum of

$$\iint_{\mathbb{R}^2} \left(\left(\frac{\partial^2 f}{\partial x^2} \right)^2 + 2 \left(\frac{\partial^2 f}{\partial x \partial y} \right)^2 + \left(\frac{\partial^2 f}{\partial y^2} \right)^2 \right).$$

This integral is proportional to $W^t H = H_k^t L_k^{-1} H_k$, where L_k^{-1}, the *bending energy matrix*, is the $k \times k$ upper left submatrix of L^{-1}, and H_k is the corresponding k-vector of "heights" $\left(h_1, h_2, \dots, h_k\right)$. For morphometric applications, this procedure is applied separately to each Cartesian coordinate H. (In 3D, $U = |r|$, and Q is $k \times 4$.)

Because the formula for f is linear in the h's, morphometricians exploit this formalism to provide a dynamic representation of interesting linear models arising within the shape coordinate hyperplane. Group mean differences, regressions, and the like can all be visualized by deformations in this wise a great deal more easily than they can be understood via tables of numbers or colors of pixel contents. Several examples appear below.

2.2.1 Implications for Functional Image Analysis

The thin-plate spline allows a visualization of the full multivariate space underlying morphometrics: any linear combination of shifts of Procrustes fit coordinates, whether a mean difference, a correlation, or a canonical component. In functional image analysis, the corresponding empirical construct is the collection of multiple peaks at different locations in the image; but your field has no equivalent for the ensuing space of linear combinations. For instance, you do not yet seem to have a formalism wherein a "neural circuit" is a vector to be extracted from data and tested for significance or localization, nor is there yet, I believe, a vectorial descriptor for "a pair of peaks" that could be tested against the alternative model of a single peak of wider halfwidth. Investigations of the variation of peaks over conditions or subjects, the covariation of hot spots with cold spots, time series of fMR activations, and the like will not go forward with sufficient rigor until suitable vector spaces are available for this purpose.

2.2.2 Analysis of Curves

The similarity group partialled out in Procrustes analysis is not the only normalization worth applying. One can also normalize within the reparametrization group of one or more curves, for instance, by matching points across the sample that correspond as jointly minimizing the bending energy of the spline that represents each as a deformation of their joint Procrustes average. In the Procrustes approach, this has the effect of replacing the original equivalence classes (shapes) by larger equivalence classes (reparametrized shapes) and thus replacing the previous set of class representatives (Procrustes fit coordinates) by a new set of representatives subject to a more extensive list of linearized constraints. For a practical algorithm, with examples, see Bookstein (1997b, 1999). Following our own advice, we have modified the procedures of statistical inference accordingly (for instance, to attend to differences only normal to average curves, not along them). Under the protection of the omnibus permutation test that results, we have procrastinated in the parallel task of erecting the corresponding noise model.

2.2.3 Implications for Functional Image Analysis

Your methods should be robust under a variety of different or concatenated normalizations, as long as each one is plausible separately. But each combination of normalizations yields geometric objects in a different space of equivalence classes; each of these spaces needs its own multivariate methods and its own noise models.

3 Partial Least Squares for Correlations of Shape

One technical transfer from morphometrics into functional image analysis has already been achieved: the technique of Partial Least Squares (PLS) for representing low-dimensional linear relationships between two or more high-dimensional measurement blocks. PLS is based on the singular-value decomposition (SVD) for cross-block covariance matrices, which becomes especially elegant when one of the blocks has a natural metric of its own. Suppose there are k landmarks or semilandmarks, and thus pk Procrustes shape coordinates, $p = 2$ or 3, in some data set of images, and also m organismal measures, such as behaviors, titres, or clinical scores. Write X_i for the ith shape coordinate variable, $i = 1, \ldots, pk$, and Y_j, $j = 1, \ldots, m$ for the jth z-scored exogenous score. PLS produces pairs (A_1, B_1), (A_2, B_2), ... of *singular vectors*, the A's, having pk elements, and the B's, having m elements. With each pair will be associated a scalar *singular value* d_i. The A's will be orthonormal (perpendicular and of unit length), and likewise the B's. When A pertains to the shape of an image, it is called a *singular warp*.

To the mathematician, these are just the components of the ordinary singular-value decomposition of the cross-covariance matrix $\sum_{l=1,N} X_{il}Y_{jl}$, $pk \times m$. Other interpretations (see Bookstein et al., 1996) are valid at the same time. The matrix $d_1 A_1 B_1^t$ is the $pk \times m$ matrix of rank one that comes closest to this full covariance matrix in the least-squares sense; the *latent variables* $LV_X = \sum A_{1i}X_i$ and $LV_Y = \sum B_{1j}Y_j$ have covariance d_1, and this is the greatest covariance of any such pair of linear combinations with coefficients of unit length; the vector A_1 is the "central tendency" of the pk columns of the covariance matrix when thought of as profiles of covariance with the outcome scores Y_j; the vector B_1 is the same for the rows of the matrix, viewed as profiles of covariance with the Procrustes shape coordinates X_i. Because the reference metric for the Procrustes block is proportional to error variance on the Mardia–Dryden distribution, the singular warps can be characterized as having greatest covariance with normalized profiles of scores from the Y-block for fixed putative variance on the null model (the ϵ's above)—this is equivalent to normalizing the "length" of these shape variables.

Successive pairs (A_2, B_2), (A_3, B_3), etc. satisfy the same properties contingent on the constraint that each A_i be perpendicular to all previous A's and each B_i to all previous B's. Note that the d's are not canonical correlations, nor are the A's and B's canonical variates in that sense. There is no matrix inversion involved in PLS, so that, in particular, computations go forward without alteration when either block has more variables than there are cases for analysis. The test for overall significance of the relation between two blocks is by reference to the permutation distribution (Good, 1994) of the first singular value d_1 when the matching of the cases of one block to the measures of the other is randomized several hundred times. The covariance structure of the blocks separately is left unchanged by this manipulation; only the tie between them is tested. When findings are significant, standard errors of the resulting descriptors (elements of the vectors A and B) may be generated by a different resampling approach, the jackknife procedure (Sampson et al., 1989). When the X-block consists of Procrustes coordinates, the vectors A can be produced as explicit landmark relocations that are then diagrammed as thin-plate spline deformations in the usual way.

3.1 Implications for Functional Image Analysis

The implications have already been drawn in your own literature (McIntosh et al., 1996, 1997, 1998, Grady et al., 1998). I can only point with pride and respectfully suggest that you imitate these exemplars. Another sort of application of this technique within morphometrics is for the correlation of shape with other shapes. Further examples can be found in Bookstein (1997a), Bookstein et al. (2000), Mardia et al. (2000).

4 Symmetry

The symmetry of the underlying Procrustes coordinate representation, as reviewed with the aid of the ϵ's in a previous section, makes for some elegant data-analytic maneuvers in a variety of scientific applications. Some deal with symmetry of method, others with symmetry as hypothesis.

4.1 Group Differences as Symmetry-Breaking

On the nothing-doing model, variation in Procrustes tangent space is expected to be spherical. Then we can test any empirical displacement in this space as long as we have an estimate of the corresponding spherical radius, postponing until later any concern for the direction of that displacement (the nature of the shape change). Consider, for instance, the following very simple but powerful test for group mean difference, due to Goodall (1991), for shapes X_{li}, $l = 1,\ldots, N_i$ from two groups $i = 1, 2$. Under the nothing-doing model, any sample average μ differs from the true mean by a set of $2k-4$ independent Gaussian vector components of the same variance σ^2/N, where N is sample size. That is, each component, squared, is $\sigma^2\chi_1^2/N$, and so, by the addition rule for independent χ^2's, the squared Procrustes distance of the sample mean shape from the template is $\sigma^2\chi_{2k-4}^2/N$ and the squared Procrustes distance $||\mu_1 - \mu_2||^2$ between two of these subsample means is distributed as $\sigma^2\chi_{2k-4}^2(N_1^{-1} + N_2^{-1})$. The total Procrustes sum of squares of either of these subsamples around its own average is, similarly, $\sigma^2\chi_{(2k-4)(N_i-1)}^2$, $i = 1, 2$, independent of any mean difference. Then, by the usual machinery of F-tests, the ratio

$$F = \frac{N_1 + N_2 - 2}{N_1^{-1} + N_2^{-1}} \frac{||\mu_1 - \mu_2||^2}{\sum_{l=1,2}\sum_{i=1,N_l} ||X_{li} - \mu_l||^2}$$
$$\sim F_{(2k-4),(2k-4)(N_1+N_2-2)}$$

—the ratio of between-group to within-group sum of squares, like any other ratio of independent χ^2's, has an F distribution on the specified degrees of freedom.

In this way one can test a pair of mean shapes for the presence of any shape difference without specifying the nature of that shape difference a priori. We manage this by recourse to that a priori spherically symmetric noise model, which permits assessment of the complete space of possible shape differences in a symmetric manner.

In practice, the assumptions of this test are often violated modestly—the perturbations at the several landmarks are often correlated, as by adjacency or symmetry—and so one would wish to estimate a number of "equivalent degrees of freedom" different from $2k - 4$ using the data rather than just counting coordinates. One can do better, however. The extremely robust *permutation version* of this statistic is the permutation distribution of the squared Procrustes distance $||\mu_1^P - \mu_2^P||^2$ over the "pseudogroups" generated when group labels are permuted over cases. An exact significance level for this omnibus hypothesis of group difference is the tail-probability of the observed $||\mu_1 - \mu_2||^2$ within this distribution, or its approximation in the usual Monte Carlo setting (not all the permutations, but only a few thousand), regardless of the null model. This agnostic form of inference is quite effective unless noise is highly anisotropic and the effect on shape spatially very concentrated, under which conditions the Procrustes distance statistic loses efficiency.

4.2 Implications for Functional Image Analysis

Permutation testing is already a familiar part of the functional imager's toolkit (cf. Holmes et al., 1996, Bullmore et al., 1996, and this volume). I need not pause to argue the desirability of distribution-free tests for any compelling summary statistic. The more important point is the explicit construction of omnibus tests *prior to any search for detailed features.* The particular question a morphometrician asks first—is there any difference anywhere between these averages?—seems equally reasonable as a candidate for the first question a functional image analyst ought to be asking: not, as at present, a question about peaks, no matter how obstreperously the neurophysiologist insists on its primacy. In the usual z-field setting for image analysis, there ought to be an equivalent of this spherically symmetric a priori setup, and thus a general procedure for asking whether two samples of normalized images differ in their contents. A convenient pivotal statistic might be the sum of squared t's, which is the first and only singular value of the corresponding PLS setup. After all, data sets need not differ merely in respect of "hot spots." They might as well differ in background gradient, center-to-periphery gradient, left-right disparity (see below), or any of a great variety of other scientifically meaningful signals. Do not think of these as mere "nuisance variables" to be normalized out. Rather, if their variation is not carefully understood beforehand, any findings about subordinate features (e.g., locations of peaks, their magnitudes, their association with causes or effects) become unreliable, entangled in those rules of normalization in an inscrutable way. The specialization of the Friston-Worsley family of tests to deal only with "excursions" (hot spots) both places blinders on the range of image-borne hypotheses that can be tested by that familiar formulary and casts universal doubt on the objective nature and reliability of findings that are produced with its aid. As Robert Adler said at the workshop, "Rejection of the null model [of the Markov random field] means that one or more of its assumptions is false; it does not tell us which ones."

4.3 Principal Components as Symmetry-Breaking

One normally thinks of principal components analysis as "eigenanalysis of a covariance matrix." Actually, it is *relative* eigenanalysis of the covariance matrix with respect to a *gauge metric* specifying what is meant by geometric orthonormality of candidate eigenvectors for the problem at hand. Different gauge metrics lead to different eigenanalyses. This is well-known in connection with "standardizing the diagonal," switching from covariances to correlations, but it applies to any other linear change of basis as well. Interpretation of any set of principal components as "dimensions of extremal variation of the sample" is an ellipsis: one has to state *with respect to what unit of length* the variation is extremal.

Principal component analyses of Procrustes shape coordinates are particularly congenial in this regard, since the gauge metric is the same Procrustes distance we have already become used to. The first principal component of a set of shapes is the dimension of largest shape variation with respect to this disk of spherically symmetric noise in the tangent space, which is, in turn, the representation within these equivalence classes of a highly plausible original model of independent identically distributed perturbations around the mean form for each original landmark coordinate separately. Because this distribution is spherical, principal components can be tested for significance in the usual textbook way regardless of the average shape. The emergence of significant components testifies to rejection of the null model (that business with ϵ's above). Reasons for this rejection might include serious anisotropy of one or a few individual perturbations (distributions that are highly elliptical or have great differences of radius) or correlated displacements of landmarks by virtue of adjacency or symmetry or as a result of joint dependence on causal factors such as growth or organismal condition.

4.4 Implications for Functional Image Analysis

In most applications of principal components (empirical orthogonal functions) in medical imaging, there is no equivalent of this theorem-based gauge metric. Without it, just as eigenmodes of vibration of a sample average shape depend on that shape, principal components of normalized brain images cannot be expected to be distributed with spherical symmetry in the space of image fields no matter what the normalization protocol. In any application of eigenanalysis to functional images, eigenvectors must be taken with respect to some plausible noise model, not to the sum of squares of normalized pixel values (the final "identity" matrix). It is particularly important to produce and understand the derived variables (geometric orthonormal components) with respect to which the noise model becomes spherical. These are known a priori for Procrustes space (cf. Bookstein, 1999, the partial warps). We need their equivalents for other normalizations that are commonly encountered.

4.5 Bilateral Symmetry

A particularly simple consequence of the a priori symmetries in the Procrustes framework is the ease with which one can separate symmetric and asymmetric parts of a given signal. Suppose there is a set of paired landmarks, left and right, in a single rigid form, and also, perhaps, some midline (unpaired) landmarks. Express the pairing by a *subscript interchange operator* S, $S^2 = I$, that exchanges "left" and "right" versions of the same landmark, but leaves midline landmarks unchanged. For each specimen z_i there is a *mirroring* $S\bar{z}_i$ involving the combination of reflection of the image and the interchanges followed by a new Procrustes fit. (The grand mean is now itself symmetric, computed from the doubled data set of all the mirrored forms together with the originals.) As notated here, reflection is in the "real axis," but since the effect of changing the axis of reflection is merely a rotation, and Procrustes analysis is invariant under rotation, it makes no difference what the axis of reflection is.

In this setup, the Procrustes tangent space can be rigidly rotated to a new set of coordinates $(z_i + S\bar{z}_i)/\sqrt{2}, (z_i - S\bar{z}_i)/\sqrt{2}$—*sum-and-difference coordinates*—that separate the symmetric and asymmetric aspects of the configuration's shape variation and covariation. The subspaces can be assessed independently for group differences or other effects, and the significance of phenomena in the asymmetric subspace per se can be assessed by the matched version of the permutation test, in which sums of Procrustes squares are observed after left and right sides are randomly permuted. (The permutation distribution arises from the pseudodatasets $(-)^{q_i}(z_i - S\bar{z}_i)$ where q is a series of independent Bernouilli coinflips.) See Mardia et al. (2000).

4.6 Implications for Functional Image Analysis

Well-designed functional experiments may hypothesize hot spots that either appear on only one side of the brain at a time, or on both. To the extent that *any* factors of interindividual variation affect the two hemispheres of the cortex symmetrically, studies that merge images over individuals gain substantially in statistical power when analysis is carried out after a rotation to sum-and-difference coordinates of this sort. This maneuver is particularly useful when images were normalized in Talairach or another similarly defective registration system not permitting adjustments for differences in anatomical symmetry as part of the normalization protocol.

5 Varieties of "Localization" for Image Phenomena

Scientific papers in functional image analysis are most often concerned with "hot spots," peaks of metabolic signal above a local background. The dominant models for these findings derive from rejections of a Markov random field null. Morphometrics has a different sort of localization to offer, one with important implications for the design and analysis of functional experiments.

Here is a recent example dealing with a group mean difference, schizophrenics vs. normals, in the shape of a set of 13 landmarks in the midplane of the brain (DeQuardo et al., 1996, Bookstein, 1997c). When each shape is fitted to the grand mean form, there result the coordinates at far left in Figure 1. The group means are as shown at center left (circles, normal mean; triangles, schizophrenic mean). The difference is significant at the 4% level by the omnibus permutation test introduced above.

From the representation by thin-plate spline, center right panel, we extract by eye a region of highest bending and a triangle of landmarks straddling the nonlinearity along it. At far right, we show the coordinates of the middle of these three landmarks in a coordinate system for which left and right endpoints have been fixed in position at $(0,1)$ and $(1,1)$, respectively—the *two-point shape coordinates* of Bookstein (1991). Schizophrenics, dots; normals, circles. The significance level of the group difference in the abscissa of this scatter is .0017, which, multiplied by the 22 degrees of freedom of the vector space, suggests the same 4% probability we had already arrived at. The feature being reported here is the ratio in which the point-landmark colliculus divides the segment from splenium to chiasm along which it approximately resides.

What is obscure here is the detection of that bulge, which was found by eye. If we were to run the transformation backwards, it would be a center of compression (Figure 2, upper left). This notion of "running a transformation backward" does not in fact refer to the inverse spline, but instead takes advantage of the happy accident that the spline is linear in the coordinates of the target form. The α-fold extrapolation of a spline $S_1 \to S_2$ is then the map $S_1 \to S_1 + \alpha(S_2 - S_1)$; for instance, the "inverse" is the map $S_1 \to 2S_1 - S_2$ for $\alpha = -1$.

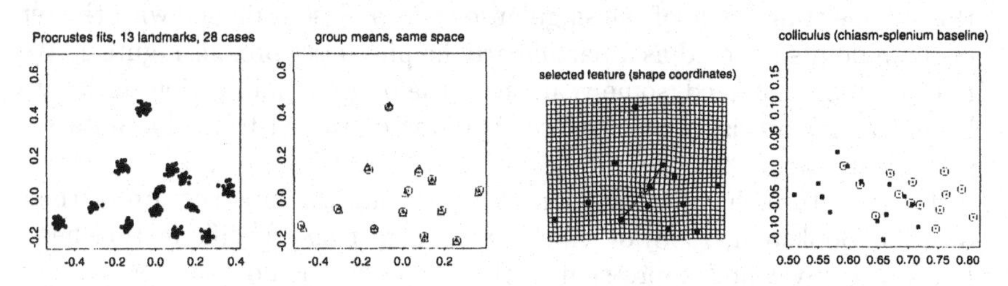

Figure 1.

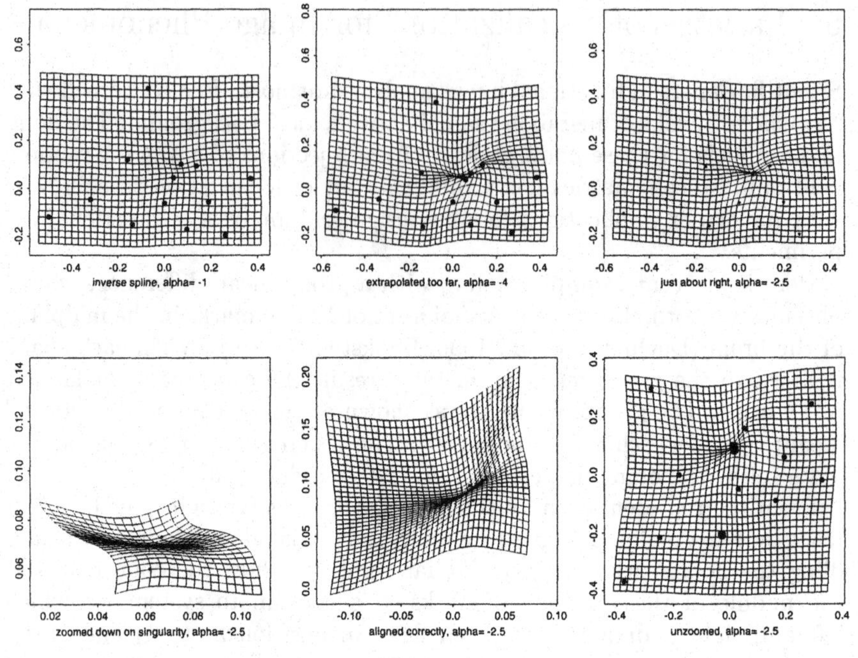

Figure 2.

When we extrapolate the transformation further backward (steadily more and more negative values of α), the compression becomes more and more severe. Eventually there emerges a region with negative Jacobian (upper center panel). The transition between these two regimes, upper right panel, generates the feature we seek, focused at the point at which the extrapolated map first exhibits an isolated zero of its Jacobian. Upon magnification (lower left), one sees that in this vicinity the (x, y) coordinate system has actually collapsed: the original x- and y-axes of the Cartesian coordinate system are mapped into curves with the same tangent. A bit of graphical exploration (lower center panel) unearths a rotation for which the image of the new y-axis traverses the singular axis at a proper corner (a piecewise smooth curve with unequal tangent directions at 0^+ and 0^-). Zooming back out to the scale of the original scene, now rotated, we see (lower right) that this unique transversal of the singular locus is precisely aligned with the set of three points selected as a particularly helpful descriptor in Figure 1. We have thereby managed, some years after the original publication, to locate by algorithm precisely the feature that was extracted from this data set some years earlier "by eye."

These features have a considerable mathematical substructure of their own (cf. Bookstein, 1998) of which I will extract one specific feature here. Lines of equivalent compression in this direction should form loci similar to concentric ellipses. Thresholding this quotient at some convenient value

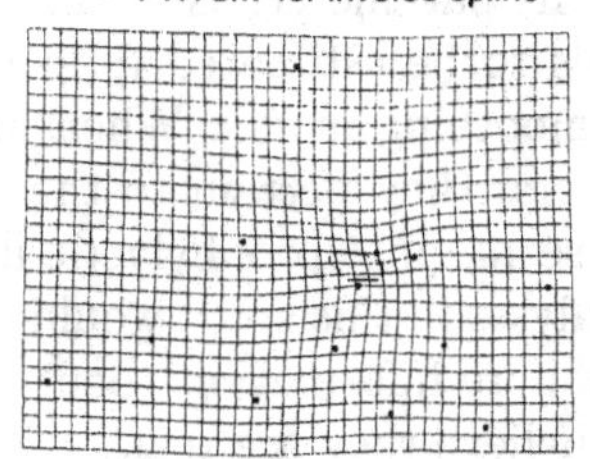

Figure 3.

(above right, for 75% of the oversmoothed "background" gradient for the data in Figure 2) supplies an oval in the preimage space that can usefully be treated by tensor-like statistics, as in Bookstein (1991). Elliptical models of this threshold (a sort of full-width-75%-maximum) track an effective "scope" of singularities in the same way that the FWHM tracks a Gaussian model of a single-peaked signal.

Significance testing of features like these can proceed in a variety of ways. In one, we set the "height" of the feature—the degree of compression of that principal traversal—and permute cases to see how often a crease emerges at or below the assigned level of extrapolation. In another, we filter the transformation first (the grid, not the original images) to take out large-scale aspects of form-variation, and carry out the permutation testing again; now the significance level is quite a bit higher, as these phenomena are usually quite localized. In practice, significant creases also seem to be stable under low-pass filtering (Bookstein, 2000), a version of smoothing (again pertaining to the grid, not the original images) in which the given shape difference $\mu_1 - \mu$ is expanded in eigenvectors of the spline's bending energy L_k^{-1} and then relaxed along a curve of forms $(\mu_1+\lambda BE\mu)/(I+\lambda BE)$, $BE = \mathrm{diag}(L_k^{-1}, L_k^{-1})$, as λ traverses the nonnegative real line.

5.1 Implications for Functional Image Analysis

Creases are the morphometric equivalent of "hot spots," but the preceding methodology does not much resemble the corresponding approaches from today's functional mapping literature, mainly because the crease is treated as a positive instance of one model rather than merely the rejection of a null. A crease will be detected only if the underlying model—here, a polynomial $(x, y) \to (x, x^2y + y^3)$—appears to fit the transformation at the feature and for some radius around it. The crease, in other words, is the organization of a report of an extended finding. If declaring a local minimum of derivative would be misleading it is not reported in this way. The functional mapping literature needs a similar protocol for deciding not only that a hot spot is improbable on a null model but that it is plausible enough on a hot-spot model. In this regard, description of a crease involves specifying a (tensor-valued) scale parameter, as in the inset above. Furthermore, scale in this sense needs to be estimated from the data, not imposed as part of some a priori smoothing model. When images are smoothed, real creases smooth with them: they pertain to the data, not the processing. Such a protocol would seem quite prophylactic if expeditiously applied in functional imaging. Analyses in this spirit would detect rather fewer peaks, but supply a good deal more information about each one: perhaps enough information to help us believe they are real.

6 Concluding Remarks

The tactics I have been sketching here, which work so well in contemporary morphometrics, seem to fall under two different rubrics. One group—the statistical geometry of Procrustes fits, the power of the thin-plate spline to visualize and localize interesting findings, the ease with which bilateral symmetry can be represented in the usual linear descriptor space—collect techniques that are at least modestly specific to the theory of data underlying this toolkit: labeled loci of Cartesian spaces governed by a model of small changes. These tools strenuously exploit the particular normalization groups that get us from image to "shape" (similarity, reparametrization) and also Kendall's 1984 powerful theorems about the classically Riemannian structure and the symmetries of these descriptor spaces per se. The methodologist of functional image analysis will not be able to apply these tools in their present form. Rather, appropriate analogues should be sought—plausible models for noise under the spatial normalization of choice, a plausible rotation of the signal space into symmetric and asymmetric subspaces, a plausible analogue for the scale of a finding that is scientifically more informative than the current "extent of an excursion set."

The other tools demonstrated above are more protean. They have nothing in particular to do with morphometrics or image analysis, but represent the sea-change in statistical computation that has taken place over the last decade or so and transformed practice in almost every application area. Permutation methods have already been introduced to your community by Holmes, Bullmore, and others. The question a permutation test asks, usually some variant of the relation of labels to cases, turns out to be, in practice, the only question about statistical inference worth asking. (As R.A. Fisher noted long ago, any parametric test is at best a surrogate for a permutation test anyway.) If one has two groups of 8 cases each, to show that the observed signal, tested without a posteriori bias (not as a peak!), has magnitude greater than any of the 12869 other subsets of 8 against 8 is to assert the evidence for the association of label and phenomenon with a crushing finality, inasmuch as *no ancillary assumptions whatsoever* have been required, and thus nothing whatever needed to be checked.

As for the other member of this pair of tools, the singular-value decomposition: our statistical models are always wrong, in particular, the Mardia–Dryden distribution of morphometrics and the Markov random field model of functional image analysis. Yet, no matter how fallible our understand of mechanism, as quantitative scientists we are in the job of making predictions. For this purpose, the singular-value decomposition, which produces optimal predictions (in the sense of explaining the most covariance) under a very broad range of circumstances, is an extraordinarily robust device for making sense when every other tool fails. Applied to covariance matrices, as we have done here, it produces quantities of three different species—coefficients, scores, and covariances—from the same standard matrix manipulation, with rules of inference and standardized graphics to match. Those of us who are comfortable with it believe it to be an almost universal panacea for crossmodal analysis, the very first analysis to be tried once a data set that measures more than one channel of reality has been laid down row by column on the screen.

The modern morphometric toolkit begins not with data but with strictly geometrical theorems. One chain of results, at the geometrical core, begins in Euclidean sums of squares, dives into the appropriate "normalization" for shape analysis (the similarity group) to arrive at the Procrustes metric, and emerges in the Kendall Riemannian manifold of shape. Another chain of theorems begins up in the tangent space of this manifold and proves that this same metric also calibrates a shrewd guess at likelihood under a plausible null model. The standard machinery of multivariate analysis is not powerful enough to cope with the dimensionality of that tangent space, but the newer, computationally intensive methods of permutation testing and singular-value decomposition can manage the linear operations. Finally, the diagram style of the thin-plate spline steps in to make visual sense of findings back in the original 2D or 3D space of the organisms supplying the original data.

There can be analogues for all of these steps in your sister literature of functional image analysis. As I read you today, however, these tactics have not yet been chained together in appropriate ways. Both of our communities need a deeper understanding of how normalizations affect image variation; we need to pass explicitly from plausible null models in the original space to the distributions over equivalence classes of images that have been normalized both spatially and (for your applications) in terms of intensity as well. Just as importantly, we need models for peaks, circuits, and the like that live in appropriate vector spaces, not graph spaces, so that they can be tested persuasively using covariance-based modeling strategies like PLS. Only such formalisms, in my view, will lend authority to the theories arrived at with their aid. Once foundations like these are set in place, and the current technological fixes of functional imaging (SPM, etc.) scrapped or modified appropriately, then the findings of this literature will reach beyond their current pictorial value and become a data resource to which the rest of us should attend in the ordinary course of applied scientific browsing. I wish your community Godspeed in producing the theorems, and the matrix manipulations, that will make your workshops, in turn, worth imitating.

Acknowledgments: This work was supported in part by USPHS grants DE-09009 and GM-37251 to Fred L. Bookstein. I appreciate the invitation from Keith Worsley to speak of these matters to a roomful of applied mathematicians and the opportunity to spread the word even further in this venue.

7 References

Bookstein, F.L. (1991). *Morphometric Tools for Landmark Data.* New York: Cambridge University Press.

Bookstein, F.L. (1997a). Analyzing shape data from multiple sources: the role of the bending energy matrix. In K. M. C. Gill and I. Dryden (Eds.), *Proceedings in the Art and Science of Bayesian Image Analysis*, pp. 106–115. Leeds University Press.

Bookstein, F.L. (1997b). Landmark methods for forms without landmarks: Localizing group differences in outline shape. *Medical Image Analysis 1*, 225–243.

Bookstein, F.L. (1997c). Shape and the information in medical images: a decade of the morphometric synthesis. *Computer Vision and Image Understanding 66*, 97–118.

Bookstein, F.L. (1998). Singularities and the features of deformation grids. In B. Vemuri (Ed.), *1998 Workshop on Biomedical Image Analysis (WBIA '98)*, Santa Barbara, CA, 1998, pp. 46–55. IEEE.

Bookstein, F.L. (1999). Linear methods for nonlinear maps: Procrustes fits, thin-plate splines, and the biometric analysis of shape variability. In A. Toga (Ed.), *Brain Warping*, pp. 157–181. Academic Press.

Bookstein, F.L. (2000). Creases as local features of deformation grids. *Medical Image Analysis 4*, 93–110.

Bookstein, F.L., A.P. Streissguth, P. Sampson, and H. Barr (1996). Exploiting redundant measurement of dose and behavioral outcome: New methods from the teratology of alcohol. *Developmental Psychology 32*, 404–415.

Bookstein, F.L., A.P. Streissguth, P. Sampson, P. Connor, and H. Barr (2000). Multivariate analysis for grouped structure—behavior studies, illustrated by the corpus callosum in fetal alcohol syndrome. *Alcoholism: Clinical and Experimental Research.* (submitted for publication).

Bullmore, E.T., M. Brammer, S. Williams, S. Rabe-Hesketh, C. Janot, A. David, J. Mellers, R. Howard, and P. Sham (1996). Statistical methods of estimation and inference for functional MR images. *Magnetic Resonance in Medicine 35*, 261–277.

DeQuardo, J.R., F.L. Bookstein, W.D.K. Green, J. Brumberg, , and R. Tandon (1996). Spatial relationships of neuroanatomic landmarks in schizophrenia. *Psychiatry Research: Neuroimaging 67*, 81–95.

Dryden, I.L. and K.V. Mardia (1998). *Statistical Shape Analysis.* Wiley Series in Probability and Statistics: Probability and Statistics. New York: Wiley.

Good, P. (1994). *Permutation Tests.* Springer Series in Statistics. New York: Springer.

Goodall, C.R. (1991). Procrustes methods in the statistical analysis of shape. *Journal of the Royal Statistical Society. Series B. Methodological 53*, 285–339.

Grady, C.L., A. McIntosh, F.L. Bookstein, B. Horwitz, S. Rapoport, and J. Haxby (1998). Age-related changes in regional cerebral blood flow during working memory for faces. *NeuroImage 8*, 409–425.

Holmes, A.P., R. Blair, J. Watson, and I. Ford (1996). Nonparametric analysis of statistical images from functional mapping experiments. *Journal of Cerebral Blood Flow and Metabolism 16*, 7–22.

Kendall, D. G. (1984). Shape-manifolds, procrustean metrics, and complex projective spaces. *Bulletin of the London Mathematical Society 16*, 81–121.

Mardia, K.V., F.L. Bookstein, and I.J. Moreton (2000). Statistical assessment of bilateral symmetry of shapes. *Biometrika 87*, 285–300.

McIntosh, A.R., F.L. Bookstein, J. Haxby, and C. Grady (1996). Multivariate analysis of functional brain images using partial least squares. *NeuroImage 3*, 143–157.

McIntosh, A.R., R. Cabeza, N. Lobaugh, F.L. Bookstein, and S. Houle (1998). Convergence of neural systems processing stimulus associations and coordinating motoric responses: neuroimaging correlates of "cause" and "effect". *Cerebral Cortex 8*, 648–659.

McIntosh, A.R., L. Nyberg, F.L. Bookstein, and E. Tulving (1997). Differential functional connectivity of prefrontal and medial temporal cortices during episodic memory retrieval. *Human Brain Mapping 5*, 323–327.

Sampson, P.D., A. Streissguth, H. Barr, and F.L. Bookstein (1989). Neurobehavioral effects of prenatal alcohol. II. Partial least squares analyses. *Neurotoxicology and Teratology 11*, 477–491.

Lecture Notes in Statistics

For information about Volumes 1 to 108, please contact Springer-Verlag.

109: Helmut Rieder (Editor), Robust Statistics, Data Analysis, and Computer Intensive Methods. xiv, 427 pp., 1996.

110: D. Bosq, Nonparametric Statistics for Stochastic Processes. xii, 169 pp., 1996.

111: Leon Willenborg and Ton de Waal, Statistical Disclosure Control in Practice. xiv, 152 pp., 1996.

112: Doug Fischer and Hans-J. Lenz (Editors), Learning from Data. xii, 450 pp., 1996.

113: Rainer Schwabe, Optimum Designs for Multi-Factor Models. viii, 124 pp., 1996.

114: C.C. Heyde, Yu. V. Prohorov, R. Pyke, and S. T. Rachev (Editors), Athens Conference on Applied Probability and Time Series Analysis, Volume I: Applied Probability In Honor of J.M. Gani. viii, 424 pp., 1996.

115: P.M. Robinson and M. Rosenblatt (Editors), Athens Conference on Applied Probability and Time Series Analysis, Volume II: Time Series Analysis In Memory of E.J. Hannan. viii, 448 pp., 1996.

116: Genshiro Kitagawa and Will Gersch, Smoothness Priors Analysis of Time Series. x, 261 pp., 1996.

117: Paul Glasserman, Karl Sigman, and David D. Yao (Editors), Stochastic Networks. xii, 298 pp., 1996.

118: Radford M. Neal, Bayesian Learning for Neural Networks. xv, 183 pp., 1996.

119: Masanao Aoki and Arthur M. Havenner, Applications of Computer Aided Time Series Modeling. ix, 329 pp., 1997.

120: Maia Berkane, Latent Variable Modeling and Applications to Causality. vi, 288 pp., 1997.

121: Constantine Gatsonis, James S. Hodges, Robert E. Kass, Robert McCulloch, Peter Rossi, and Nozer D. Singpurwalla (Editors), Case Studies in Bayesian Statistics, Volume III. xvi, 487 pp., 1997.

122: Timothy G. Gregoire, David R. Brillinger, Peter J. Diggle, Estelle Russek-Cohen, William G. Warren, and Russell D. Wolfinger (Editors), Modeling Longitudinal and Spatially Correlated Data. x, 402 pp., 1997.

123: D. Y. Lin and T. R. Fleming (Editors), Proceedings of the First Seattle Symposium in Biostatistics: Survival Analysis. xiii, 308 pp., 1997.

124: Christine H. Müller, Robust Planning and Analysis of Experiments. x, 234 pp., 1997.

125: Valerii V. Fedorov and Peter Hackl, Model-Oriented Design of Experiments. viii, 117 pp., 1997.

126: Geert Verbeke and Geert Molenberghs, Linear Mixed Models in Practice: A SAS-Oriented Approach. xiii, 306 pp., 1997.

127: Harald Niederreiter, Peter Hellekalek, Gerhard Larcher, and Peter Zinterhof (Editors), Monte Carlo and Quasi-Monte Carlo Methods 1996. xii, 448 pp., 1997.

128: L. Accardi and C.C. Heyde (Editors), Probability Towards 2000. x, 356 pp., 1998.

129: Wolfgang Härdle, Gerard Kerkyacharian, Dominique Picard, and Alexander Tsybakov, Wavelets, Approximation, and Statistical Applications. xvi, 265 pp., 1998.

130: Bo-Cheng Wei, Exponential Family Nonlinear Models. ix, 240 pp., 1998.

131: Joel L. Horowitz, Semiparametric Methods in Econometrics. ix, 204 pp., 1998.

132: Douglas Nychka, Walter W. Piegorsch, and Lawrence H. Cox (Editors), Case Studies in Environmental Statistics. viii, 200 pp., 1998.

133: Dipak Dey, Peter Müller, and Debajyoti Sinha (Editors), Practical Nonparametric and Semiparametric Bayesian Statistics. xv, 408 pp., 1998.

134: Yu. A. Kutoyants, Statistical Inference For Spatial Poisson Processes. vii, 284 pp., 1998.

135: Christian P. Robert, Discretization and MCMC Convergence Assessment. x, 192 pp., 1998.

136: Gregory C. Reinsel, Raja P. Velu, Multivariate Reduced-Rank Regression. xiii, 272 pp., 1998.

137: V. Seshadri, The Inverse Gaussian Distribution: Statistical Theory and Applications. xii, 360 pp., 1998.

138: Peter Hellekalek and Gerhard Larcher (Editors), Random and Quasi-Random Point Sets. xi, 352 pp., 1998.

139: Roger B. Nelsen, An Introduction to Copulas. xi, 232 pp., 1999.

140: Constantine Gatsonis, Robert E. Kass, Bradley Carlin, Alicia Carriquiry, Andrew Gelman, Isabella Verdinelli, and Mike West (Editors), Case Studies in Bayesian Statistics, Volume IV. xvi, 456 pp., 1999.

141: Peter Müller and Brani Vidakovic (Editors), Bayesian Inference in Wavelet Based Models. xiii, 394 pp., 1999.

142: György Terdik, Bilinear Stochastic Models and Related Problems of Nonlinear Time Series Analysis: A Frequency Domain Approach. xi, 258 pp., 1999.

143: Russell Barton, Graphical Methods for the Design of Experiments. x, 208 pp., 1999.

144: L. Mark Berliner, Douglas Nychka, and Timothy Hoar (Editors), Case Studies in Statistics and the Atmospheric Sciences. x, 208 pp., 2000.

145: James H. Matis and Thomas R. Kiffe, Stochastic Population Models. viii, 220 pp., 2000.

146: Wim Schoutens, Stochastic Processes and Orthogonal Polynomials. xiv, 163 pp., 2000.

147: Jürgen Franke, Wolfgang Härdle, and Gerhard Stahl, Measuring Risk in Complex Stochastic Systems. xvi, 272 pp., 2000.

148: S.E. Ahmed and Nancy Reid, Empirical Bayes and Likelihood Inference. x, 200 pp., 2000.

149: D. Bosq, Linear Processes in Function Spaces: Theory and Applications. xv, 296 pp., 2000.

150: Tadeusz Caliński and Sanpei Kageyama, Block Designs: A Randomization Approach, Volume I: Analysis. ix, 313 pp., 2000.

151: Håkan Andersson and Tom Britton, Stochastic Epidemic Models and Their Statistical Analysis. ix, 152 pp., 2000.

152: David Ríos Insua and Fabrizio Ruggeri, Robust Bayesian Analysis. xiii, 435 pp., 2000.

153: Parimal Mukhopadhyay, Topics in Survey Sampling. x, 303 pp., 2000.

154: Regina Kaiser and Agustín Maravall, Measuring Business Cycles in Economic Time Series. vi, 190 pp., 2000.

155: Leon Willenborg and Ton de Waal, Elements of Statistical Disclosure Control. xvii, 289 pp., 2000.

156: Gordon Willmot and X. Sheldon Lin, Lundberg Approximations for Compound Distributions with Insurance Applications. xi, 272 pp., 2000.

157: Anne Boomsma, Marijtje A.J. van Duijn, and Tom A.B. Snijders (Editors), Essays on Item Response Theory. xv, 448 pp., 2000.

158: Dominique Ladiray and Benoît Quenneville, Seasonal Adjustment with the X-11 Method. xxii, 220 pp., 2001.

159: Marc Moore (Editor), Spatial Statistics: Methodological Aspects and Applications. xvi, 282 pp., 2001.

9 780387 952406